COMFORTABLE QUARTERS FOR LABORATORY ANIMALS

Animal Welfare Institute

EDITED BY
CATHY LISS,
KENNETH LITWAK,
DAVE TILFORD,
AND VIKTOR REINHARDT
TENTH EDITION

The Animal Welfare Institute is a nonprofit charitable organization founded in the United States in 1951 and dedicated to reducing animal suffering caused by people. AWI engages policymakers, scientists, industry, and the public to achieve better treatment of animals everywhere—in the laboratory, on the farm, in commerce, at home, and in the wild.

Published by Animal Welfare Institute
900 Pennsylvania Avenue, SE, Washington, DC 20003
www.awionline.org

Comfortable Quarters for Laboratory Animals / by Animal Welfare Institute

ISBN 978-0-938414-79-7
LCN 2015947590

Tenth Edition, 2015

Printed in the United States
Edited by Cathy Liss, Kenneth Litwak, Dave Tilford, and Viktor Reinhardt
Design by Ava Rinehart and Alexandra Alberg

References to Internet websites (URLs) were accurate at the time of writing. The Animal Welfare Institute is not responsible for URLs that may have expired or changed since the book was printed.

Cover photograph by Dr. Brianna Gaskill

This book is devoted to those who recognize the need to improve the welfare of animals in research and have the will to make it happen.

COMFORTABLE QUARTERS FOR LABORATORY ANIMALS

Foreword

More than 30 years ago, I began conducting visits at registered research facilities for the Animal Welfare Institute (AWI). I observed a wide range of different species and their housing, handling and care, along with surgical sites, storage rooms, offices, and libraries. I also spent time meeting with various staff at research laboratories across the United States. Most laboratories appeared familiar with AWI's reputation and were willing to open their doors to me, if only a little. While I was not always allowed to see all of the animals or facilities, the interactions I observed and had with the technicians, investigators, and management staff told me much about their animal care programs. Perhaps most telling was the interaction between the people and the animals. How did staff members behave as they entered an animal room and how did the animals respond?

What was nearly universal in the facilities I visited for many years was that the animals were kept in small, barren cages with feeders and waterers—and nothing more. Enrichment was unthinkable, uncharted territory, viewed by labs as both costly and a source of extraneous variables that would threaten research results. Nonhuman primates (with the exception of breeding animals), dogs, and other social species were individually housed. I routinely witnessed primates engaging in a range of stereotypies, including hair plucking and self-mutilating. I saw dogs cowering and shaking in the backs of their tiered cages, while others were circling round and round in their small confines. I observed rabbits sitting in the middle of their cages, not moving because there was no room or reason to do so. I saw many rodents being kept in wire-bottom cages, while those in shoebox cages only had a bit of litter on the floor.

Now, it appears that a reversal in perspective is underway. There is increasing recognition of the need to keep animals physically and psychologically healthy to *reduce* extraneous variables—and this is done by providing them with species-appropriate housing and enrichment and reducing pain and distress when possible. The eighth edition of the *Guide for the Care and Use of Laboratory Animals*, published in 2011, embraces this view (albeit without the regulatory force of the Animal Welfare Act). Similarly, the Institute of Medicine and the Working Group of the Council of Councils (an advisory body to the NIH) recommend significant improvements in the manner in which chimpanzees are housed, with the Working Group calling for "ethologically appropriate physical and social environments." This recommendation was later accepted in large part by the NIH.

Spending significant moneys on shiny new cages and commercially available enrichment devices may not be necessary, though; simple steps such as reconfiguring old cages and making your own enrichment can have a positive impact on animal welfare, while also minimizing research confounds. Teaching animals to cooperate with positive reinforcement instead of forcing them into compliance is increasingly recognized as beneficial to the animals as well as to research outcomes.

Agreeing with the change in perspective is one thing, but implementing it for all animals is quite another. While the situation for animals in research is changing, and improvements have been made, there is still much that needs to be done. The vast majority of animals used in research—rats, mice and fish—are not covered under the Act, nor are birds, amphibians and other cold-blooded species. The requirements under the Act sorely need to be updated and expanded. All animals deserve an environment adequate to promote their well-being.

We at AWI hope you will take inspiration from this book to go well beyond the minimum standards in seeking to ensure the best possible welfare of the animals who are completely dependent upon you. This book is intended for anyone involved with animals in laboratories—technicians, veterinarians, scientists, institutional officials, enrichment specialists, IACUC members, and inspectors. Thank you to those who are already moving the bar ever higher. To those who aren't there yet, we don't underestimate the task before you in trying to facilitate change, but such change is warranted, and we hope this book will be helpful to you.

It has been 13 years since the previous edition of this book was published and this new edition is more than twice the length, in part because research on improved housing and enrichment has been, and continues to be, conducted. There are 14 chapters on different animals in research (including new chapters on ferrets and zebrafish, as well as chapters on extraneous variables and the human-animal bond) that describe species-specific needs and offer recommendations on how to address them in the laboratory.

My deepest appreciation goes to Vera Baumans, Kaile Bennett, David Cawston, Joanna Cruden, Michele Cunneen, Louis DiVincenti, Jr., Marcie Donnelly, Alvaro Duque, Christian Lawrence, Jennifer Lofgren, Pascalle van Loo, Kathleen Pritchett-Corning, Viktor Reinhardt, Irene Rochlitz, Jodi Scholz, Evelyn Skoumbourdis, and Russell Yothers for the chapters you have written. Thank you to the editors, Viktor Reinhardt, Dave Tilford and Kenneth Litwak, and to Ava Rinehart and Alex Alberg for the beautiful design and layout. Thanks to the photographers for sharing your images. And finally, although they prefer to remain anonymous, a big thank-you to our donors; your support has made it possible for the Animal Welfare Institute to produce and widely distribute this book.

Cathy Liss
President, Animal Welfare Institute

Mice

Mice

Pascalle LP van Loo, PhD
NETHERLANDS ORGANISATION FOR APPLIED SCIENTIFIC RESEARCH TNO

Vera Baumans, DVM, PhD, DipECLAM
DEPARTMENT OF ANIMALS IN SCIENCE AND SOCIETY, LABORATORY ANIMAL SCIENCE DIVISION,
UTRECHT UNIVERSITY

In the previous edition of *Comfortable Quarters*, the chapter on housing mice focused on what was known from wild mice and how we can take that into account when providing them with optimal housing conditions. Since our knowledge of wild mice is the basis for our understanding the needs of laboratory mice, these issues will certainly be addressed again in the majority of paragraphs of this sequel, supplemented with information from more recent literature. Our understanding of the needs of mice is not only a prerequisite when designing comfortable quarters, but also when performing procedures with minimal stress. Thus, in the last paragraphs we will address several procedures that are commonly performed on laboratory mice and provide the reader with tips and tricks as to how these procedures can be performed in a way that is least stressful for both the animal and the caretaker.

Mice in the wild

The house mouse, *Mus musculus*, is one of the most common mammals on earth. This animal is believed to have originated in Central Asia and subsequently dispersed widely around the globe, where it displays remarkable variations in color and size as an environmental adaptation (Marshall, 1978). There is evidence that the house mouse has lived in close proximity with humans since the end of the last glacial period (about 12,000 years ago; Hedrich, 2012, chapter 1). Mice inhabit most parts of the world where humans live. Perhaps because mice live in such close proximity with humans and are so successful in outsmarting people when it comes to their survival, the house mouse has often been the subject of study. Specifically advisable readings are "Mice all over" (Crowcroft, 1973) and the chapter on the laboratory mouse in the *UFAW Handbook*, which includes an elaborate table on mouse standard biological parameters (Baumans, 2010). In the next paragraphs we will provide a brief

outline of house mouse biology and behavior to increase the reader's understanding of the precarious balance between their extraordinary adaptability to extreme circumstances and the boundaries thereof in captivity.

Life cycle

Typically, wild female mice begin breeding at 6–7 weeks of age, and have their first litter around the age of 9–10 weeks, after a 20-day gestation (on average). Newborn mouse pups are born hairless (with the exception of whiskers), deaf, and blind, and are completely dependent on their mother for survival. Over the course of 2 to 3 weeks, they subsequently start growing fur, their incisors erupt, their ears open, and finally their eyes open (Baumans, 2010). These life events happen in a strict order at set times, as does the onset of their motor skills, starting as rooting and circling in the first week and presenting as refined activities and sensory responses in the third week, before weaning (Fox, 1965). Around 3 weeks of age, the pups start to explore the immediate surroundings of their nest and increasingly become more independent of their mother for food (Latham & Mason, 2004). This strictly ordered development is very useful to detect any developmental effects in laboratory mice and is used on a wide scale as part of phenotyping the large number of transgenic mouse lines (Van der Meer, 2001).

The young mice grow up in demes (territories populated by family groups), usually consisting of a dominant male, several females with their progeny, and subordinate males. Female juveniles may stay or disperse as adults to neighboring demes. Unfamiliar males are chased out of the territory while juvenile and subordinate adult males are, to a certain extent, tolerated by the dominant male within the boundaries of the deme. Evidence suggests that the level of tolerance depends on population numbers and food availability, with highest tolerance in densely populated areas with an abundance of food (Crowcroft, 1973; Mackintosh, 1970; Mackintosh, 1973; Poole & Morgan, 1976; Hurst et al., 1993). When population density increases further, subordinate (juvenile) males may disperse out of the territory and live in bachelor groups (Busser et al., 1974).

Wild mice have an average life span of 9–18 months, although some can live up to 2 or even 3 years in captivity or protected environments. The oldest mouse ever to have lived, as far as we know, was kept in captivity in an enriched environment, but did not receive any genetic, pharmacological, or dietary treatment and lived for 1,551 days (about 50 months; Than, 2006). Usually, wild mice live less than a year, due to the high level of predation, pest control, and exposure to harsh living conditions.

Senses and behavior

In the wild, mice are most active from dawn to dusk and then seek shelter from bright light during the daytime. Thus, it is not surprising that mice have excellent senses of smell, taste, touch, and hearing, but have generally poor vision. However, they do have good peripheral vision that allows them to detect movement.

The mouse retina consists of rods and two varieties of cones, one serving the traditional green-yellow region of the vision spectrum and another serving the ultra-violet (UV) region, essentially invisible to humans and many other mammals. These UV cones are more concentrated in the ventral retina, which may reflect the background spectrum of the sky at times of the day or twilight when mice are most active. UV sensitivity and the presence of two varieties of cones (UV and green-yellow) that are wide apart in the spectrum increase contrast, and thus help mice distinguish form and movement achromatically (Gouras & Ekesten, 2004).

The mouse auditory and vocalization apparatus has evolved to hearing and emitting frequencies well beyond the human auditory range. Their hearing is especially sensitive to sounds in the 5–20 kHz range and around 50 kHz (Baumans, 2010), which is beyond the human auditory range (20 Hz to 20 kHz). Both audible and ultrasonic calls are used for communication. Pups emit wriggling calls (<10 kHz) to invoke nursing behavior of the mother and distress calls (50–70 kHz) when separated from their mothers (Latham & Mason, 2004). Male mice grunt and squeak audibly (1–2 kHz) during aggressive encounters and sing ultrasonic courtship songs to females. Female mice can discriminate between the characteristics of male songs and prefer the songs of males of different strains. It is probable that male songs contribute to kin recognition by females, thus avoiding inbreeding and resulting in greater heterozygosity of offspring (Kikusui & Koide, 2011).

Not only the auditory and vocal apparatus, but also the olfactory sense plays an important role in social communication between wild mice. Pheromones from urinary and plantar glands are used to mark territories and to recognize which individuals are familiar. They may invoke or suppress aggression between males and influence mating behavior (Hurst et al., 1993; Hurst et al., 1998; Humphries et al., 1999; Nevison et al., 2000). Mice use their olfactory sense to receive information on edible foods by smelling the breath of other mice (Munger et al., 2010) and use social cues in choosing feeding sites (Baker, 1985). Mice nibble whatever food is available, eating small portions during many feeding bouts throughout both day and night (Baumans, 2010). These olfactory and behavioral strategies are an excellent way to find out which novel foods are edible and which are not. Mice are omnivorous and are able to chew through almost everything. Their diet, typically 10–15% of their body weight per diem, typically consists of seeds and grains, but they also eat roots, leaves and stems, and insects such as beetle larvae, caterpillars, and cockroaches. Mice, like other rodents, also engage in coprophagy, typically consuming up to 10% of fecal matter as a means of nutritional supplementation (Heinrichs, 2001).

The mouse is a very agile creature. When not eating, mice spend their time exploring their surroundings and engage in running, climbing, digging, and even swimming when they need to. Their ability to climb vertical walls, jump as high as 30 cm and squeeze through cracks as small as 0.5 cm allows them to reach almost any area. Since mice have generally poor eyesight, they explore their surroundings by moving along barriers, touching the walls with their whiskers and sides of their bodies. Although they generally avoid open spaces (thigmotaxis), they are curious and explore their surroundings continuously, memorizing pathways, obstacles, food, water, shelter, and other elements in their habitats.

Mice sleep in several short and long bouts throughout the day, with the longer sleeping bouts typically occurring during the daylight phase (Van de Weerd et al., 1997). Each sleeping bout is preceded by an elaborate amount of nesting behavior. The nests,

DOME SHAPE NEST, BUILT BY A BALB/C MOUSE

CUP SHAPE NEST, BUILT BY A C57BL/6 MOUSE

equally formed by female and male mice, are constructed of any soft material the mice can shred and form to their liking. Mice originating from surface nesters (e.g., commensals of some field mice, as well as the laboratory strain BALB/c) typically build dome-shape nests with single tiny openings. Burrow nesters such as C57BL/6, on the other hand, build more cup-shape nests. This difference in nest-building behavior is believed to be genetically determined and can still be found in laboratory strains today (Van de Weerd et al., 1997; Sluyter & Van Oortmerssen, 2000). Nest building behavior is an excellent way to check the health of mice (Deacon, 2012; Jirkof et al., 2013). Mice who are subclinically ill build increasingly frumpy nests or fail to build a nest at all. This behavior has been successfully applied in several mouse strains to establish subclinical disease (Jirkof et al., 2010; Deacon, 2012).

Essentially, wild mice spend their days (or nights) eating, sleeping, reproducing, and exploring—sensing the world in a way considerably different from humans.

Emergence of the laboratory mouse

The question as to why the house mouse was initially chosen (and remains) the most popular laboratory animal is not difficult to answer; it is due to their success in the wild. They are small, immensely adaptive in almost all circumstances, and fast breeders. Above that, they are mammals, sharing about 90% of their genetic makeup with humans (Shakespeare, 2013). In the early 19th century, people—fascinated by this agile little creature with such bewildering polymorphism—started to breed them as pets with specific characteristics such as coat color, and mice from around the world were exchanged to create new lines. Since experimental science started to develop in the late 19th century, the use of these specifically bred pet mice was a logical step, and many of the strains of mice that were bred then are still used today; the most well-known example being the strain C57BL/6, established in the United States by Lathrop as an intercross between the black progeny of female 57. Genetic analysis of most common laboratory mice reveals that they all originate from an intercross between three subspecies of wild mice: *Mus musculus domesticus*, *M. m. musculus* and *M. m. castaneus*. An extensive discussion of the emergence of laboratory mice can be found in Part 1 of *The Laboratory Mouse* (Hedrich, 2012).

Meshing human and mouse needs

With the choice of mice as laboratory animals, came the question how best to house and care for them. The number of laboratory animals has rapidly increased since the mid-20th century, posing constraints on economically feasible housing. At the same time, there has been much attention paid to the importance of standardization to reduce intra- and inter-experimental variability and increase reproducibility of results within and between laboratories (Olsson et al., 2003). This has led to such an intertwining of economically feasible housing and standardization, that the two are routinely viewed as equal, even as the definition of standard housing has evolved. There has been a gradual change from a jar containing some sawdust, to a shoebox-shaped cage with a wire mesh

bottom for individual mice, to a shoebox-shaped open cage with sawdust, to the technically ingenious present-day individually ventilated cage (IVC) systems.

Nevertheless, it has been shown that despite rigorous efforts to equalize conditions, inbred mouse strains, which originated simultaneously from three well-recommended laboratories, have significant site-based effects for nearly all variables examined (Crabbe et al., 1999; Van de Weerd et al., 2002; Wahlsten et al., 2003). Further, increased complexity of housing conditions does not necessarily increase variation between animals. It may even be argued that, since it is variation between the animals that we wish to decrease, housing conditions should be designed with respect to individual needs of animals, much like we nowadays treat human beings through personalized health care (Snyderman, 2012).

Boundaries to adaptability

Captivity changes behavior of animals. Animals who have been domesticated generally become more tame, possibly through artificial selection for traits such as ease of handling and decreased aggression. Evidence for this has been clearly demonstrated in pets, zoo and farm animals, and laboratory animals (Jones et al., 2011). These changes in behavior enable animals to better cope with the circumstances. Nevertheless, there is compelling evidence that feral animals (once domesticated and released back into the wild) are still able to exhibit the behaviors of their wild counterparts that are necessary for survival. In general, changes in behavior brought about by domestication are quantitative rather than qualitative (Howard et al., 2010). Berdoy (2003) provided what is perhaps the most illustrative and enjoyable example of this wild behavior that is still present in laboratory animals. Berdoy created a semi-natural farm environment into which he released laboratory rats from two strains: Wistar and Lister hooded rats; here, they had to compete, like their wild cousins, for food, shelter and mates for a period of 6 months. The resulting film, *The Laboratory Rat: A Natural History*, reviews the range of behaviors and needs that, despite generations of domestication, remain innate and ready to be expressed when given the opportunity. The film is highly recommended for both students and teachers in the field of ethology and laboratory animal science. Although a similar experiment has not been performed with mice, it is likely that adaptation of laboratory mice in feral conditions would be similar.

These examples pose the question as to what the boundaries are to adaptability of animals in captivity, especially with regard to behaviors that animals are motivated to perform per se (behavioral needs), even if the physiological need that the behavior serves is fulfilled (see, e.g., Jensen & Toates, 1993).

For laboratory mice, despite their extraordinary capabilities to adapt to changing circumstances, evidence is accumulating that the present-day standardized housing and management is still a long way from an environment meeting their behavioral needs, leading to behavioral problems indicative of decreased

ABNORMAL BEHAVIOR: WHISKER TRIMMING

welfare—such as stereotypies, aggression, and whisker and fur trimming. The underlying mechanisms for development of these detrimental behavioral patterns are only partly understood and some strains of mice appear to be more prone to developing them than others. There is, however, general consensus that rearing mice in barren, restricted cages, lacking appropriate stimuli, is a precursor for the development of abnormal behavior (Latham & Mason, 2004; Würbel et al., 1996). Garner et al. (2004) found that cage height in the animal room and cage material were factors influencing severity of barbering and stereotypic behaviors, while Nevison (1999) linked them to repetitive and futile attempts to flee from the cage. Aggression, especially between male mice, may be the result of inbreeding, environmentally disturbed social behavior, frustration, or lack of control (Van Loo et al., 2003).

There is an increasing awareness of the importance of an environment meeting the mice's needs both for animal welfare and for the quality of research, clearly articulated by Poole (1997): "Happy animals make good science." This awareness is evident in what is considered standard today: shoebox-shaped cages with bedding material, nesting material or nest boxes, gnawing blocks, and social housing. Environmental refinement is an ongoing process and we should aim to provide stimuli beyond the satisfaction of the basic needs normally accommodated in standard housing conditions (Baumans & Van Loo, 2013). The notion of catering to behavioral needs has been embedded in present legislation and guidelines around the world, all with more or less similar import: that animals should be (a) allowed adequate space to express a wide behavioral repertoire, (b) socially housed wherever possible, and (c) provided with an adequately complex environment within the animal enclosure to enable them to carry out a range of normal behaviors (see, e.g., European Parliament and Council of the European Union, 2010; National Research Council [of the United States], 2011; National Health and Medical Research Council [of Australia], 2013).

Making use of our knowledge of how mice perceive the world

With proper management of the wide range of stimuli in the laboratory mouse environment—both physically (such as climate, cage furniture and procedures) and socially (other mice, human caretakers)—and through knowledge of the mice's behavior in the wild and knowledge of and empathy for the way laboratory mice experience the world around them, a lot can be achieved, sometimes with only minor adaptations (Van de Weerd & Baumans, 1995).

In the following paragraphs, both scientific and anecdotal evidence on welfare-enhancing adaptations in housing, management and experimental procedures commonly performed on mice are discussed in relation to what we know of their wild counterparts.

Housing and husbandry

Temperature: The climate in animal rooms is usually kept as stable as possible, with temperatures between 20–24°C (68–75°F). Several studies have shown that mice prefer temperatures quite above the temperatures usually provided (Blom et al., 1993; Gaskill et al., 2009). This does not mean that room temperature needs to be increased to levels uncomfortable to work in for animal caretakers. Mice prove to be perfectly able to adapt their microclimate to their needs. Housing the mice socially and providing them with nesting material enables them to create nests with temperatures around 30–32°C (86–90°F; Gaskill et al., 2011; Gaskill, 2013). Under special circumstances, however, it is advisable to increase cage temperature—for example, when mice are housed individually

in metabolic cages or when they are recovering from general anesthesia. Again, in these circumstances, it is advisable to give the mice a choice. When providing heating mats after surgery, placing half of the cages on the mat ensures that mice can move their nest away from the heat during recovery.

Ventilation: In our attempt to standardize lab conditions—especially with regard to our desire to keep SPF (specific pathogen free) and immunocompromised animals, as well as to minimize allergen load for animal caretakers—individually ventilated cages have been designed. These cages, although in appearance no different from the standard shoebox cages, provide a microclimate that can be carefully regulated. Cages can be ventilated with a rate up to 120 air changes per hour, reducing the need for frequent cage cleaning. However, health monitoring and inspection of the animals may be difficult, procedures and cage cleaning might be more time-consuming, and the high intra-cage ventilation rate could induce chronic stress and heat loss due to the draft (Baumans et al., 2002; Krohn, 2002).

In a personal communication on CompMed several years ago, a researcher asked advice on unexplained death of his mice who were housed in recently purchased IVC cages. Further inquiry revealed that the IVC air inlet was situated around the drinking nipple, forcing the mice to drink in a constant draft. The mice, it turned out, suffered from dehydration, either from avoiding the nipple due to the draft, or the constant drying airflow in the cage. Humans consider air speed greater than 0.2 m/s to be drafty and this is generally agreed to be an upper limit for rodents, as well (Lipman, 1999). This can be ameliorated by moving the air inlet to a different location in the cage. When the air inlet was located at the top of the cage and nesting material was provided, air changes of up to 60 per hour were tolerated, with no

adverse effects on the physiology or behavior of the mice (Baumans et al., 2002). This means that the location of the air supply to the cage (from the side or from the top), the ventilation rate, and the presence of nesting material are important considerations when assessing the impact of IVC housing on the well-being of mice. Evidence that animals are reacting to draft could be a change of location of the nest and the building of barriers of bedding.

Lights and sounds: Mouse rooms usually have a 12-hour light-dark cycle with lights off during the night and light intensity during the day as high as 300 lux. Light has damaging effects on the retina, but can be particularly damaging in nocturnal species such as mice, and is even more detrimental in albino strains, which lack pigment protection (Lanum, 1978). Therefore, it is of utmost importance that light intensity be kept to a minimum. In general, for albino mice, light intensity should not exceed 60 lux at the cage level. This can be achieved by covering the highest cage shelves and by providing mice with structures, such as nesting material, that enable them to shelter from bright light.

Mouse rooms are a constant source of sounds emitted by the mice themselves, animal technicians and caretakers, and equipment. Some of these sounds, such as equipment producing ultrasound, or sudden noises, such as doors and cages opening or closing, may be a source of stress for the animals. Chronic and/or loud noises may induce impaired behavior, cognition, and immune function in mice (Cheng et al., 2011; Pascuan et al., 2014; Tamura et al., 2012). Playing background music in animal rooms may help mask stressful sounds (Van Loo & Baumans, 2004; Alworth & Buerkle, 2013).

Structuring and size of the cages: Appropriate structuring of the cage environment is typically more beneficial than provision of a larger floor area; however, a minimum floor area is necessary to provide a structured space. This enables mice to use the vertical cage dimension as well. It is difficult to scientifically specify the minimal sizes of cages for maintaining laboratory mice, as much depends on the strain, group size and age of the animals, their familiarity with each other, and their reproductive condition (see Whittaker et al., 2012, for an excellent review). In terms of structure, the home cage can be furnished with, for example, nest boxes, tubes, partitions, and nesting material (Latham & Mason, 2004; Sherwin & Nicol, 1997; Baumans, 2005).

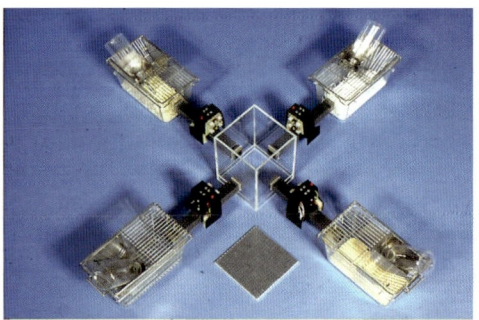

PREFERENCE TEST SYSTEM

However, provision of environmental refinement should not be a process of randomly applying objects that staff consider attractive for the animals; instead, environmental refinement should be regarded as an essential component of the overall animal care program, and equally important as nutrition and veterinary care. It is critical to evaluate environmental refinement in terms of the benefit to the animal—assessing the use of and preference for certain refinements, the effect on behavior (in particular, species-typical behavior), and the effect on physiological parameters. At the same time, it is necessary to evaluate the impact on scientific outcome, how the refinement influences the scientific study, and whether and how the statistical power is affected. Communication and teamwork among animal welfare scientists, animal research scientists, institutional animal welfare officers, veterinarians, animal ethics committees, and animal facility management and personnel is a key to success (Ottesen et al., 2004; Weed & Raber, 2005; Baumans et al., 2006). Many experiments

Comfortable Quarters for Laboratory Animals

have been performed that reveal the mice's preferences for cage type, bedding and environmental refinement, and how this affects the animals' well-being and quality of research (see, e.g., Blom et al., 1996; Garner et al., 2004; Kirchner et al., 2012; Van Loo et al., 2005; Van de Weerd et al., 1997; Nicol et al., 2008). These studies support the now-accepted notion that controllability and predictability of the environment are highly important factors in enhancing the mice's physical and psychological well-being, by providing stimuli meeting the species-specific needs (Poole, 1997; Baumans, 2005; Van Loo et al., 2005). These needs include social contact, nest building, hiding, exploration, foraging, gnawing, and resting. Mice are highly susceptible to predation and are likely to show strong fear responses in unfamiliar situations if they cannot shelter. These responses include attempts to flee, biting when handled, or sudden immobility to avoid being detected. For this reason, cages should be provided with a shelter or hiding places. Security can be achieved via manipulable nesting material, hiding places, and compatible cage mates. Even simple environmental refinement induces a robust and replicable anxiolytic-like effect in mice (Sztainberg & Chen, 2010). Moreover, providing nesting material helps mice keep their nests clean, thus always providing them with a feces-free resting area (Godbey et al., 2011; Boivin, 2013).

Food and water provision: About 15% of the time that mice are awake is spent eating food, in numerous small bouts of feeding behavior (Van de Weerd et al., 1997). Mice feed in social bouts and learn from each other with regard to the type of food and drink that can be consumed, and how it is consumed (Baker, 1985). In the laboratory, mice may be confronted with novel ways to obtain food and water several times in their lives; for example, a drinking bottle, an automated watering system, and water-containing substances such as potatoes or commercially available gels (e.g., Solid Drink) during transport. In our experience, acclimatization within a social group to these novel ways considerably speeds up the learning curve. In studies where mice were individually housed in cages with an atypical food dispenser for metabolic measurements, they typically lost weight for up to 3 days. Through acclimatization to the food dispenser in groups prior to experiment, mice learned to eat from the novel dispenser within a day (A. M. Van den Hoek, personal communication, May 2014). Food can also be used as environmental refinement. With ad libitum food pellets readily available, foraging behavior cannot be expressed

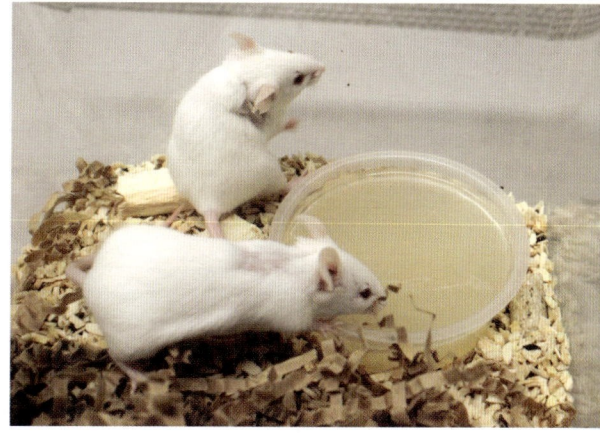

EASILY ACCESSIBLE LIQUID-CONTAINING GELS ENSURE DEBILITATED MICE HAVE CONSTANT ACCESS TO WATER

fully. This may lead to stereotypies such as food grinding behavior. Scattering grain, as refinement, has been shown to decrease this behavior (Pritchett-Corning et al., 2013). If mice are not well—for example, when used for studying progressive disease models—providing readily available food and water via such things as lengthened drinking nipples, food pellets in the cage, and glucose-containing substances such as Solid Drink is a prerequisite not only to reducing discomfort for the animals, but also to ensuring that experimental data reflect the disease under study, rather than dehydration or famine.

Social contact: For gregarious species, such as the mouse, social contact is an important part of their environment and should only be denied in exceptional cases, e.g., extreme aggression or for scientific reasons. Provided that the group composition is harmonious, social interactions are important contributors to animal welfare. Group-housed mice are able to engage in social exploration, and the behavioral activities of one animal—such as scent marking or digging—may also be a valuable source of novelty that elicits exploration by the other individuals (Olsson & Westlund, 2007). More importantly, group-housed mice provide each other with social support (or "social buffering") when encountering a stressful situation (Hennessy et al., 2009), and several studies have suggested that mice benefit from being socially housed with respect to postoperative recovery and the need for pain relief (Van Loo et al., 2007; Pham et al., 2010; Jirkof et al., 2012).

The successful establishment of harmonious single-sex groups requires the grouping of individuals who are compatible, a task that is especially challenging with male mice. Compatibility is strongly influenced by internal factors such as age, sex and hierarchical rank, and external factors such as availability and distribution of resources, and availability of space (Van Loo et al., 2003; Akre et al., 2011). The effects of space availability on the welfare of mice are not consistent. In general, aggression between male mice seems to decrease with increased crowding, however, other studies indicate that crowding increases stress-related parameters (see Olsson & Westlund, 2007, for a review). Some factors, however, can be managed by good husbandry practices, including housing mice in small, socially stable groups of three males (Van Loo et al., 2001), transferring

nesting material, but not dirty bedding, during cage cleaning (Van Loo et al., 2000), and avoiding exposure of male mice to (unfamiliar) male urine (Lacey et al., 2007). Anecdotal evidence from our lab shows that housing male and female mice in different rooms, handling males before females, and generally keeping disturbances to a minimum clearly helps in reducing aggression further. This intuitively makes sense, since stressful events are known to trigger aggression (Pant & Nath, 1993). Another very interesting observation was made by P. Y. Wielinga (personal communication, September 2011). In a study in which a high fat diet was tested for its effect on metabolic parameters in male BALB/c mice, the mice fed the control diet all had to be separated due to high levels of aggression, while the mice fed the high fat diet lived peacefully together. Unfortunately, this finding has never been investigated further. It could be worthwhile investigating whether the nutritional balance for laboratory mice needs to be re-evaluated.

If social housing is not possible due to experimental restraints or excessive aggression, several other options may be worth considering. Co-housing aggressive males with an ovariectomized female as companion may be a solution when the animals have to be kept for long periods, although whether the negative impact of the surgical procedure is outweighed by the benefit of social housing needs to be taken into account. In studies with instrumented animals, non-instrumented buddies may be an option. If the instrumentation does not allow this, some authors have suggested the provision of mirrors (Sherwin, 2004; Fuss et al., 2013) or cohousing animals with a divider in between (Van Loo et al., 2007). The latter two solutions have not proven as yet to alleviate the detrimental effects of individual housing.

Experimental procedures

When evaluating the way in which different techniques are performed on mice, knowledge of their natural behavior and the way they see the world, together with a little common sense, can lead to minor adaptations in technique with decidedly positive consequences for the mice.

Training, conditioning and reward: A rather unexplored area of investigation in mice is the use of training, conditioning and reward. These are very common means of refinement for larger animals, such as dogs or primates. Training and conditioning help the animal to predict what is coming, thereby decreasing the stress response (Weiss, 1972). Adding a reward may help in associating stressful stimuli with overall positive events. Dogs and primates, for example, can be taught to readily offer their paw or arm for blood removal. For mice, the use of training, conditioning and reward is less commonly done, even though several studies show promising results with substantial effects on stress response to minor procedures, such as restraint (Meijer, 2006). The use of training, conditioning and reward for mice, therefore, is certainly worth promoting.

Handling and restraint: Handling is by far the most frequently performed technique on mice. Virtually all procedures involve picking up the mouse from the cage (i.e., cage cleaning, health check, and other invasive procedures). The most common method for taking mice from cages is to pick them up by their tails, usually by hand, or sometimes even with forceps. This, of course, is very unnatural for the mouse (or, to the extent it is "natural," it is like being caught by a predator). By making use of natural responses of the mouse to climb in or on things, Hurst & West (2010) have shown that picking up mice by use of tunnels or open hands led to voluntary approach, low

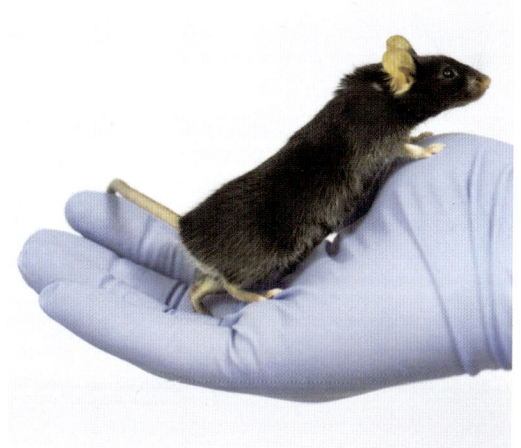

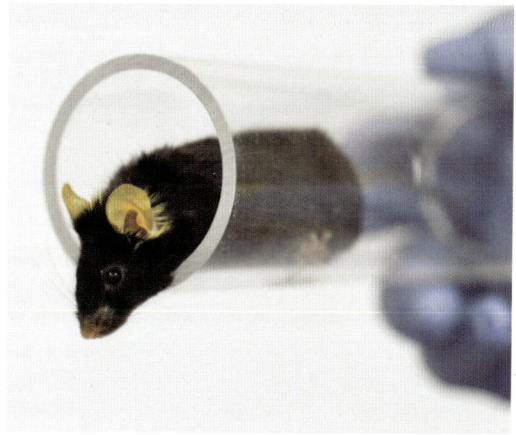

HANDLING OF MICE: (TOP) CUPPING IN A HAND, (BOTTOM) VOLUNTARY ENTRY INTO A TUBE

anxiety, and acceptance of physical restraint. An important aspect here is that the mice continue to keep their feet on a base, rather than hanging from their tails, staring into an abyss. Similarly, other natural responses from mice can be used to aid restraint procedures with less stress. For example, a mouse who needs to be restrained in a tube for blood removal or otherwise could be expected to more readily do so if the tube is a dark, apparently safe haven.

Oral dosing: In studies that require daily oral dosage, voluntary ingestion via water or

VOLUNTARY CONSUMPTION OF A TEST SUBSTANCE AS AN ALTERNATIVE TO ORAL GAVAGE

a tasty gel or paste has proven to be a valuable alternative to oral gavage in mice (Zhang, 2011). Voluntary dosing is a positive rather than a stressful event. Further, the risk of mishaps related to oral gavage is absent. Mice appear to have a specific appetite for savory tastes such as peanut butter, cheese and bacon grease (Witmer et al., 2014). The use of voluntary ingestion of daily dosage is increasingly used for administration of analgesics (Van Loo et al., 1997; Pham et al., 2010; Kalliokoski et al., 2011; Abelson et al., 2012; Molina-Cimadevila et al., 2014), as well as dosing of test substances (Walker et al., 2012; Gonzales et al., 2014).

Several other procedures may be candidates for replacement with less stressful ones. For example, keeping mice in metabolic cages overnight is not necessary if single urine or feces samples are needed. Simply transferring the mouse into a plastic bucket for a short period of time will usually trigger voiding of urine and feces (Van Loo et al., 2001).

Concluding remarks

In preparing this chapter, we have tried to guide the reader into the world as perceived by mice. Being humans, we realize that in no way can we be certain that we are correct on all counts. Nevertheless, we hope that we have provided the reader with some new knowledge and a huge amount of empathy with the way laboratory mice experience the world around them. The examples we have given on ways to improve the life of laboratory mice are by no means exhaustive. Instead, we invite the reader to consider them and to continuously be aware of what we can do to make lives easier for our mice.

REFERENCES

Abelson KSP, Jacobsen KR, Sundbom R, Kalliokoski O and Hau J 2012 Voluntary ingestion of nut paste for administration of buprenorphine in rats and mice. *Laboratory Animals 46*: 349–351

Akre AK, Bakken M, Hovland AL, Palme R and Mason G 2011 Clustered environmental enrichments induce more aggression and stereotypic behavior than do dispersed enrichments in female mice. *Applied Animal Behavior Science 131*: 145–152

Alworth LC and Buerkle SC 2013 The effects of music on animal physiology, behavior and welfare. *Lab Animal Europe 13*(2): 54–61

Baker AEM 1985 Distribution of feeding and drinking by groups of captive house mice. *Behavioural Processes 10*(3): 297–308

Baumans V 2005 Environmental enrichment for laboratory rodents and rabbits: Requirements of rodents, rabbits and research. *Institute for Laboratory Animal Research Journal 46*: 162–170. http://dels-old.nas.edu/ilar_n/ilarjournal/46_2/pdfs/v4602baumans.pdf

Baumans V 2010 The laboratory mouse. In: Hubrecht RC and Kirkwood J (eds) *The UFAW Handbook on the Care and Management of Laboratory Animals, Eighth Edition* pp 276–310. Wiley-Blackwell: Oxford, UK

Baumans V and Van Loo PLP 2013 How to improve housing conditions of laboratory animals: The possibilities of environmental refinement. *The Veterinary Journal 195*: 24–32

Baumans V, Schlingmann F, Vonck M and Van Lith HA 2002 Individually ventilated cages: Beneficial for mice and man? *Contemporary Topics in Laboratory Animal Science 41*: 13–19

Baumans V, Clausing P, Hubrecht R, Reber R, Vitale A and Wyffels E 2006 *FELASA Working Group Standardization of Enrichment: Working Group Report.* http://www.felasa.eu/media/uploads/WG_Enrichment_2006_Report-Final.pdf

Berdoy M (producer) 2003 *The Laboratory Rat: A Natural History* [film]. Oxford University: Oxford, UK. http://www.ratlife.org

Blom HJM 1993 *Evaluation of housing conditions for laboratory mice and rats: The use of preference tests for studying choice behavior* [thesis]. Utrecht University: Utrecht, Netherlands

Blom HJM, Van Tintelen G, Van Vorstenbosch CJAHV, Baumans V and Beynen AC 1996 Preferences of mice and rats for type of bedding material. *Laboratory Animals 30*: 234–244

Boivin GP 2013 Availability of feces-free areas in rodent shoebox cages. *Lab Animal Europe 13*(5): 13–21

Busser J, Zweep A and Van Oortmerssen GA Variability in the aggressive behavior of *Mus musculus domesticus*, its possible role in population structure. In: Van Abeelen JNF (ed) *The Genetics of Behavior* pp 185–199. North-Holland Publishing Company: Amsterdam, Netherlands

Cheng L, Wang SH, Chen QC and Liao XM 2011 Moderate noise induced cognition impairment of mice and its underlying mechanisms. *Physiology & Behavior 104*: 981–988

Crabbe JC, Wahlsten D and Dudek BC 1999 Genetics of mouse behavior: Interactions with the laboratory environment. *Science 284*: 1670–1672

Crowcroft P 1973 *Mice All Over.* Chicago Zoological Society: Brookfield, IL

Deacon R 2012 Assessing burrowing, nest construction, and hoarding in mice. *Journal of Visualized Experiments 59*: 1–10

European Parliament and Council of the European Union 2010 Directive 2010/63/EU of the European Parliament and of the Council of 22 September 2010 on the protection of animals used for scientific purposes. *Official Journal of the European Union*: L 276/33. http://eur-lex.europa.eu/legal-content/EN/TXT/?uri=CELEX:32010L0063

Fox WM 1965 Reflex-ontogeny and behavioral development of the mouse. *Animal Behaviour 13*: 234–241

Fuss J, Richter SH, Steinle J, Deubert G, Hellweg R and Gass P 2013 Are you real? Visual simulation of social housing by mirror image stimulation in single housed mice. *Behavioral Brain Research 243*: 191–198

Garner JP, Dufour B, Gregg LE, Weisker SM and Mench JA 2004 Social and husbandry factors affecting the prevalence and severity of barbering ('whisker trimming') by laboratory mice. *Applied Animal Behaviour Science 89*: 263–282

Gaskill BN, Rohr SA, Pajor EA, Lucas JR and Garner JP 2009 Some like it hot: Mouse temperature preferences in laboratory housing. *Applied Animal Behavior Science 116*: 279–285

Gaskill BN, Rohr SA, Pajor EA, Lucas JR and Garner JP 2011 Working with what you've got: Changes in thermal preference and behavior in mice with or without nesting material. *Journal of Thermal Biology 36*: 193–199

Gaskill BN, Gordon CJ, Pajor EA, Lucas JR, Davis JK and Garner JP 2013 Impact of nesting material on mouse body temperature and physiology. *Physiology & Behavior 110–111*: 87-95

Godbey T, Gray G and Jeffery D 2011 Cage-change interval preference in mice. *Lab Animal Europe 11*(8): 15–22

Gonzales C, Zaleska MM, Riddell DR, Atchison KP, Robshaw A, Zhou H and Sukoff Rizzo SJ 2014 Alternative method of oral administration by peanut butter pellet formulation results in target engagement of BACE1 and attenuation of gavage-induced stress responses in mice. *Pharmacology Biochemistry and Behavior 126*: 28–35

Gouras P and Ekesten B 2004 Why do mice have ultra-violet vision? *Experimental Eye Research 79*: 887–892

Hedrich HJ 2012 *The Laboratory Mouse, Second Edition.* Academic Press: New York, NY

Heinrichs SC 2001 Mouse feeding behavior: ethology, regulatory mechanisms and utility for mutant phenotyping. *Behavioral Brain Research 125*: 81–88

Hennessy MB, Kaiser S and Sachser N 2009 Social buffering of the stress response: Diversity, mechanisms, and functions. *Frontiers in Neuroendocrinology 30*: 470–482

Howard B, Nevalainen T and Perretta G (eds) 2010 *The COST Manual of Laboratory Animal Care and Use: Refinement, Reduction, and Research.* CRC Press: Boca Raton, FL. http://www.crcpress.com/product/isbn/9781439824924

Humphries RE, Robertson DHL, Beynon RJ and Hurst JL 1999 Unravelling the chemical basis of competitive scent marking in house mice. *Animal Behaviour 58:* 1177–1190

Hurst JL, Fang J and Barnard CJ 1993. The role of substrate odours in maintaining social tolerance between male house mice, *Mus musculus domesticus. Animal Behaviour 45:* 997–1006

Hurst JL, Robertson DHL, Tolladay U and Beynon RJ 1998 Proteins in urine scent marks of male house mice extend the longevity of olfactory signals. *Animal Behaviour 55:* 1289–1297

Hurst JL and West RS 2010 Taming anxiety in laboratory mice. *Nature Methods 7:* 825–826

Jensen P and Toates FM 1993 Who needs 'behavioral needs'? Motivational aspects of the needs of animals. *Applied Animal Behaviour Science 37:* 161–181

Jirkof P, Cesarovic N, Rettich A, Nicholls F, Seifert B and Arras M 2010 Burrowing behavior as an indicator of post-laparotomy pain in mice. *Frontiers in Behavioral Neuroscience 4:* 165

Jirkof P, Cesarovic N, Rettich A, Fleischmann T and Arras M 2012 Individual housing of female mice: Influence on postsurgical behavior and recovery. *Laboratory Animals 46:* 325–334

Jirkof P, Fleischmann T, Cesarovic N, Rettich A, Vogel J and Arras M 2013 Assessment of postsurgical distress and pain in laboratory mice by nest complexity scoring. *Laboratory Animals 47:* 153–161

Jones MA, Mason GJ and Pillay N 2011 Correlates of birth origin effects on the development of stereotypic behavior in striped mice, *Rhabdomys. Animal Behaviour 82:* 149–159

Kalliokoski O, Jacobsen KR, Hau J and Abelson KSP 2011 Serum concentrations of buprenorphine after oral and parenteral administration in male mice. *The Veterinary Journal 187:* 251–254

Kikusui T and Koide T 2011 Ultrasound communication in mice. *Neuroscience Research 71:* 11

Kirchner J, Hackbarth H, Selzer HD and Tsai PP 2012 Preferences of group-housed female mice regarding structure of softwood bedding. *Laboratory Animals 46:* 95–100

Krohn TC 2002 *Method Developments and Assessments of Animal Welfare in IVC Systems* [thesis]. The Royal Veterinary and Agricultural University: Frederiksberg, Denmark

Lacey JC, Beynon RJ and Hurst JL 2007 The importance of exposure to other male scents in determining competitive behavior among inbred male mice. *Applied Animal Behaviour Science 104:* 130–142

Lanum J 1978 The damaging effects of light on the retina. Empirical findings, theoretical and practical implications. *Survey of Ophthalmology 22:* 221–249

Latham N and Mason G 2004 From house mouse to mouse house: the behavioral biology of free-living *Mus musculus* and its implications in the laboratory. *Applied Animal Behaviour Science 86:* 261–289

Lipman NS 1999 Isolator rodent caging systems (state of the art): A critical view. *Contemporary Topics in Laboratory Animal Science 38:* 9–17

Mackintosh JH 1970 Territory formation by laboratory mice. *Animal Behaviour 18:* 177–183

Mackintosh JH 1973 Factors affecting the recognition of territory boundaries by mice (*Mus musculus*). *Animal Behavior 21:* 464–470

Marshall JT 1978 Brief review of European house mice. In: Morse III HC (ed) *Origins of Inbred Mice* pp 511–518. Academic Press: New York, NY. http://www.informatics.jax.org/morsebook/index.shtml

Meijer MK 2006 *Neglected impact of routine: Refinement of experimental procedures in laboratory mice* [thesis]. Utrecht University: Utrecht, Netherlands

Molina-Cimadevila MJ, Segura S, Merino C, Ruiz-Reig N, Andrés B and de Madaria E 2014 Oral self-administration of buprenorphine in the diet for analgesia in mice. *Laboratory Animals 48:* 216–224

Munger SD, Leinders-Zufall T, McDougall LM, Cockerham RE, Schmid A, Wandernoth P, ... Kelliher KR 2010 An olfactory subsystem that detects carbon disulfide and mediates food-related social learning. *Current Biology 20:* 1438–1444

National Health and Medical Research Council 2013 *Australian Code of Practice for the Care and Use of Animals for Scientific Purposes, Eighth Edition.* Australian Government Publishing Service: Canberra, Australia. https://www.nhmrc.gov.au/guidelines/publications/ea28

National Research Council 2011 *Guide for the Care and Use of Laboratory Animals, Eighth Edition.* National Academies Press: Washington, DC http://www.nap.edu/catalog.php?record_id=12910

Nevison CM, Hurst JL and Barnard CJ 1999 Why do male ICR (CD-1) mice perform bar-related (stereotypic) behavior? *Behavioural Processes 47:* 95–111

Nevison CM, Barnard CJ, Beynon RL and Hurst JL 2000 The consequences of inbreeding for recognizing competitors. *Proceedings of the Royal Society London, Series B 267:* 687–694

Newberry RC 1995 Environmental enrichment: Increasing the biological relevance of captive environments. *Applied Animal Behaviour Science 44:* 229–243

Nicol CJ, Brocklebank S, Mendl M and Sherwin CM 2008 A targeted approach to developing environmental enrichment for two strains of laboratory mice. *Applied Animal Behaviour Science 110:* 341–353

Olsson IAS, Nevison CM, Patterson-Kane EG, Sherwin CM, Van de Weerd HA and Würbel H 2003 Understanding behavior: The relevance of ethological approaches in laboratory animal science. *Applied Animal Behaviour Science 81:* 245–264

Olsson IAS and Westlund K 2007 More than numbers matter: The effect of social factors on behavior and welfare of laboratory rodents and non-human primates. *Applied Animal Behaviour Science 103:* 229–254

Ottesen JL, Weber A, Gürtler H and Mikkelsen LF 2004 New housing conditions: Improving the welfare of experimental animals. *Alternatives to Laboratory Animals 32*(Suppl. 1B): 397–404

Pant KK and Nath C 1993 Dopaminergic involvement in the effects of piracetam on foot shock induced aggression in mice. *Indian Journal of Medical Research 98:* 155–9

Pascuan CG, Uran SL, Gonzalez-Murano MR, Wald MR, Guelman LR and Genaro AM 2014 Immune alterations induced by chronic noise exposure: Comparison with restraint stress in BALB/c and C57Bl/6 mice. *Journal of Immunotoxicology 11:* 78–83

Pham TM, Hagman B, Codita A, Van Loo PLP, Strömmer L and Baumans V 2010 Housing environment influences the need for pain relief during post-operative recovery in mice. *Physiology & Behavior 99:* 663–668

Poole TB and Morgan HDR 1976 Social and territorial behavior of laboratory mice (*Mus musculus L.*) in small complex areas. *Animal Behaviour 24:* 476–480.

Poole TB 1997 Happy animals make good science. *Laboratory Animals 31:* 116

Pritchett-Corning KR, Keefe R, Garner JP and Gaskill BN 2013 Can seeds help mice with the daily grind? *Laboratory Animals 47:* 312–315.

Shakespeare T 2013 A point of view: Fly, fish, mouse and worm. Retrieved from http://www.bbc.com/news/magazine-22904931

Sherwin CM 2004 Mirrors as potential environmental enrichment for individually housed laboratory mice. *Applied Animal Behaviour Science 87:* 95–103

Sherwin CM and Nicol C J 1997 Behavioral demand functions of caged laboratory mice for additional space. *Animal Behaviour 53:* 67–74

Sluyter F and Van Oortmerssen GA 2000 A mouse is not just a mouse *Animal Welfare 9:* 193–205

Snyderman R 2012 Personalized health care: From theory to practice. *Biotechnology Journal 7:* 973–979

Sztainberg Y and Chen A 2010 An environmental enrichment model for mice. *Nature Protocols 5:* 1535–1539

Tamura H, Ohgami N, Yajima I, Iida M, Ohgami K, Fujii N, ... Kato M 2012 Chronic exposure to low frequency noise at moderate levels causes impaired balance in mice. *PLoS ONE 7*(6): e39807

Than K 2006 Anti-aging competitions go head-to-head. Retrieved from http://www.livescience.com/10479-anti-aging-competitions-head-head.html

Van Loo PLP, Croes IAA, Baumans V 2004 Music for mice: Does it affect behavior and physiology? Abstract, Telemetry Workshop. FELASA meeting, Nantes, France

Van Loo PLP, Everse LA, Bernsen MR, Baumans V, Hellebrekers LJ, Kruitwagen CLJJ and den Otter W 1997 Analgesics in mice used in cancer research: reduction of discomfort? *Laboratory Animals 31:* 318–325

Van Loo PLP, Kruitwagen CLJJ, Van Zutphen LFM, Koolhaas, JM and Baumans V 2000 Modulation of aggression in male mice: influence of cage cleaning regime and scent marks. *Animal Welfare 9:* 281–295

Van Loo PLP, Mol JA, Koolhaas JM, Van Zutphen LFM, and Baumans V 2001 Modulation of aggression in male mice: Influence of group size and cage size. *Physiology & Behavior 72:* 675–683

Van Loo PLP, Van Zutphen LFM and Baumans V 2003 Male management: Coping with aggression problems in male laboratory mice. *Laboratory Animals 37:* 300–313

Van Loo PLP, Blom HJM, Meijer MK and Baumans V 2005 Assessment of the use of two commercially available environmental enrichments by laboratory mice by preference testing. *Laboratory Animals 39:* 58–67

Van Loo PLP, Kuin N, Sommer R, Avsaroglu H, Pham T and Baumans V 2007 Impact of 'living apart together' on post-operative recovery of mice compared to social and individual housing. *Laboratory Animals 41:* 441–455

Van der Meer M 2001 *Transgenesis and animal welfare: Implications of transgenic procedures for the well-being of the laboratory mouse* [thesis]. Labor Grafimedia BV: Utrecht, Netherlands

Van de Weerd HA and Baumans V 1995 Environmental enrichment in rodents. In: Environmental Enrichment Information Resources for Laboratory Animals. *AWIC Resource Series 2:* 145–149.

Van de Weerd HA, Van Loo PL, Van Zutphen LF, Koolhaas JM and Baumans V 1997 Preferences for nesting material as environmental enrichment for laboratory mice. *Laboratory Animals 31:*133–143

Van de Weerd HA, Aarsen EL, Mulder A, Kruitwagen CLJJ, Hendriksen CFM and Baumans V 2002 Effects of environmental enrichment for mice on variation in experimental results. *Journal of Applied Animal Welfare Science 5:* 87–108

Wahlsten D, Metten P, Phillips TJ, Boehm II SL, Burkhart-Kasch S, Dorow J, ... Crabbe JC 2003 Different data from different labs: Lessons from studies in gene-environment interaction. *Journal of Neurobiology 54:* 283–311

Walker MK, Boberg JR, Walsh MT, Wolf V, Trujillo A, Skelton Duke M, ... Felton LA 2012 A less stressful alternative to oral gavage for pharmacological and toxicological studies in mice. *Toxicology and Applied Pharmacology 260:* 65–69

Weed JL and Raber JM 2005 Balancing animal research with animal well-being: Establishment of goals and harmonization of approaches. *ILAR Journal 46:* 118–128

Weiss JM 1972 Psychological factors in stress and disease. *Scientific American 226*(6): 104–113

Whittaker AL, Howarth GS and Hickman DL 2012 Effects of space allocation and housing density on measures of wellbeing in laboratory mice: a review. *Laboratory Animals 46:* 3–13

Witmer GW, Snow NP and Moulton RS 2014 Responses by wild house mice (*Mus musculus*) to various stimuli in a novel environment. *Applied Animal Behaviour Science 19:* 99–106

Zhang L 2011 Voluntary oral administration of drugs in mice. *Protocol Exchange.* doi:10.1038/protex.2011.236

Rats

Rats

Kathleen Pritchett-Corning, DVM, DACLAM

OFFICE OF ANIMAL RESOURCES, FACULTY OF ARTS AND SCIENCES,
HARVARD UNIVERSITY

The brown rat, or Norway rat (*Rattus norvegicus*) probably originated in Asia, near the Caspian Sea, and has spread throughout the world as a human commensal (Donaldson, 1912; Hedrich, 2000). With their spread from Asia to Europe and places beyond in the Middle Ages, brown rats were quickly recognized not only as pests that had an economic impact through competition with humans for limited resources, but also as vectors of disease. Rat-catching has existed as a trade since at least medieval times; typically, these early exterminators were paid a per-rat bounty, although sometimes payment per job was arranged (Matthews, 1898). Rats were also captured for the rat-baiting trade, which rose to prominence after other popular blood sports, such as bull and bear baiting, were banned in the early 19th century. Rat-catchers may also have bred rats to increase their financial security, rather than relying solely on what nature provided, and this practice may have led to the rat fancy and rat domestication, both of which arose in Europe during Victorian times. Unusual color variants captured for rat baiting were instead saved and tamed, then sold. Although some claim that the rat was the first mammal domesticated solely for research, it is unclear if research or the rat fancy was the primary driver (Lindsey & Baker, 2006). Many books have been written on the natural history, behavior, biology, and research uses of the rat (Calhoun, 1963; Cooley & Vanderwolf, 2005; Krinke, 2000; Sharp & Villano, 2013; Suckow et al., 2006; Sullivan, 2005; Waynforth & Flecknell, 1992; Wishaw & Kolb, 2004). Brown rats are clever, social, physically robust rodents (Ben-Ami Bartal et al., 2011), who are used extensively in every aspect of teaching and biomedical research, from psychology to genetics to safety testing to infectious disease to neuroscience.

Species-typical characteristics of laboratory rats

Origin: Laboratory rats are domesticated wild brown rats with no genetic evidence of crossing with black rats (*Rattus rattus*) (Hedrich, 2006), unlike laboratory mice, which carry genes from several closely related species and subspecies (Didion & de Villena, 2013). For the rest of this work, "rat" will be used to refer, specifically, to the brown rat rather than any other rat species in use in research or found in the wild. Inbred strains and outbred stocks of laboratory rats originated in the early 20th century in the laboratories of H. H. Donaldson and W. E. Castle (Castle, 1947; Donaldson, 1912). Rats have been selected and bred by humans for various characteristics such as coat color, tumor

susceptibility, disease development, or responses to compounds (Castle, 1947). Domestic rats differ from their wild ancestors in several important ways, including greater docility, larger body size, and increased fecundity. However, if released into the wild, laboratory rats will readily revert to ancestral behaviors (Berdoy, 2003), so discussion of wild rat characteristics is appropriate when considering laboratory rats.

Behavior: Rats are social animals, living in kin groups in the wild. The kin group maintains and defends a home range that varies in size, depending on resources and availability of cover (Calhoun, 1963; Davis et al., 1948; Taylor, 1978). Above-ground movement trails are readily apparent in the environment. Residents of a home range excavate an extensive burrow system with many chambers and exits (Berdoy, 2003; Calhoun, 1963). A burrow system is typically occupied by a dominant male and related or familiar females with offspring. Low-status animals, often juveniles, are crowded into territory not claimed by a group (Calhoun, 1963). Although low-status females may breed, the matings are usually unsuccessful in producing weaned pups unless the female has a territory and burrow (Calhoun, 1963). Typically, related or familiar females will communally rear pups (Schultz & Lore, 1993). Female rats are aggressive in defense of the young and burrow. Gestation lasts 21–24 days, with larger litters generally having shorter gestation periods. Stress may lengthen the gestation period, probably through embryonic diapause (Pritchett-Corning et al., 2013).

Prior to parturition, a female typically isolates herself in a burrow and builds a nest. Rats nurse in various positions, depending on security, experience, and demands of pups. Pups play a great deal once they are mobile at approximately 12–14 days of age, and their play consists of chasing, wrestling, and pouncing (Pellis & Pellis, 1997; Vanderschuren et al., 1997; see also other work by Pellis & Pellis), all while emitting chirps that some have characterized

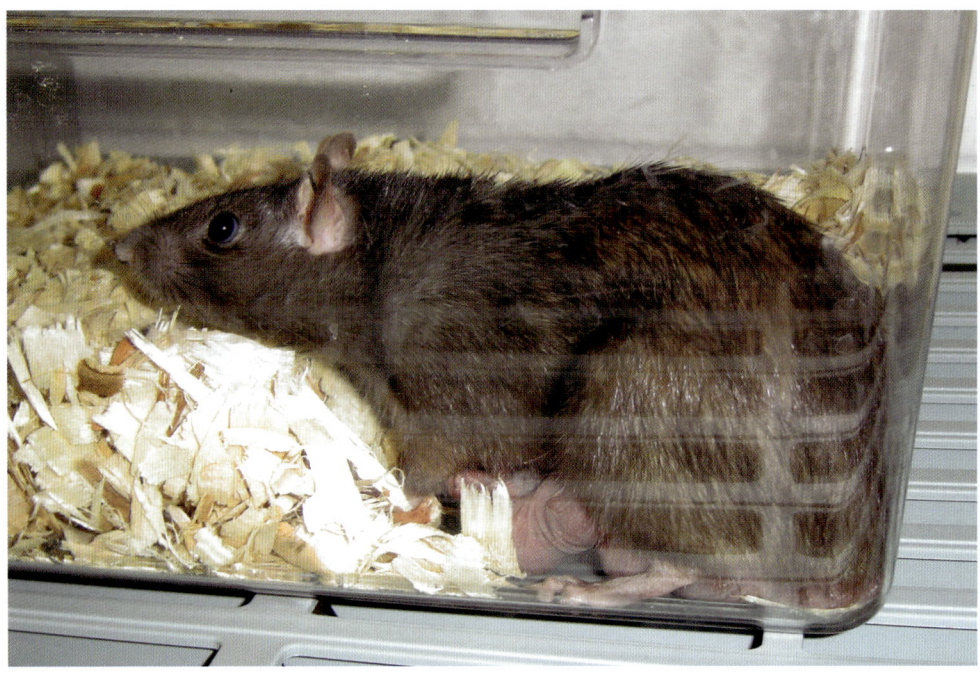

TWO OF THE MANY NURSING POSTURES ADOPTED BY RAT DAMS: (TOP) BROWN NORWAY RAT NURSING IN THE ACTIVE COVER POSITION, (BOTTOM) LONG EVANS RAT NURSING IN A RELAXED POSTURE

as laughter (Panksepp, 2007). Juvenile rats are weaned by the mother in a gradual process occurring between 3 and 4 weeks of age (Cramer et al., 1990) and reach sexual maturity between 4 and 7 weeks of age (Kennedy & Mitra, 1963). Reproduction may occur year round, although it declines substantially in the winter months (Andrews et al., 1972). Some laboratory rats are photoresponsive, indicating that

wild rats may also have reproductive and other somatic responses to shorter days (Heideman et al., 2000; Lorincz et al., 2001; Shoemaker & Heideman, 2002).

Rats are both nocturnal and crepuscular animals. Wild rats seen above ground during daylight hours are typically under duress, either social or environmental. The mainstay of their diet is plant material, although they are omnivorous and will kill and consume insects, birds, amphibians, and other mammals (Bandler & Moyer, 1970; Kemble et al., 1985). The early evening is the peak time for feeding and foraging but there is a second peak right before sunrise. In areas of poor resource availability, they may hoard food, but in urban environments, food hoarding is apparently rare (Takahashi & Lore, 1980). Rats will transport food from exposed sources to protected areas for consumption (Calhoun, 1963; Thompson, 1948). Rat feeding patterns involve tentative approaches and repeated small samples of unfamiliar items (Barnett, 1956) until they determine that the items are safe to

Comfortable Quarters for Laboratory Animals

consume. Rats urinate and defecate near safe food sites to signal other rats (Galef & Beck, 1985; Laland & Plotkin, 1991; Laland & Plotkin, 1993). Juveniles learn foraging skills and about feeding sites from nearby adults (Galef & Clark, 1971). Adult rats modify their food choices by smelling food on the fur, whiskers, and breath of other rats, but do not learn to avoid poisoned food that way, as the preference extends to food smelled on the breath of ill rats (Galef & Wigmore, 1983; Galef et al., 1983). All rats exhibit some neophobia when exposed to unfamiliar foods (Modlinska et al., 2015), although this is readily overcome by hunger.

Norway rats are strong swimmers (Galef, 1980) but poor climbers (when compared to black rats) (Foster et al., 2011). They often assume a temporary bipedal position, stabilized by their tails, to investigate changes in their environment. Rat gaits include a walk, a trot, and a gallop, gaits common to quadrupeds, although the gallop is rarely seen in the laboratory due to the lack of predators and cage-size constraints (Gillis & Biewener, 2001). Various postures adopted by rats during intraspecies interactions are well illustrated by Grant and Mackintosh (1963) and Barnett et al. (1982).

When not investigating their environment or foraging for food, rats spend a great deal of time grooming themselves (autogrooming) and each other (allogrooming; Bolles, 1960). Autogrooming typically occurs upon awakening, as a displacement activity or when anxious, and after eating (Komorowska & Pellis, 2004). Organized bouts of autogrooming proceed from rostral to caudal, beginning with forepaw wiping of the face and finishing with cleaning the tail (Sachs, 1988). Allogrooming is often directed from mother to pup, as well as from pup to pup; in adults, it can serve to reinforce social hierarchies as well as promote affiliative behavior (Pellis & Pellis, 1997).

To delve deeper into the natural behavior of the rat, the author recommends Calhoun (1963), an extensive examination of the behavior of rats in the wild as well as in a semi-natural enclosure. To view wild rats in a semi-natural enclosure in action, the film *The Laboratory Rat: A Natural History* is recommended (Berdoy, 2003). Another overview of the ethology of the wild rat and its laboratory counterparts may be found in Würbel et al. (2009).

Senses: As with other rodents, the primary sensory modalities of the rat differ from humans, which can make it difficult for caretakers to detect environmental issues that may be disturbing rats. Burn (2008) provides an excellent review of the rat's senses. The two primary sensory modalities of rats are olfaction and touch, with hearing and vision taking a less dominant role. Olfaction is one of the primary ways in which they gain sensory input. Rats use pheromones to communicate basic information, such as gender, health, and relatedness, as well as more complex emotional states, such as anxiety (Inagaki et al., 2014). In rats, sites of concentrated pheromonal emission include the perianal region (produced by the anal sacs and feces), the preputial or urinary papillar region (produced by the preputial or clitoral glands, as well as by urine itself), the face (produced by the sebaceous glands in the whisker pads), and the pads of the feet (mediated by the plantar glands). The action of pheromones is often mediated through the vomeronasal organ, which communicates directly with the amygdala, while a typical airborne odor is recognized by the olfactory cortex. In other words, pheromones act on emotions and responses "beneath" conscious thought. Olfaction contributes to the sensation of taste as well, and rats have the same complement of taste receptors that other rodents have: sweet, sour, umami, bitter, fat, and salt (Gilbertson & Khan, 2014; Ma et al., 2007).

Rat tactile sensory input is present throughout the body, but focused on the vibrissae. Vibrissae, also known as whiskers, are specialized hair cells with a large, blood-filled sinus and representation in the somatosensory cortex. Whiskers are primarily found on the head, although some may be found on the carpus as well. Rats have two types of facial vibrissae, the macrovibrissae, the large whiskers arranged in parallel rows on the snout, as well as the microvibrissae, which are found under the nostrils and around the lips (see photo, page 21). Rats explore their environment with active whisking movements of the vibrissae (Welker, 1964), as well as head movements that help them to determine the shape, size, and texture of objects in their environment (Hartmann, 2001; Hartmann, 2011). Whiskers on the face and feet also provide information on speed and foot placement when running (Niederschuh et al., 2015; Thé et al., 2013). Whisking is consciously controlled by the animal; it is not a reflexive response to obstacles in the environment (Berg & Kleinfeld, 2003). The amount of rat cortical space devoted to input from the whiskers is roughly equivalent to human cortical space devoted to hand and finger input. In addition to the whisker inputs, rats have touch-sensitive guard hairs that detect the presence of surfaces against the body and, like other mammals, have a subset of neurons that are sensitive to stroking (Vrontou et al., 2013).

Rats are dichromats with a rod-dominated retina, as is often found with nocturnal animals. Their rods have the typical mammalian sensitivity to light, while their two types of cone cells have peak sensitivities at 359 nm (UV light) and 510 nm ("green") (Jacobs et al., 2001). Practically, this means that rats cannot perceive "red" and behave as though red objects are opaque. Although facilities do not routinely expose their animals to UV light, both LED and fluorescent light act similarly on rats' circadian systems (Syrkin, 1999). Unlike laboratory mice, where retinal degeneration genes are relatively common, there is only one strain of rats known to be blind, the RCS rat (D'Cruz et al., 2000). Rat vision would be classed as "nearsighted" by humans (distant objects are blurry), but the severity varies by strain/stock (Prusky et al., 2002). Their degree of nearsightedness would render most markers hung on walls as navigation cues for behavioral tasks useless (Prusky et al., 2002). Rat vision is also sensitive to motion, with a sensitivity 2–3 times higher than that of humans (Douglas et al., 2006).

The hearing range of rats overlaps with that of humans, although rats can hear frequencies that humans cannot. Their hearing range, as defined by sounds audible at 60 dB, is from 500 Hz to 64 kHz, with the peak sensitivity (sounds detectible at 10 dB) at approximately 4–32 kHz (Heffner & Heffner, 2007). In comparison, human peak hearing sensitivity is from 250 Hz to 8 kHz. Many of the sounds made by rats are inaudible to humans without an ultrasonic frequency converter. In addition, noise in the ultrasonic range, rarely audible to humans, can cause stress to rats and disrupt communication.

Rats apparently have one other sense that humans may not have (or may not be able to easily access)—that of magnetoreception. Based on work with mice and other rodents, it is likely that rats have the ability to detect the Earth's magnetic field (Wiltschko & Wiltschko, 2005). This may help them to navigate or to orient their nests with the Earth's axis, the importance of which is unknown.

Comfortable Quarters for Laboratory Animals

Addressing the species-typical characteristics of rats in the research laboratory

Caging: The size of the basic home cage recommended differs slightly between the US and the EU and has changed through time.

Evolution of cage space requirements for rats through versions of the *Guide for the Care and Use of Laboratory Animals* (National Research Council, 2011)

Version of the Guide	Number/weight of rats (g)	Housing area per animal (cm²)	Height (cm)
1963	1–3/250 4–10/250	185.8–650.3/animal 185.8–464.5/animal	20.3
1972 and 1974	Up to 100 100–200 201–300 Over 300	110 148 187 258	17.8
1978 and 1980	<100 100–200 201–300 >300	110 148 187 258	17.8
1985 and 1996	<100 100–200 200–300 300–400 400–500 >500	109.68 148.40 187.11 258.08 387.12 451.64	17.8
2011	<100 100–200 201–300 301–400 401–500 >501 Mother and litter	109.6 148.35 187.05 258.0 387.0 >451.5 800	17.8

Although rats can successfully reproduce in much smaller cages than recommended (Gaskill & Pritchett-Corning, in press; Horn et al., 2012), this may be related to the fact that they are domesticated animals who have been selected for successful reproduction under various conditions and stressors. Rats show a strong preference for a larger cage, but desires for space are subservient to the desire to have conspecifics present (Patterson-Kane, 2002). Juvenile rats exhibit a great deal of active play behavior and benefit from access to more space. Basic caging should also be sized so the rat has the ability to express all natural postures. Cages are often too short for rats to fully extend vertically.

Cages with solid bottoms are preferred to those with wire floors (Manser et al., 1995; van de Weerd et al., 1996), although wire floors with resting platforms are seemingly well tolerated. When preference testing was used to determine the strength of the preference for solid-bottomed floors, results were inconclusive, as animals would work just as hard to access space to explore as to access solid-bottomed caging for resting (Manser et al., 1996). However, when animals were not asked to work for access, a clear preference for solid-bottomed caging readily emerged (van de Weerd et al., 1996). Rats exhibit signs of stress in cages with wire floors, whether large or small, if no enrichment is provided (Foulkes, 2004). One reason rats are still housed on wire-bottomed cages is to prevent coprophagy from interfering with certain types of scientific endeavor. Coprophagy may occur both through ingestion of feces found on the cage floor as well as directly from the anus (Ebino, 1993), so the utility of wire-bottomed cages in preventing all coprophagy is questionable. Transitions from solid-bottomed to wire-bottomed cages are likely to stress rats (Giral et al., 2011). Metabolism cages, with their wire floors, lack of cover, and social isolation are likely very stressful for rats (Gil et al., 1989)

Cage space recommendations for rats found in 2010/63/EU (Table 1.2 in Annex—unchanged from Table A.2 in 86/609/EEC)

	Body weight (g)	Minimum enclosure size (cm²)	Floor area per animal (cm²)	Number of animals that can be housed in minimum enclosure	Minimum enclosure height (cm)
In stock and during procedures	≤200 ≥201 to 300 ≥301 to 400 ≥401 to 600 ≥601	800 800 800 800 1,500	200 250 350 450 600	4 3 2 1 2	18
Breeding		800 Mother and litter. For each additional adult animal permanently added to the enclosure, add 400 cm²			18
Stock at breeders in 1,500 cm² cages	≤50 ≥51 to 100 ≥101 to 150 ≥151 to 200	1,500 1,500 1,500 1,500	100 125 150 175	15 12 10 8	18
Stock at breeders in 2,500 cm² cages	≤100 ≥101 to 150 ≥151 to 200	2,500 2,500 2,500	100 125 150	25 20 16	18

and data from experiments using metabolism cages should be interpreted through that filter. The observation of clinical signs in rats in toxicologic studies is not impaired by solid-bottom caging (Van Vleet et al., 2008); in fact, the only impairment in human observation of deliberately induced mild clinical signs in rats was found in wire-bottomed cages. Cage placement on a rack may also affect rat behavior and physiology (Cloutier & Newberry, 2010) and should be considered in experimental design and analysis.

Typical laboratory rat beddings include: wood shavings, wood chips, corncob processed to various diameters, cellulose, and wood pulp. Rats prefer wood-based bedding with a larger particle size (Blom et al., 1996; Ras et al., 2002). Aspen shavings were associated with a greater rate of lung pathology when compared to a cellulose-based bedding (Burn, Peters, Day, et al., 2006). Corncob bedding has been shown to affect rats' physiology with changes in estrous cyclicity associated with corn's natural estrogenic compounds, as well as disruption of slow-wave sleep (Leys et al., 2012; Markaverich et al., 2005). Being reared on corncob bedding has also been shown to reduce measures of anxiety in male rats (Sakhai et al., 2013). Cellulose-based bedding is well tolerated by rats but does not provide the absorption of some other types of bedding (Burn & Mason, 2005). Facility-wide bedding changes may be difficult to implement, since the choice of bedding is often dictated by cost or disposal concerns.

Cage cleaning affects rats by placing them in a new environment from which all pheromonal markers have been removed. Bind et al. (2013) provides a review of how lab procedures may disrupt pheromonal communication in rodents. An additional disruption is that this usually takes place during the day, when a nocturnal animal is resting (Abou-Ismail et al., 2008). If animals are kept on a reverse day-night schedule, this is of less concern. Schedules of cage cleaning are reliant on types of caging used, with frequencies varying from once every 2 weeks to three times per week for various types of solid-bottomed cages. Rat behavior is disrupted for about an hour after cage change (Burn, Peters & Mason, 2006; Duke et al., 2001; Saibaba et al., 1996), although this disruption may be related to novelty and handling rather than disruption of pheromones, as nonbreeding rats show no preference for scent-marked cages (Burn &

Mason, 2008b). Changing the cage of a rat close to parturition or with newborn pups may result in cannibalism (Burn & Mason, 2008a). Rats close to parturition and rats with new litters should be left undisturbed for as long as is feasible. It should also be noted that frequent cage changes may have an additional follow-on effect of accustoming rats to human contact and, thus, positively affecting handleability (Burn, Peters, Day, et al., 2006).

Enrichment: Before addressing recommendations, a distinction should be drawn between Enrichment (capital E) and standard enrichment. Enrichment is typically seen as part of neurobiology or psychology projects and usually involves very large cages, training or habituation to handling, multiple manipulanda, and a constant changing or refreshment of offered objects. In contrast, enrichment entails objects or interactions that should be readily provided in a standard home cage. The figure at above right shows a cage currently being used in an Enrichment study, while the figure at below right shows a standard enriched cage. Although advising widespread Enrichment would likely benefit rats in some ways, the practicalities of research make the appropriate use of rat-relevant standard enrichment more likely to benefit a greater number of rats overall (Abou-Ismail et al., 2010; Baumans et al., 2010; Patterson-Kane, 2010; Patterson-Kane, 2004). It is worth noting that spatially and socially enriched environments were once considered unreasonable for rabbits, but the shift of many institutions to larger pens and group housing has been relatively rapid. A similar shift in perspective and practice may occur with rats now that larger caging is more readily commercially available. Consideration of the relevance of the enrichment to rats is important; things humans find enriching, rats may not (Krohn et al., 2011).

Rats are social creatures and the most highly-valued enrichment is a compatible conspecific

TOP: HOUSING USED FOR ENRICHMENT STUDY (NOTE SIZE, MULTIPLE LEVELS, MANY MANIPULANDA, AND SOCIAL ASPECTS BOTTOM: STANDARD ENRICHMENT (TWO RATS, ONE TUBE, ONE BONE, AND LONG-FIBER NESTING MATERIAL)

(Patterson-Kane et al., 2002), although a physically enriched cage for a singly housed rat may be more beneficial than a barren cage with social partners (Abou-Ismail et al., 2014). If rats are housed in stable groups, removing animals results in signs of stress in the remainder (Burman et al., 2008), illustrating that interactions with cagemates are important. It is generally agreed that rats

Comfortable Quarters for Laboratory Animals

benefit from social interactions with other rats, and that housing rats singly is stressful, although some feel that definitive data that support the stress of single housing are lacking (Krohn et al., 2006). In some cases, physiologic or behavioral differences between singly and group-housed rats are difficult to interpret (Azar et al., 2011), but in other cases, they seem to support that singly housed animals are stressed (Kruegel et al., 2014). For example, adult male rats typically maintain a breeding territory, shared only with females and their offspring, so only subdominant male rats are found in groups in the wild. Which is more stressful, being the sole dominant animal in a territory or interacting with another animal to establish a dominance hierarchy? If it is stressful to be housed with another animal(s), is it eustress or distress, and does the eustress of the dominant animal outweigh the potential distress of subordinates (Abou-Ismail, 2011b; Hurst et al., 1996)? Regardless, social animals should be allowed the opportunity to socialize and caging should be sized appropriately for housing rats in groups of two or more. Rats with implants or other modifications that might make group housing dangerous should be given extra enrichment and the ability to hide, and may also benefit somewhat from limited contact—either visual or tactile—with conspecifics (Angermeier, 1960; Hurst et al., 1997; Hurst et al., 1998; Walton et al., 1972). Some investigators are group housing rats with head implants successfully and this should be attempted when possible (Schwarz et al., 2010).

Other changes to caging are possible and are being investigated by researchers. Rats may prefer opaque caging (Cloutier et al., 2010), but the necessity of daily animal health examinations have resulted in most institutions moving entirely to clear caging. Caging that appears opaque to the rat, such as red-tinted caging, may reduce stress in rats and this is being examined by researchers; although, as with any change from "normal," changes in "normal" physiologic values may occur (Dauchy et al., 2013). Multi-level caging may be another way of increasing welfare in both singly housed and breeding rats (Wheeler et al., 2015) since rats are motivated to climb onto objects (Williams et al., 2009). For breeding pairs of rats, dams will spend time away from their pups if this is made possible by cage configuration (Cramer et al., 1990). Providing a way for lactating females to temporarily

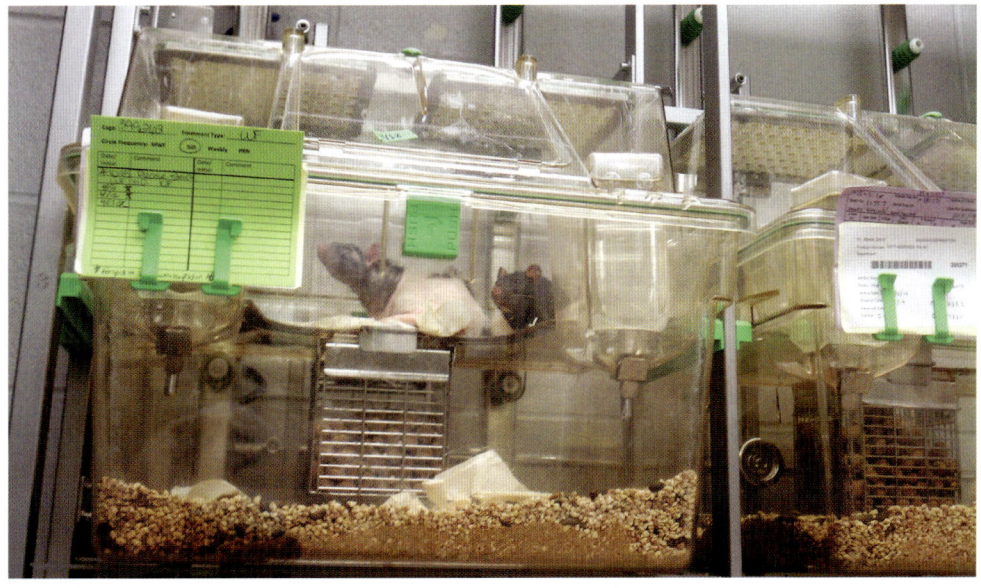

escape pups has proven to be beneficial in other species (Buob et al., 2013; Cloutier et al., 2013; Dawson et al., 2013).

Rats have been shown to have a preference for cages with increased interior complexity (Anzaldo et al., 1994). Also highly valued by rats is a source of cover such as a shelter, hut, or box. A nest box is valued more than nesting material (Manser et al., 1998b), but if both are offered, both will be used. The type of box preferred is an opaque, thermoplastic box fully enclosed on at least four sides, with a fifth side containing a small opening (Patterson-Kane, 2003). Any shelter provided will be used, however, with the rats both climbing on top of it (if vertical space allows) and going inside. Rats prefer long paper strips for nesting (Manser et al., 1998a; Ras et al., 2002) but will also nest with paper towels or facial tissue (Bradshaw et al., 1991; Van Loo et al., 2004). Female rats are motivated to seek out nesting material and build nests as they near parturition (Kinder, 1927; Price et al., 1977),

but the response of male rats to nesting material may vary by strain/stock (Jegstrup et al., 2005). Virgin rats may need to be exposed to nesting material as youngsters in order to use it effectively (Van Loo et al., 2004). Nesting material is not commonly used as standard enrichment; however it has been shown to improve rat physiology (Vitalo et al., 2009; Vitalo et al., 2012).

Some means of enriching rat cages have become standard, such as providing rats with gnawing items made of nylon, wood, or plastic (Abou-Ismail, 2011a). Although their incisors wear mainly on the occlusal surfaces, rats are motivated to gnaw and will chew objects placed in their cage for that purpose. They will also gnaw at shelters, food crocks, or other objects placed in their cages. Although rats rarely injure themselves on sharp edges they create, objects that have sharp edges from gnawing should be removed.
Other factors may need to be considered before implementing enrichments such as

foraging or running wheels. Providing rats with a foraging enrichment (food hidden under gravel in a metal dish) decreased aggression and allowed rats to perform species-specific feeding behaviors, but also increased rates of obesity (Johnson et al., 2004). Rats will spontaneously use running wheels if provided, and the frequency of use and effects on the rat and research will differ by sex, strain, and age (Novak et al., 2012). Standard housing results in sedentary rats with poorer performance on tests of agility and strength than rats housed in large pens (Spangenberg et al., 2005); running wheels may be one way to manage rats' metabolic abnormalities (Martin et al., 2010).

Refining husbandry and research procedures for rats

When considering the rat in research, it is important to acknowledge that vendors differ, transport differs, labs differ, housing differs, husbandry differs, and individual rats differ. Few of these variables can be completely controlled for, so it is important to recognize that all these aspects can affect research outcomes (Nevalainen, 2014). If thorough information is provided in supplemental materials and methods sections of published work, it may be easier to identify some of these effects so they may be examined in the future (Prager et al., 2011).

Relatively few rats are bred at institutions; most are purchased from vendors. This means that rats used for research arrive and must adjust to completely different housing types, social interactions, husbandry schedules, enrichment, and food, among other things. Recent work indicates that rats may need longer acclimation periods (up to 2 weeks) after transport than previously thought (Arts et al., 2014). Once they have arrived, animals may be identified through tail markings, tattoos, or microchips. Marking the tail with a permanent marker

was found to change behavior in rats (Burn et al., 2008). Behavioral changes associated with other methods of identification have not been studied, or the results have not been published.

Sometimes immediately upon arrival, and definitely after acclimation, rats undergo research procedures such as being handled, weighed, injected, or having blood sampled. Common procedures that occur in the laboratory or animal housing room such as cage changes and weighing stress rats, but returning rats to group housing decreases the effects of this stress (Sharp et al., 2003b; Sharp et al., 2002b). Watching most research-related procedures does not disturb rats, but observing (or more likely smelling) decapitation is stressful (Sharp et al., 2002a; Sharp et al., 2003a). Rats will react negatively to certain conspecific residues such as blood or muscle, while ignoring others such as brain (Stevens et al., 1977; Stevens et al., 1973). Alarm pheromones that researchers cannot smell and alarm calls that researchers cannot hear should be considered when performing techniques in close contact with other rats. Cleaning equipment such as behavioral apparatuses used by multiple rats should include both water and alcohol-based cleaners so that scent marks and pheromones are removed.

Despite all these sources of stress, researchers and husbandry staff can also help rats acclimate to the research environment and tolerate research-related procedures. Rats can be easily accustomed to human handling, especially when young (Maurer et al., 2008). If humans interact with rats in a way similar to the way young rats interact with each other while playing (called either tickling or playful handling, as opposed to stroking), rats will also be less fearful of humans compared to rats that were not handled (Cloutier et al., 2012). Handling by humans has been investigated as a means of reducing stress associated with common research procedures. Tickling is not necessarily a better reward when compared to food or stroking after intraperitoneal injection (Cloutier et al., 2008) but if rats are accustomed to playful handling, they show less aversion to repeated intraperitoneal injections (Cloutier et al., 2014). Human interaction may be considered enrichment for singly housed rats, and group-housed rats benefit also (Cloutier et al., 2013). Rats are readily trained using operant conditioning methods (numerous online videos show pet rats performing all sorts of feats) but training of rats to perform research-related tasks, as is undertaken with monkeys and dogs, is rarely attempted. For example, rats can learn to accept oral dosing of some compounds via syringe feeding rather than oral gavage (Atcha et al., 2010). Human handling may also decrease the effects of social isolation in rats, decrease anxiety, and improve learning skills (Costa et al., 2012; Pritchard et al., 2013).

Conclusion

During the process of domestication, we have selected and bred the rats who thrived and reproduced in the limited environment provided to them in captivity. Although laboratory rats are domesticated animals, they retain behaviors exhibited by their wild ancestors; working with, rather than against, those behavioral patterns is a good starting point. Humans must also realize that the way rats perceive their environment is foreign to the way we do, and account for this difference. The square centimeters of variance in size of typically available commercial caging is probably of little importance to rats since, in all cases, it is so much less than what would be available in the wild. Rats should have solid-bottomed cages with a wood-product bedding. If animals must be kept on wire flooring, resting platforms must be provided. Caging should be large enough to allow animals to fully extend their bodies vertically. Cages should also be large enough to allow rats of any size to be socially housed, and consideration should be given to the fact that rats will willingly tolerate less space for conspecific contact. Some retreat from human view should also be available, and rats have a clear preference for opaque, enclosed nest boxes. Nesting material will also be used by many rats, as will gnawing items, and provision of those will also enhance animal welfare. Finally, gentle, considerate, consistent handling by humans will significantly decrease stress on both sides and result in both better research subjects and better research results.

Acknowledgements

The author would like to acknowledge the staff of AWI, including Dr. Kenneth Litwak, and Dave Tilford for their edits and comments, and Dr. Emily Patterson-Kane for her review.

REFERENCES

Abou-Ismail UA 2011a Are the effects of enrichment due to the presence of multiple items or a particular item in the cages of laboratory rat? *Applied Animal Behaviour Science 134*(1–2), 72–82

Abou-Ismail UA 2011b The effects of cage enrichment on agonistic behaviour and dominance in male laboratory rats (*Rattus norvegicus*). *Research in Veterinary Science 90*(2), 346–351. doi: 10.1016/j.rvsc.2010.06.010

Abou-Ismail UA, Burman OH, Nicol CJ and Mendl M 2010 The effects of enhancing cage complexity on the behaviour and welfare of laboratory rats. *Behavioural Processes 85*(2), 172–180. doi: 10.1016/j.beproc.2010.07.002

Abou-Ismail UA, Burman OHP Nicol CJ and Mendl M 2008 Let sleeping rats lie: Does the timing of husbandry procedures affect laboratory rat behaviour, physiology and welfare? *Applied Animal Behaviour Science 111*(3–4), 329–341. doi: 10.1016/j.applanim.2007.06.019

Abou-Ismail UA, Darwish RA and Ramadan SG 2014 Should cages of laboratory rats be enriched physically or socially? *Global Veterinaria 13*(4), 570–582.

Andrews RV, Belknap RW, Southard J, Lorincz M and Hess S 1972 Physiological, demographic and pathological changes in wild Norway rat populations over an annual cycle. *Comparative Biochemistry and Physiology Part A: Physiology 41*(1), 149–165. doi: 10.1016/0300-9629(72)90043-6

Angermeier WF 1960 Some basic aspects of social reinforcements in albino rats. *Journal of Comparative and Physiological Psychology 53*(3), 364–367. doi: 10.1037/h0047435

Anzaldo AJ, Harrison PC, Riskowski GL, Sebek LA, Maghirang R, Stricklin WR and Gonyou HW 1994 Increasing welfare of laboratory rats with the help of spatially enhanced cages. *AWIC Newsletter 5*(3), 1–2, 5.

Arts JWM, Oosterhuis NR, Kramer K and Ohl F 2014 Effects of transfer from breeding to research facility on the welfare of rats. *Animals 4*, 721-728. doi:10.3390/ani4040712

Atcha Z, Rourke C, Neo AH, Goh CW, Lim JS, Aw CC, Pemberton DJ 2010 Alternative method of oral dosing for rats. *Journal of the American Association for Laboratory Animal Science 49*(3), 335-343.

Azar T, Sharp J and Lawson D 2011 Heart rates of male and female sprague-dawley and spontaneously hypertensive rats housed singly or in groups. *Journal of the American Association for Laboratory Animal Science 50*(2), 175–184.

Bandler R and Moyer KE 1970 Animals spontaneously attacked by rats. *Communications in Behavioral Biology 5*, 177–182.

Barnett S, Fox I and Hocking W 1982 Some social postures of five species of *Rattus. Australian Journal of Zoolog 30*(4), 581–601. doi: 10.1071/ZO9820581

Barnett SA 1956 Behaviour components in the feeding of wild and laboratory rats. *Behaviour 9*(1), 24–43

Baumans V, Van Loo PLP and Pham TM 2010 Standardisation of environmental enrichment for laboratory mice and rats: Utilisation, practicality and variation in experimental results. *Scandinavian Journal of Laboratory Animal Science 37*(2), 101–114

Ben-Ami Bartal I, Decety J and Mason P 2011 Empathy and pro-social behavior in rats. *Science 334*(6061), 1427–1430. doi: 10.1126/science.1210789

Berdoy M (producer) 2003 The Laboratory Rat: A Natural History [film]. Oxford University: Oxford, UK. http://www.ratlife.org

Berg RW and Kleinfeld D 2003 Rhythmic whisking by rat: Retraction as well as protraction of the vibrissae is under active muscular control. *Journal of Neurophysiology 89*(1), 104–117. doi: 10.1152/jn.00600.2002

Bind RH, Minney SM, Rosenfeld S and Hallock RM 2013 The role of pheromonal responses in rodent behavior: Future directions for the development of laboratory protocols. *Journal of the American Association for Laboratory Animal Science 52*(2), 124–129

Blom HJM, VanTintelen G, VanVorstenbosch C, Baumans V and Beynen AC 1996 Preferences of mice and rats for types of bedding material. *Laboratory Animals 30*(3), 234–244. doi: 10.1258/002367796780684890

Bolles RC 1960 Grooming behavior in the rat. *Journal of Comparative and Physiological Psychology 53*(3), 306–310. doi: 10.1037/h0045421

Bradshaw AL and Poling A 1991 Choice by rats for enriched versus standard home cages: Plastic pipes, wood platforms, wood chips, and paper towels as enrichment items. *Journal of the Experimental Analysis of Behavior 55*(2), 245–250. doi: 10.1901/jeab.1991.55-245

Buob M, Meagher R, Dawson L, Palme R, Haley D and Mason G 2013 Providing 'get-away bunks' and other enrichments to primiparous adult female mink improves their reproductive productivity. *Applied Animal Behaviour Science 147*(1–2), 194–204. doi: 10.1016/j.applanim.2013.05.004

Burman O, Owen D, Aboulsmail U and Mendl M 2008 Removing individual rats affects indicators of welfare in the remaining group members. *Physiology & Behavior 93*(1–2), 89–96. doi: 10.1016/j.physbeh.2007.08.001

Burn CC 2008 What is it like to be a rat? Rat sensory perception and its implications for experimental design and rat welfare. *Applied Animal Behaviour Science 112*, 1–32

Burn CC, Deacon RM and Mason GJ 2008 Marked for life? Effects of early cage-cleaning frequency, delivery batch, and identification tail-marking on rat anxiety profiles. *Developmental Psychobiology 50*(3), 266–277. doi: 10.1002/dev.20279

Burn CC and Mason GJ 2005 Absorbencies of six different rodent beddings: Commercially advertised absorbencies are potentially misleading. *Laboratory Animals 39*(1), 68–74. doi: 10.1258/0023677052886592

Burn CC and Mason GJ 2008a Effects of cage-cleaning frequency on laboratory rat reproduction, cannibalism, and welfare. *Applied Animal Behaviour Science 114*(1–2), 235–247. doi: 10.1016/j.applanim.2008.02.005

Burn CC and Mason GJ 2008b Rats seem indifferent between their own scent-marked homecages and clean cages. *Applied Animal Behaviour Science 115*(3–4), 201–210. doi: 10.1016/j.applanim.2008.06.002

Burn CC, Peters A, Day MJ and Mason GJ 2006 Long-term effects of cage-cleaning frequency and bedding type on laboratory rat health, welfare, and handleability: A cross-laboratory study. *Laboratory Animals 40*(4), 353–370. doi: 10.1258/002367706778476460

Burn CC, Peters A and Mason GJ 2006 Acute effects of cage cleaning at different frequencies on laboratory rat behaviour and welfare. *Animal Welfare 15*(2), 161–171

Calhoun JB 1963 *The Ecology and Sociology of the Norway Rat.* US Deptartment of Health, Education, and Welfare, Public Health Service: Bethesda, MD

Castle WE 1947 The domestication of the rat. *Proceedings of the National Academy of Sciences of the United States of America 33*(5), 109–117. doi: 10.1073/pnas.33.5.109

Cloutier S, Baker C, Wahl, K, Panksepp J and Newberry RC 2013 Playful handling as social enrichment for individually- and group-housed laboratory rats. *Applied Animal Behaviour Science 143*(2–4), 85–95. doi: 10.1016/j.applanim.2012.10.006

Cloutier S and Newberry RC 2008 Use of a conditioning technique to reduce stress associated with repeated intra-peritoneal injections in laboratory rats. *Applied Animal Behaviour Science 112*, 158–173

Cloutier S and Newberry RC 2010 Physiological and behavioural responses of laboratory rats housed at different tier levels and levels of visual contact with conspecifics and humans. *Applied Animal Behaviour Science 125*(1–2), 69–79. doi: 10.1016/j.applanim.2010.03.003

Cloutier S, Panksepp J and Newberry RC 2012 Playful handling by caretakers reduces fear of humans in the laboratory rat. *Applied Animal Behaviour Science 140*(3–4), 161–171. doi: 10.1016/j.applanim.2012.06.001

Cloutier S, Wahl K, Baker C and Newberry RC 2014 The social buffering effect of playful handling on responses to repeated intraperitoneal injections in laboratory rats. *Journal of the American Association for Laboratory Animal Science 53*(2), 168–173

Cooley RK and Vanderwolf CH 2005 *Stereotaxic Surgery in the Rat: A Photographic Series, Second Edition.* A J Kirby Co: London, Canada

Costa R, Tamascia ML, Nogueira MD, Casarini DE and Marcondes FK 2012 Handling of adolescent rats improves learning and memory and decreases anxiety. *Journal of the American Association for Laboratory Animal Science 51*(5), 548–553

Cramer CP, Thiels E and Alberts JR 1990 Weaning in rats: I. Maternal behavior. *Developmental Psychobiology 23*, 479–493

D'Cruz PM, Yasumura D, Weir J, Matthes MT, Abderrahim H, LaVail MM and Vollrath D 2000 Mutation of the receptor tyrosine kinase gene Mertk in the retinal dystrophic RCS rat. *Human Molecular Genetics 9*(4), 645–651. doi: 10.1093/Hmg/9.4.645

Dauchy RT, Wren MA, Dauchy EM, Hanifin JP, Jablonski MR, Warfield B, ... Blask DE 2013 Effect of spectral transmittance through red-tinted rodent cages on circadian metabolism and physiology in nude rats. *Journal of the American Association for Laboratory Animal Science 52*(6), 745–755

Davis DE, Emlen JT and Stokes AW 1948 Studies on home range in the brown rat. *Journal of Mammalogy 29*(3), 207–225. doi: 10.2307/1375387

Dawson, L, Buob, M, Haley D, Miller S, Stryker J, Quinton M and Mason G 2013 Providing elevated 'getaway bunks' to nursing mink dams improves their health and welfare. *Applied Animal Behaviour Science 147*(1–2), 224–234. doi: 10.1016/j.applanim.2013.04.001

Diamond ME and Arabzadeh E 2013 Whisker sensory system - From receptor to decision. *Progress in Neurobiology 103*, 28–40. doi: 10.1016/j.pneurobio.2012.05.013

Didion JP and de Villena FP 2013 Deconstructing Mus gemischus: Advances in understanding ancestry, structure, and variation in the genome of the laboratory mouse. *Mamm Genome 24*(1–2), 1–20. doi: 10.1007/s00335-012-9441-z

Donaldson HH 1912 The history and zoological position of the albino rat. *Journal of the Academy of Natural Sciences of Philadelphia 15*, 363–369

Douglas RM, Neve A, Quittenbaum JP, Alam NM and Prusky GT 2006 Perception of visual motion coherence by rats and mice. *Vision Research 46*(18), 2842–2847. doi: 10.1016/j.visres.2006.02.025

Duke JL, Zammit TG and Lawson DM 2001 The effects of routine cage-changing on cardiovascular and behavioral parameters in male Sprague-Dawley rats. *Contemporary Topics in Laboratory Animal Science 40*(1), 17–20

Ebino KY 1993 Studies on coprophagy in experimental animals. *Jikken Dobutsu 42*(1), 1–9

Foster S, King C, Patty B and Miller S 2011 Tree-climbing capabilities of Norway and ship rats. *New Zealand Journal of Zoology 38*(4), 285–296. doi: 10.1080/03014223.2011.599400

Foulkes A 2004 Do laboratory rats benefit from more cage space? *AWI Quarterly.* Retrieved from http://www.awionline.org/pubs/Quarterly/04-53-3/533p18.htm

Galef BG 1980 Diving for food: Analysis of a possible case of social learning in wild rats (*Rattus norvegicus*). *Journal of Comparative and Physiological Psychology 94*(3), 416

Galef BG and Beck M 1985 Aversive and attractive marking of toxic and safe foods by Norway rats. *Behavioral and Neural Biology 43*(3), 298–310. doi: 10.1016/s0163-1047(85)91645-0

Galef BG and Clark MM 1971 Social factors in poison avoidance and feeding behavior of wild and domesticated rat pups. *Journal of Comparative and Physiological Psychology 75*(3), 341–357. doi: 10.1037/h0030937

Galef BG and Wigmore SW 1983 Transfer of information concerning distant foods: A laboratory investigation of the 'information-center' hypothesis. *Animal Behaviour 31*(3), 748–758. doi: 10.1016/s0003-3472(83)80232-2

Galef BG, Wigmore SW and Kennett DJ 1983 A failure to find socially mediated taste-aversion learning in Norway rats (*R. norvegicus*). *Journal of Comparative Psychology 97*(4), 358–363. doi: 10.1037//0735-7036.97.4.358

Gaskill BN and Pritchett-Corning KR (in press) The effect of cage space on behavior and reproduction in Crl:CD(SD) and BN/Crl laboratory rats. *Journal of the American Association for Laboratory Animal Science*

Gilbertson TA and Khan NA 2014 Cell signaling mechanisms of oro-gustatory detection of dietary fat: Advances and challenges. *Progress in Lipid Research 53*(0), 82–92. doi: 10.1016/j.plipres.2013.11.001

Gill TJ, Smith GJ, Wissler RW and Kunz HW 1989 The Rat as an Experimental Animal. *Science 245*(4915), 269–276. doi: 10.1126/science.2665079

Gillis GB and Biewener AA 2001 Hindlimb muscle function in relation to speed and gait: In vivo patterns of strain and activation in a hip and knee extensor of the rat (*Rattus norvegicus*). *Journal of Experimental Biology 204*(15), 2717–2731

Giral M, Garcia-Olmo DC and Kramer K 2011 Effects of wire-bottom caging on heart rate, activity and body temperature in telemetry-implanted rats. *Laboratory Animals 45*(4), 247–253. doi: 10.1258/la.2011.010071

Grant EC and Mackintosh JH 1963 A comparison of the social postures of some common laboratory rodents. *Behaviour 21*, 246–259

Hartmann M 2001 Active sensing capabilities of the rat whisker system. *Autonomous Robots 11*(3), 249–254. doi: 10.1023/A:1012439023425

Hartmann MJ 2011 A night in the life of a rat: Vibrissal mechanics and tactile exploration. *Annals of the New York Academy of Sciences 1225*, 110–118. doi: 10.1111/j.1749-6632.2011.06007.x

Hedrich HJ 2000 History, strains and models. In: Krinke GJ (ed) *The Laboratory Rat, First Edition* pp 3–16. Academic Press: San Diego, CA

Hedrich HJ 2006 Taxonomy and stocks and strains. In: Suckow MA, Weisbroth SH and Franklin CL (eds), *The Laboratory Rat, Second Edition*. pp 71–92. Academic Press: Burlington, MA

Heffner HE and Heffner RS 2007 Hearing ranges of laboratory animals. *Journal of the American Association for Laboratory Animal Science 46*(1), 20–22

Heideman PD, Bierl CK and Galvez ME 2000 Inhibition of reproductive maturation and somatic growth of Fischer 344 rats by photoperiods shorter than L14:D10 and by gradually decreasing photoperiod. *Biology of Reproduction 63*(5), 1525–1530

Horn MJ, Hudson SV, Bostrom LA and Cooper DM 2012 Effects of cage density, sanitation frequency, and bedding type on animal wellbeing and health and cage environment in mice and rats. *Journal of the American Association for Laboratory Animal Science 51*(6), 781–788

Hurst JL, Barnard CJ, Hare R, Wheeldon EB and West CD 1996 Housing and welfare in laboratory rats: Time-budgeting and pathophysiology in single-sex groups. *Animal Behaviour 52*(2), 335–360

Hurst JL, Barnard CJ, Nevison CM and West CD 1997 Housing and welfare in laboratory rats: Welfare implications of isolation and social contact among caged males. *Animal Welfare 6*, 329–347

Hurst JL, Barnard CJ, Nevison CM and West CD 1998 Housing and welfare in laboratory rats: The welfare implications of social isolation and social contact among females. *Animal Welfare 7*, 121–136

Inagaki H, Kiyokawa Y, Tamogami S, Watanabe H, Takeuchi Y and Mori Y 2014 Identification of a pheromone that increases anxiety in rats. *Proceedings of the National Academy of Sciences 111*(52), 18751–18756. doi: 10.1073/pnas.1414710112

Jacobs GH, Fenwick JA and Williams GA 2001 Cone-based vision of rats for ultraviolet and visible lights. *Journal of Experimental Biology 204*(14), 2439–2446

Jegstrup IM, Vestergaard R, Vach W and Ritskes-Hoitinga M 2005 Nest-building behaviour in male rats from three inbred strains: BN/HsdCpb, BDIX/OrlIco and LEW/Mol. *Animal Welfare 14*(2), 149–156

Johnson SR, Patterson-Kane EG and Niel L 2004 Foraging enrichment for laboratory rats. *Animal Welfare 13*(3), 305–312

Kemble ED, Flannelly KJ, Salley H and Blanchard RJ 1985 Mouse killing, insect predation, and conspecific attack by rats with differing prior aggressive experience. *Physiol Behav 34*(4), 645–648

Kennedy GC and Mitra J 1963 Body weight and food intake as initiating factors for puberty in the rat. *The Journal of Physiology 166*(2), 408–418. doi: 10.1113/jphysiol.1963.sp007112

Kinder EF 1927 A study of the nest-building activity of the albino rat. *Journal of Experimental Zoology 47*(2), 117–161. doi: 10.1002/jez.1400470202

Komorowska J and Pellis SM 2004 Regulatory mechanisms underlying novelty-induced grooming in the laboratory rat. *Behavioural Processes 67*(2), 287–293. doi: 10.1016/j.beproc.2004.05.001

Krinke GJ (ed) 2000 *The Laboratory Rat, First Edition*. Academic Press: San Diego, CA

Krohn TC, Salling B and Hansen AK 2011 How do rats respond to playing radio in the animal facility? *Laboratory Animals 45*(3), 141–144. doi: 10.1258/la.2011.010067

Krohn TC, Sorensen DB, Ottesen JL and Hansen AK 2006 The effects of individual housing on mice and rats: A review. *Animal Welfare 15*(4), 343–352

Kruegel U, Fischer J, Bauer K, Sack U and Himmerich H 2014 The impact of social isolation on immunological parameters in rats. *Archives of Toxicology 88*(3), 853–855. doi: 10.1007/s00204-014-1203-0

Laland KN and Plotkin HC. 1991 Excretory deposits surrounding food sites facilitate social-learning of food preferences in norway rats. *Animal Behaviour 41*, 997–1005. doi: 10.1016/S0003-3472(05)80638-4

Laland KN and Plotkin HC 1993 Social transmission of food preferences among Norway rats by marking of food sites and by gustatory contact. *Animal Learning & Behavior 21*(1), 35–41. doi: 10.3758/Bf03197974

Leys LJ, McGaraughty S and Radek RJ 2012 Rats housed on corncob bedding show less slow wave sleep. *Journal of the American Association for Laboratory Animal Science 51*(6), 764–768

Lindsey JR and Baker HJ 2006 Historical Foundations. In Suckow MA, Weisbroth SH and Franklin CL (eds), *The Laboratory Rat, Second Edition*, pp 1–52. Academic Press: Burlington, MA

Lorincz AM, Shoemaker MB and Heideman PD 2001 Genetic variation in photoperiodism among naturally photoperiodic rat strains. *Am J Physiol Regul Integr Comp Physiol 281*(6), R1817–1824

Ma H, Yang R, Thomas SM and Kinnamon JC 2007 Qualitative and quantitative differences between taste buds of the rat and mouse. *BMC Neuroscience 8*(1), 1–13. doi: 10.1186/1471-2202-8-5

Manser CE, Broom DM, Overend P and Morris TH 1998a Investigations into the preferences of laboratory rats for nest-boxes and nesting materials. *Laboratory Animals 32*(1), 23–35. doi: 10.1258/002367798780559365

Manser CE, Broom DM, Overend P and Morris TH 1998b Operant studies to determine the strength of preference in laboratory rats for nest-boxes and nesting materials. *Laboratory Animals 32*(1), 36–41. doi: 10.1258/002367798780559473

Manser CE, Elliott H, Morris TH and Broom DM 1996 The use of a novel operant test to determine the strength of preference for flooring in laboratory rats. *Laboratory Animals 30*(1), 1–6. doi: 10.1258/002367796780744974

Manser CE, Morris TH and Broom DM 1995 An investigation into the effects of solid or grid cage flooring on the welfare of laboratory rats. *Laboratory Animals 29*(4), 353–363. doi: 10.1258/002367795780740023

Markaverich BM, Crowley JR, Alejandro MA, Shoulars K, Casajuna N, Mani S, ... Sharp J 2005 Leukotoxin diols from ground corncob bedding disrupt estrous cyclicity in rats and stimulate MCF-7 breast cancer cell proliferation. *Environmental Health Perspectives 113*(12), 1698–1704

Martin B, Ji S, Maudsley S and Mattson MP 2010 "Control" laboratory rodents are metabolically morbid: Why it matters. *Proceedings of the National Academy of Sciences of the United States of America 107*(14), 6127–6133. doi: 10.1073/pnas.0912955107

Matthews I 1898 *Full Revelations of a Professional Rat-Catcher After 25 Years' Experience.* The Friendly Society Printing Company, Manchester, UK

Maurer BM, Doring D, Scheipl F, Kuchenhoff H and Erhard MH 2008 Effects of a gentling programme on the behaviour of laboratory rats towards humans. *Applied Animal Behaviour Science 114*(3–4), 554–571. doi: 10.1016/j.applanim.2008.04.013

Modlinska K, Stryjek R and Pisula W 2015 Food neophobia in wild and laboratory rats (multi-strain comparison). *Behavioural Processes 113*(0), 41–50. doi: 10.1016/j.beproc.2014.12.005

National Research Council 2011 *Guide for the Care and Use of Laboratory Animals, Eighth Edition.* The National Academies Press: Washington, DC

Nevalainen T 2014 Animal husbandry and experimental design. *ILAR Journal 55*(3), 392–398. doi: 10.1093/ilar/ilu035

Niederschuh SJ, Witte H and Schmidt M 2015 The role of vibrissal sensing in forelimb position control during travelling locomotion in the rat (*Rattus norvegicus*, Rodentia) *Zoology 118*(1), 51–62. doi: 10.1016/j.zool.2014.09.003

Novak CM, Burghardt PR and Levine JA 2012 The use of a running wheel to measure activity in rodents: Relationship to energy balance, general activity, and reward. *Neuroscience & Biobehavioral Reviews 36*(3), 1001–1014. doi: 10.1016/j.neubiorev.2011.12.012

Panksepp J 2007 Neuroevolutionary sources of laughter and social joy: Modeling primal human laughter in laboratory rats. *Behavioural Brain Research 182*(2), 231–244

Patterson-Kane E 2010 Thinking outside our cages. *Journal of Applied Animal Welfare Science 13*(1), 96–99. doi: 10.1080/10888700903372283

Patterson-Kane EG 2002 Cage size preference in rats in the laboratory. *Journal of Applied Animal Welfare Science 5*(1), 63–72

Patterson-Kane EG 2003 Shelter enrichment for rats. *Contemporary Topics in Laboratory Animal Science 42*(2), 46–48

Patterson-Kane EG 2004 Enrichment of laboratory caging for rats: A review. *Animal Welfare 13*, 209–214

Patterson-Kane EG, Hunt M and Harper D 2002 Rats demand social contact. *Animal Welfare 11*, 327–332

Pellis SM and Pellis VC 1997 The prejuvenile onset of play fighting in laboratory rats (*Rattus norvegicus*) *Developmental Psychobiology 31*(3), 193–205

Prager EM, Bergstrom HC, Grunberg NE and Johnson LR 2011 The importance of reporting housing and husbandry in rat research. *Frontiers in Behavioral Neuroscience 5*. doi: 10.3389/fnbeh.2011.00038

Prescott TJ, Mitchinson B and Grant RA 2011 Vibrissal behavior and function. *Scholarpedia 6*. doi: 10.4249/scholarpedia.6642

Price EO and Belanger PL 1977 Maternal behavior of wild and domestic stocks of Norway rats. *Behavioral Biology 20*(1), 60–69. doi: 10.1016/S0091-6773(77)90511-9

Pritchard LM, Van Kempen TA and Zimmerberg B 2013 Behavioral effects of repeated handling differ in rats reared in social isolation and environmental enrichment. *Neuroscience Letters 536*, 47–51. doi: 10.1016/j.neulet.2012.12.048

Pritchett-Corning KR, Clifford CB and Festing MF 2013 The effects of shipping on early pregnancy in laboratory rats. *Birth Defects Research Part B: Developmental and Reproductive Toxicology 98*(2), 200–205. doi: 10.1002/bdrb.21056

Prusky GT, Harker KT, Douglas RM and Whishaw IQ 2002 Variation in visual acuity within pigmented, and between pigmented and albino rat strains. *Behavioural Brain Research 136*(2), 339–348. doi: S0166432802001262

Ras T, van de Ven M, Patterson-Kane EG and Nelson K 2002 Rats' preferences for corn versus wood-based bedding and nesting materials. *Laboratory Animals 36*(4), 420–425. doi: 10.1258/002367702320389080

Sachs BD 1988 The development of grooming and its expression in adult animals. *Annals of the New York Academy of Sciences 525*(1), 1–17. doi: 10.1111/j.1749-6632.1988.tb38591.x

Saibaba P, Sales GD, Stodulski G and Hau J 1996 Behaviour of rats in their home cages: Daytime variations and effects of routine husbandry procedures analysed by time sampling techniques. *Laboratory Animals 30*(1), 13–21. doi: 10.1258/002367796780744875

Sakhai SA, Preslik J and Francis DD 2013 Influence of housing variables on the development of stress-sensitive behaviors in the rat. *Physiology & Behavior 120*, 156–163. doi: 10.1016/j.physbeh.2013.08.003

Schultz LA and Lore RK 1993 Communal reproductive success in rats (*Rattus norvegicus*): Effects of group composition and prior social experience. *Journal of Comparative Psychology 107*(2), 216–222

Schwarz C, Hentschke H, Butovas S, Haiss F, Stüttgen MC, Gerdjikov TV, ... Waiblinger C 2010 The head-fixed behaving rat—Procedures and pitfalls. *Somatosensory & Motor Research 27*(4), 131–148. doi: 10.3109/08990220.2010.513111

Sharp J, Zammit T, Azar T and Lawson D 2002a Does witnessing experimental procedures produce stress in male rats? *Contemporary Topics in Laboratory Animal Science 41*(5), 8–12

Sharp J, Zammit T, Azar T and Lawson D 2002b Stress-like responses to common procedures in male rats housed alone or with other rats. *Contemporary Topics in Laboratory Animal Science 41*(4), 8–14

Sharp J, Zammit T, Azar T and Lawson D 2003a Are "by-stander" female Sprague-Dawley rats affected by experimental procedures? *Contemporary Topics in Laboratory Animal Science 42*(1), 19–27

Sharp J, Zammit T, Azar T and Lawson D 2003b Stress-like responses to common procedures in individually and group-housed female rats. *Contemporary Topics in Laboratory Animal Science 42*(1), 9–18

Sharp P and Villano J 2013 *The Laboratory Rat, Second Edition* [Laboratory Animal Pocket Reference Series]. CRC Press: Boca Raton, FL

Shoemaker MB and Heideman PD 2002 Reduced body mass, food intake, and testis size in response to short photoperiod in adult F344 rats. *BMC Physiology 2*, 11

Spangenberg EM, Augustsson H, Dahlborn K, Essén-Gustavsson B and Cvek K 2005 Housing-related activity in rats: Effects on body weight, urinary corticosterone levels, muscle properties and performance. *Laboratory Animals 39*, 45–57

Stevens DA and Gerzogthomas DA 1977 Fright Reactions in Rats to Conspecific Tissue. *Physiology & Behavior 18*(1), 47–51. doi: 10.1016/0031-9384(77)90092-0

Stevens DA and Saplikos NJ 1973 Rats reactions to conspecific muscle and blood: Evidence for an alarm substance. *Behavioral Biology 8*(1), 75–82. doi: 10.1016/S0091-6773(73)80008-2

Suckow MA, Weisbroth SH and Franklin CL (eds) 2006 *The Laboratory Rat, Second Edition.* Academic Press: New York, NY

Sullivan R 2005 *Rats: Observations on the history and habitat of the city's most unwanted inhabitants.* Bloomsbury USA: New York, NY

Syrkin NJ 1999 *LED & fluorescent light have similar effects on the circadian system of the rat* [thesis]. San Jose State University: San Jose, CA

Takahashi LK and Lore RK 1980 Foraging and food hoarding of wild *Rattus norvegicus* in an urban environment. *Behavioral and Neural Biology 29*(4), 527–531. doi: 10.1016/S0163-1047(80)92863-0

Taylor KD 1978 Range of movement and activity of common rats (*Rattus norvegicus*) on agricultural land. *Journal of Applied Ecology 15*(3), 663–677. doi: 10.2307/2402767

Thé L, Wallace ML, Chen CH, Chorev E and Brecht M 2013 Structure, function, and cortical representation of the rat submandibular whisker trident. *The Journal of Neuroscience 33*(11), 4815–4824. doi: 10.1523/jneurosci.4770-12.2013

Thompson HV 1948 Studies of the behaviour of the common brown rat (*Rattus norvegicus* Berkenhout.) I. Watching marked rats taking plain and poison bait. *Bulletin of Animal Behaviour 6*, 26–40

van de Weerd HA, van den Broek FAR and Baumans F 1996 Preference for different types of flooring in two rat strains. *Applied Animal Behaviour Science 46*(3–4), 251–261. doi: 10.1016/0168-1591(95)00654-0

Van Loo PL and Baumans V 2004 The importance of learning young: The use of nesting material in laboratory rats. *Laboratory Animals 38*(1), 17–24. doi: 10.1258/00236770460734353

Van Vleet TR, Rhodes JW, Waites CR, Schilling BE, Nelson DR and Jackson TA 2008 Comparison of technicians' ability to detect clinical signs in rats housed in wire-bottom versus solid-bottom cages with bedding. *Journal of the American Association for Laboratory Animal Science 47*(2), 71–75

Vanderschuren LJ, Niesink RJ and Van Pee JM 1997 The neurobiology of social play behavior in rats. *Neuroscience & Biobehavioral Reviews 21*(3), 309–326

Vitalo A, Fricchione J, Casali M, Berdichevsky Y, Hoge EA, Rauch SL, … Levine JB 2009 Nest making and oxytocin comparably promote wound healing in isolation reared rats. *PLoS One 4*(5), e5523. doi: 10.1371/journal.pone.0005523

Vitalo AG, Gorantla S, Fricchione JG, Scichilone JM, Camacho J, Niemi SM, … Levine JB 2012 Environmental enrichment with nesting material accelerates wound healing in isolation-reared rats. *Behavioural Brain Research 226*(2), 606–612. doi: 10.1016/j.bbr.2011.09.038

Vrontou S, Wong AM, Rau KK, Koerber HR and Anderson DJ 2013 Genetic identification of C fibres that detect massage-like stroking of hairy skin in vivo. *Nature 493*(7434), 669–673. doi: 10.1038/nature11810

Walton D and Latané B 1972 Visual vs physical social deprivation and affiliation in rats. *Psychonomic Science 26*(1), 4–6. doi: 10.3758/BF03337865

Waynforth HB and Flecknell PA 1992 *Experimental and Surgical Technique in the Rat, Second Edition.* Academic Press: London, UK

Welker WI 1964 Analysis of sniffing of the albino rat. *Behaviour 22*(3), 223–244. doi: 10.1163/156853964X00030

Wheeler RR, Swan MP and Hickman DL 2015 Effect of multilevel laboratory rat caging system on the well-being of the singly-housed Sprague Dawley rat. *Laboratory Animals 49*(1), 10–19. doi: 10.1177/0023677214547404

Williams CM, Hanmer LA and Riddell PM 2009 The effect of the functional attributes of objects within the caged environment on interaction time in laboratory rats. *Applied Animal Behaviour Science 120*(3–4), 208–215. doi: 10.1016/j.applanim.2009.06.004

Wiltschko W and Wiltschko R 2005 Magnetic orientation and magnetoreception in birds and other animals. *Journal of Comparative Physiology A 191*(8), 675–693. doi: 10.1007/s00359-005-0627-7

Wishaw IQ and Kolb B 2004 *The Behavior of the Laboratory Rat: A Handbook With Tests.* Oxford University Press: Oxford, UK

Würbel H, Burn C and Latham N 2009 *The behaviour of laboratory mice and rats.* In: Jensen P (ed) *Ethology of Domestic Animals: An Introductory Text, Second Edition,* pp 217–233. CAB International: Wallingford, UK

Guinea Pigs

Guinea Pigs

Marcie Donnelly, BS, LATg, SRA

BRISTOL-MYERS SQUIBB

Originally from South America, guinea pigs are a diurnal crepuscular species, being active in early morning and evening with intermittent periods of rest, activity, and nibbling of food during the day and night. Grass is the natural diet of guinea pigs. In the Andes, their natural habitat, they live in herds or small groups of 5 to 10 animals and exhibit a definitive social hierarchy with a dominant male and female (Berryman, 1976). They are very alert for predators and frequently seek shelter in the burrows of other animals, as well as in crevices and tunnels formed by vegetation. Guinea pigs typically live an average of 4–5 years, but may live as long as 8 years.

They are members of the rodent suborder *Hystricomorpha*, characterized by their

relatively long gestation periods, the precocious state of development of their young at birth, and the membrane covering the vaginal orifice except during estrus and parturition (Weir & Rowlands, 1974). The exact time when guinea pigs were domesticated is unknown. Through domestication, guinea pigs have become less aggressive, exhibit increased social tolerance, and are less attentive to their surrounding environment than their wild counterparts (Berryman, 1976). Scent marking with urine or secretions from the perineal and supracaudal glands rubbed on the substratum reflect the animals' social status and social roles within the group. Strange individuals are identified by the absence of group characteristic scents (Reinhardt, 1971). Although they do not groom each other, they do seek out bodily contact during times of rest.

Domesticated guinea pigs are nonaggressive, docile animals, and with frequent gentle handling and petting are extremely responsive to attention. They will get to know their human caretakers and readily respond by whistling when such caretakers enter the room. Guinea pigs frequently lick human caretakers, which is often seen as a sign of affection and acceptance (Berryman, 1976).

Comfortable Quarters for Laboratory Animals

The young are born after a relatively long gestation period of about 66 days. Unlike other rodents, guinea pigs do not build nests, as the young are born precocial. They look like small adults and begin to consume solid food the day they are born. Young guinea pigs are very active, often enjoying games of running and jumping alone or with peers, or "popcorning" as it's often called. This period of development is very brief. Females (known as sows) can successfully breed as early as 3 weeks old and give birth at the age of 3 months. The young sow is best mated at approximately 2.5 to 3 months of age. Breeding should always occur before the age of 6 months for females, as after that age, the pubic symphysis becomes more rigid, causing issues with parturition.

Males (known as boars) also engage in sexual courtship activities when they are just 3 weeks old. When the little males start courting females, they inevitably become targets of the dominant boar, who will persistently chase them. Young males will gradually become sexually and behaviorally inhibited unless they are removed from the group (Reinhardt, 1971).

The guinea pig is an extremely social species, and bonding has been shown to be very important. Both males and females placed in challenging situations show lower cortisol levels when supported by a familiar conspecific (Kaiser et al., 2003).

Females rarely engage in fighting. They have little to fight over as they neither hoard nor compete for food. Females are so tolerant of each other that they may even nurse each other's young (Reinhardt, 1971). The mothers seem to treat all newborns equally and the young will suckle off any available female, although once nursing has started, the mother will butt away any other infants that approach her. The mothers set the nursing timetable and when ready to nurse they will pace back and forth attracting the infants. Nursing ends

abruptly after approximately 10 minutes, as the nursing mother will walk away. The lactation period ends after just 3 weeks. Although females do not appear to develop a bond with their offspring, the presence of suckling young causes them to become aggressive toward strange females (Reinhardt, 1971). It is best to introduce new females to a group when there are no lactating females or suckling young present and they do not have the scent of another group on them.

Males fight viciously in the presence of females in estrus. Generally one boar takes the role as the dominant male who then monopolizes the females. This dominance will result in all other males in the group acting more like females and even cause them to stop emitting male-typical pheromones; this prevents fights between them. The dominant male will even display courtship behaviors, rather than aggression toward these males (Reinhardt, 1971). To prevent frustration and stress from being under constant inhibition from the dominant male, it is best to separate subordinate males from the group and form bachelor groups or other harems with these males.

When housed in solid-bottom pens or cages, guinea pigs do well within a temperature range of 16–24°C (61–75°F); their preferred temperature is 20°C (68°F). High temperatures of 32°C (90°F) should be avoided, as this species does not dissipate heat well and is subject to hyperthermia and heat stroke (Canadian Council on Animal Care, 1984).

Being prey animals, guinea pigs easily panic when an unfamiliar or unseen person comes into their room. Cages or pens with open sides of metal wire are recommended so that the animals have good visual contact with their environment. They will not panic when familiar caretakers enter their room, if they are able to see them. Theses enclosure

types also provide the caretakers with easy observation of the animals and better ventilation. Being heavy rodents, weighing close to 1 kg, solid-bottom caging is highly preferred to help prevent pressure sores and pododermatitis that can develop if housed on wire bottom cages (Fullerton & Gilliatt, 1967).

Adult guinea pigs measure up to approximately 30 cm in length and require at least 3 cm additional horizontal space to allow for free expression of the stretching posture. They are poor jumpers and diggers but greatly enjoy burrowing in hay. The hay provided should be soft to avoid eye injuries.

Vocalization plays an important role in the social and sexual behavior of guinea pigs. The animals have quite a repertoire of sounds; one will always hear lots of purring, squeaking, chirping, whistling, or teeth chattering in a guinea pig room. A favorite caretaker is greeted with a noisy welcome, especially if she or he brings produce or treats! Like most other rodents, though, guinea pigs are susceptible to noise stress; sudden loud noises and other stressful sounds should be minimized (Anthony & Harclerode, 1959).

"Rodents appear to prefer sheltered areas of the cage, especially if those areas have decreased light and height. Providing such a confined space within a cage might be one way to enrich the environment of rodents" (National Research Council, 1996). Guinea pigs are very easily startled. A protected, safe refuge is a basic necessity to buffer stress in guinea pigs and assure that the data collected are not compromised by stress-related factors. Guinea pigs tend to keep close to the outer cage or pen walls, as they instinctively avoid open surfaces that would expose them to potential predators (White et al., 1989).

Addressing guinea pig-specific characteristics in the research institution

When it comes to enrichment, guinea pigs present a special challenge. They do not welcome changes and react negatively toward new food types, feeders, and water containers. Enrichment may be met with skepticism or even fear. However, guinea pigs do seem to enjoy having their common furnishing moved around to different places of the cage. Positive reactions to such changes include normal play behavior and excitement when caretakers enter, comfortable appearance, and exploration of the "new" or different cage space. Conversely, if guinea pigs do become stressed by new items or the rearranging of their furnishings, they may stop eating.

Guinea pigs do well on a commercial pelleted diet supplemented daily with hay and fresh produce. Acceptable food enrichment such as hay, hay cubes, and dried corn on the cob all allow the animals to graze throughout the day, keeping them busy and offering variety. "When good quality hay is supplied the consumption of the more expensive pelleted diet is reduced and, by their vocalization when they realize that the hay is about to be replenished, the animals clearly indicate the great pleasure they obtain from eating it and burrowing in it" (Sutherland & Festing, 1987). A daily supply of hay and other preferred fresh produce is very important, as guinea pigs forage continuously and may develop habits such as chewing and eating their own hair (trichophagia) if fresh or dry grass is not available (Sutherland & Festing, 1987; Gerold et al., 1997).

When exposed to enrichment items from a young age, guinea pigs enjoy tasting everything and generally welcome things to chew on, such as treat sticks, wood blocks, or wood sticks. Guinea pigs' front teeth continue to grow throughout their lifetime, so chewing on hard items is essential. Hard pelleted diets, as well as wooden sticks or blocks, help to prevent overgrowth of their front teeth. Fresh produce is a welcome treat, and may include greens such as kale or romaine, carrots, apple, strawberries, or other such fruits and vegetables. Regular distribution of these food items help to foster a positive human-animal relationship. It is good practice to maintain a consistent

standard food selection, as guinea pigs can be rather fickle eaters who may stop eating and starve rather than accept any new food stuff.

Guinea pigs need a social environment to maintain physiological and behavioral well-being (Sachser & Lick, 1991; Fenske, 1992). Being housed with other conspecifics also helps them cope with living in confinement (Olfert et al., 1993). Compatible group- or pair-housing should be standard practice in a research laboratory. Group-housed animals should be provided a floor space of no less than 750 cm^2 for weaned, nonbreeding guinea pigs and no less than 1,200 cm^2 for breeding females. A shelter should be provided as a refuge for the animals. Such areas serve as a comfortable sleeping area or an area to give birth. Places of refuge not only provide a safe haven but also increase the usable floor space when placed in the middle of the cage where animals typically would not go. A large box with a sliding door provides an excellent way to capture an entire group for cage cleaning; the box can simply be lifted out of the cage or pen. If the box is equipped with a removable top, this can allow easy handling of a single animal for a procedure or veterinary observation (Gray, 1988).

Guinea pigs must never be kept isolated. If the research protocol requires single-housed animals, they should always have visual, auditory, and olfactory contact with their own kind (Fenske, 1992; Olfert et al., 1993). Floor space for a research protocol requiring single-housed animals should consist of an area of at least 35 x 70 cm (2,450 cm^2) so that an adult animal can stretch and turn around freely, and a refuge box can be provided. If a medical event arises that requires temporary single housing of an animal, the minimum space needed is 35 x 35 cm (1,225 cm^2) to provide normal free movement of an adult

animal and locomotor play behaviors of a young animal.

With the exception of short-term experimental protocols, guinea pigs should be kept on solid-bottom caging with bedding (e.g., National Research Council, 1996). "When grid or perforated floors are used, a solid resting area must be provided" (Council of Europe, 2006) that is large enough to allow all animals to lie on it simultaneously. To maintain a hygienic cage environment, bedding should be dust-free, seasoned soft wood and changed at least twice per week. Minimum environmental and feeding enrichment should include the bedding, a hide box, and fresh, high-quality hay given daily.

Social tensions often arise from keeping several mature males together or from overcrowding; to minimize this, one mature male should be kept with three to six females and their young. At the weaning age of about 3 weeks, the young guinea pigs should be removed and kept in same-sex groups. Adolescents housed in same-sex groups do well together, provided no females are kept within visual or olfactory contact with groups of males. Exposure to the smell of female urine will turn even the most compatible males into fractious enemies who will no longer tolerate each other (Reinhardt, 1971). A mature male can be removed from a group and replaced by another male with no problem. The females will accept him with no aggression. Strange females can be introduced to a new group without causing turmoil as long as there are no nursing females present (Raje & Stewart, 2000). Thus, it is recommended that new females be introduced to a group only when young are no longer present. Individual animals can be returned to their group without overt aggression, provided they have not been scent marked by another conspecific from a different group.

Uniform lighting should be provided for all animals, and is a fundamental condition of scientifically valid research methodology (American Medical Association, 1992), assuring that no more than the minimum numbers of guinea pigs are used to obtain statistically significant research results. Multi-tier caging systems should be avoided, as the top tier casts shadows on the bottom tiers, making it impossible to assure that the lighting provides "uniformly distributed illumination" (United States Department of Agriculture, 1995; Bellhorn, 1980; Clough, 1982).

Regular distribution of food treats such as hay, fresh produce, and yogurt drops, as well as gentle handling, help guinea pigs overcome their fear of personnel. Guinea pigs should be handled "as expeditiously and carefully as possible in a manner that does not cause trauma, overheating, ... behavioral stress, ... or unnecessary discomfort" (United States Department of Agriculture, 1995). The animals should be handled gently with both hands, one firmly around the shoulder and the other supporting the hindquarters. Nervous or impatient investigators can startle and distress guinea pigs, rendering research data collected from such animals highly suspect.

Proper handling depends on the investigator rather than on the subject. "Animal care staff are expected, at all times, to have a caring and respectful attitude towards animals in their care, and to be proficient in their handling" (Council of Europe, 2000). "Unless the contrary is established, investigators should consider that procedures that cause pain or distress in human beings may cause pain or distress in other animals" (Interagency Research Animal Committee, 1996). "All who care for or use animals in research, teaching, or testing must assure responsibility for their well being. ... A good management program provides the environment, housing, and care that ... minimizes variations that can affect research results" (National Research Council, 1996) and hence, reduces the number of research subjects needed to achieve statistically significant results.

REFERENCES

American Medical Association 1992 *Use of Animals in Biomedical Research: The Challenge and Response, An American Medical Association White Paper*. American Medical Association: Chicago, IL

Anthony A and Harclerode JE 1959 Noise stress in laboratory rodents. II: Effects of chronic noise exposure on sexual performance and reproductive function of guinea pigs. *Journal of the Acoustical Society of America 31*: 1437–1440

Bellhorn RW 1980 Lighting in the animal environment. *Laboratory Animal Science 30*: 440–450

Berryman JC 1976 Guinea-pig vocalizations: Their structure, causation and function. *Zeitschrift für Tierpsychologie 41*: 80–106. http://dx.doi. org/10.1111/j.1439-0310.1976.tb00471.x

Canadian Council on Animal Care 1984 *Guide to the Care and Use of Experimental Animals, Volume 2*. Canadian Council on Animal Care: Ottawa, Canada. http://www.ccac.ca/en_/standards/guidelines/additional/vol2_guinea_pigs

Clough G 1982 Environmental effects on animals used in biomedical research. *Biological Reviews 57*: 487–523

Council of Europe 2006 *Appendix A of the European Convention for the Protection of Vertebrate Animals Used for Experimental and Other Scientific Purposes (ETS No. 123) enacted June 15, 2007*. Council of Europe: Strasbourg, France. http://conventions.coe.int/Treaty/EN/Treaties/PDF/123-Arev.pdf

Fenske M 1992 Body weight and water intake of guinea pigs: influence of single caging and an unfamiliar new room. *Journal of Experimental Animal Science 35*: 71–79

Fullerton PM and Gilliatt RW 1967 Pressure neurophathy in the hind foot of the guinea pig. *Journal of Neurology, Neurosurgery and Psychiatry 30*: 18–25

Gerold S, Huisinga E, Iglauer F, Kurzawa A, Morankic A and Reimers S 1997 Influence of feeding hay on the alopecia of breeding guinea pigs. *Zentralblatt für Veterinärmedizin 44*: 341–348

Gray G 1988 Guinea pigs. *Humane Innovations and Alternatives in Animal Experimentation 2*: 48–49. http://www.awionline.org/lab_animals/biblio/hiaa-88.html

Interagency Research Animal Committee 1996 U.S. Government principles for the utilization and care of vertebrate animals used in testing, research, and training. In: National Research Council *Guide for the Care and Use of Laboratory Animals, Seventh Edition* pp 117–118. National Academy Press: Washington, DC

Kaiser S, Kirtzeck M, Hornschuh G and Sachser N 2003 Sex specific difference in social support: A study in female guinea pigs. *Physiology and Behavior 79*: 297–303

National Research Council 1996 *Laboratory Animal Management: Rodents*. National Academy Press: Washington, DC

Olfert ED, Cross BM and McWilliam AA (eds) 1993 *Guide to the Care and Use of Experimental Animals, Volume 1, Second Edition*. Canadian Council on Animal Care: Ottawa, Canada http://www.ccac.ca/Documents/Standards/Guidelines/Experimental_Animals_Vol1.pdf

Raje SS and Stewart KL 2000 Group housing female guinea pigs. *Lab Animal 29*(8): 31–32. http://www.awionline.org/lab_animals/biblio/la29-8gp.html

Reinhardt V 1971 *Soziale Verhaltensweisen und soziale Rollen des Hausmeerschweinchens* [Social behavior and social roles of guinea pigs]. Dissertationsdruck Novotny: Starnberg, Germany

Sachser N and Lick C 1991 Social experience, behavior and stress in guinea pigs. *Physiology and Behavior 50*: 83–90

Sutherland SD and Festing MFW 1987 The guinea-pig. In: Poole TB (ed) *The UFAW Handbook on the Care and Management of Laboratory Animals, Sixth Edition* pp 393–410. Churchill Livingstone: New York, NY

United States Department of Agriculture 1995 *Code of Federal Regulations, Title 9, Chapter 1, Subchapter A: Animal Welfare*. US Government Printing Office: Washington, DC

Weir BJ and Rowlands IW 1974 Functional anatomy of the hystricomorph ovary. *Symposia of the Zoological Society of London 34*: 303–332

White WJ, Balk MW and Lang CM 1989 Use of cage space by guinea pigs. *Laboratory Animals 23*: 208–214

Hamsters

Hamsters

Michele Cunneen, BA, LATg
ANIMAL RESEARCH CONSULTING LLC

Hamsters represent a tiny fraction of the overall number of animals used in research. Most that are used are Syrian (or "golden") hamsters (*Mesocricetus auratus*). They are descended almost entirely from a single female and 12 offspring, captured in Syria in the 1930s, with only a few additional individuals added to the lineage in 1965 and 1971 (van Hoosier & McPherson, 1987; Laber-Laird et al., 1996). The other common research hamster, the Armenian hamster (*Cricetulus migratorius*) was introduced as a research animal in 1963 as part of the USSR-USA cultural exchange program. Less commonly used species include the Chinese hamster (*Cricetulus griseus*), brought to Harvard in 1948 (Gad, 2007), the European hamster (*Cricetus cricetus*), the Siberian dwarf hamster (*Phodopus sungorus*), and the Russian dwarf hamster (*Phodopus campbelli*).

In the wild, hamsters inhabit a wide variety of climate zones and environments, from hot deserts to more temperate European zones to frigid areas of Russia. (EU Wildlife and Sustainable Farming Project, 2009). Hamsters maintain a constant environmental temperature by digging deeper burrows in the extreme climates. Burrows are almost always dug and lived in by a single animal, with no overlap between burrows.

Except during estrus, female hamsters are very aggressive and will attack other animals entering their burrow. Conflict is avoided using urine and scent marking at tunnel entrances, delineating the females' receptiveness.

All hamsters share certain characteristics. Their solitary burrows contain sleeping, pantry, and bathroom areas. Typically, hamsters are active at night and can cover more than 6 miles as they search for grains, grasses, and insects (Willows Veterinary Centre and Referral Service, n.d.). Some studies of specific wild hamster populations suggest a larger diurnal component to their lifestyle (Gattermann, 2008). Regardless of time of day, behaviors outside the burrow are focused on locating food, storing it in their distensible cheek pouches, and returning the food to the burrow. In hard times more food is collected (Day et al., 1999; Garretson & Bartness, 2014). If food is plentiful, the hamster may return will only high value items in their cheek pouches to deposit in their pantry (Garretson & Bartness, 2014).

Although considered nocturnal, hamsters eat throughout the day, awakening to a state of semi-arousal to eat nearby food (Anderson & Shettleworth, 1977; Hoosier

& McPherson, 1987). This behavior can be disrupted in the research setting if a traditional food hopper (wire lid or basket) is used. This type of food storage requires the hamster to fully awaken and stand to eat, presenting a source of stress. Items on the cage floor or in a J-type hopper will allow a hamster to more easily maintain a sleeping position and not require full waking.

The carrying and manipulation of food are so ingrained, that they should be considered imperative activities in hamsters. Hamsters in the research setting, when deprived of this ability, show signs of stress and demonstrate more hoarding behavior when provided access to food (Anderson & Shettleworth, 1977).

In their natural habitat, hamsters continuously renovate the size and shape of their burrows, as they dig out new bathroom, sleeping, and eating areas. Feces are typically brought out of the burrow and deposited on the ground. Within the nesting and sleeping areas, the contents are widely varied and can include grass, their own fur and that of other animals, feathers, paper waste, bark, etc.

These species-typical behaviors bring unique challenges to maintaining hamsters in a research facility, particularly when they are housed in ventilated rack units. This technology can make providing species-appropriate housing quite difficult. The following sections will examine each component of the cage, offering suggestions to make for better housing for hamsters. While it is recognized that institutions may not be able to make all changes, many are very simple and can provide major dividends in improved enrichment and housing for the hamsters.

Cage construction and physical size

The eighth edition of the *Guide for the Care and Use of Laboratory Animals* (*Guide*) specifies that a hamster weighing more than 100 grams should be provided with a cage space of 19 inches square and a minimum height of 6 inches (National Research Council, 2011). Most ventilated mouse and rat cages are compliant with these suggestions, although ultra-high density mouse cages may be less than 6 inches in height.

The standard ventilated mouse cage, while technically compliant, is not an ideal habitat for hamsters. After deep bedding is added for burrowing, there may not be adequate head room. Further, small cages have been linked to chronic stress in the hamster (Kuhnen, 1999). If any structures are added to the cage, the floor space can become quite limited, particularly if the caging includes a wire-basket food hopper. If a ventilated mouse cage is the only option, studies have shown that hamsters prefer large quantities of nesting material, which they can shape into a nest and burrow-like structure (Hauzenberger et al., 2006). When provided with sufficient nesting material, the bedding depth appears to be less important. Examples of ample nesting material include large handfuls of shredded paper, several paper towels plus hay and straw, and a cardboard (chewable) shack plus additional nesting material. One study suggested at least 15–25 grams worth of material per hamster (Richards, 1969).

The larger, ventilated rat cages provide much more appropriate housing for hamsters. Some of these are 8 inches tall and will accommodate a larger shelter in the cage without it interfering with food and water dispensers. They also leave room for a thicker layer of bedding and possibly a J-type feeder instead of a wire-basket feeder. Since rat cages are at least twice the size of mouse cages (~150 vs 75 in²), they provide space for clear separation of bathroom, pantry and sleeping areas, even with multiple animals in a cage.

The rat cages can also accommodate a running wheel. Since wild hamsters can range more than 6 miles per day and many studies show them to be prodigious wheel runners, a wheel can provide highly valuable enrichment for them.

Shoebox-style, non-ventilated guinea pig cages offer even more room, typically having over 200 square inches of floor space and a height of 9 or 9.5 inches. These cages offer a variety of better features for the hamster. A guinea pig food hopper is usually a J-type feeder, which allows hamsters to carry away food. The extended height offers space for a wheel. It also allows for deeper bedding and more nesting material without incurring additional risk of flooding. Structures could have two levels and the hamster could use the roof or top for additional useable space.

Other larger rodent or rabbit cages may also work well, as they will accommodate more environmental variety. Some may require modification, as hamsters caging should be solid bottomed, with no gaps that might entrap a foot. The sides should also be solid to prevent climbing and escape. These cages may be tall enough for placement of tube-style pet systems (e.g., Habitrail OVO brand), with multiple levels and designated play, eating, and rest areas.

Feeder type and location

Conflicting opinions exist regarding feeding methods for laboratory hamsters. The current wording in the *Guide* states "feeders should be designed and placed to allow easy access to food and to minimize contamination with urine and feces" (National Research Council, 2011). At the same time, there are many citations in the literature that state that provision of food on the floor is essential (Harkness et al., 2010; Phillips, 1966). For example, Charles River, Inc., the largest breeder of the Syrian hamster for

research, states in their laboratory animal guide, "Hoarding is an important behavior in hamsters. If food is only provided via food hopper, then the ability to collect and hoard it in a store is lost. ... Providing food inside the cage, such as on the cage floor, allows the animal to display this natural hoarding behavior" (Winnicker, 2012). Hamsters whose cache is removed will lose weight and hoard more when loose food is offered again, suggesting that maintenance of body condition is dependent on the presence of the food cache (Phillips, 1966). Another study notes that dams with litters should receive their food directly on the cage floor to prevent preoccupation with collecting food, at the expense of the litter (Harkness et al., 2010). It is also important to remember that hamsters are coprophagic and will eat feces directly from the anus throughout the day (Harkness et al., 2010).

A viable compromise can be found with a J-shaped feeder or a food bowl placed on the bottom of the cage. Both offer the hamster the opportunity to carry entire pellets to make a pantry area and hoard food. Since this behavior can result in emptied feeders, but not require additional food, the veterinarian, animal care staff, and IACUC should develop an appropriate standard to monitor the food quality and quantity in hamster cages.

If the J-feeder or food bowl is not an option at a given facility, wire-bar hoppers that are rat or hamster width (7/16 in.) are also commonly used with hamsters. These wide spaces do allow them to bite off small pieces of food, which can be carried around. When combined with placement of food on the cage floor, this arrangement can provide hamsters with adequate conditions.

Feeders that are externally mounted, so no dust or parts of pellets can fall into the cage, are not suitable for hamsters. These

Comfortable Quarters for Laboratory Animals

feeders deprive the animal of all tactile stimuli when feeding and force head-extended gnawing, maximizing stress.

Shelters, houses and tunnels (shreddable and reuseable)

An appropriate shelter or material to construct a shelter is one of the most important components of the hamster cage. Unlike mice, hamsters prefer shelters above familiar, old (dirty) bedding (Veillette & Reebs, 2010). If only a single shelter is used in a cage, it must be large enough for all hamsters to fit inside. When placing the shelter in the cage, one should ensure it does not contact the water source (to prevent flooding) and that it does not prevent free movement through the cage. If a solid shelter is provided, it should ideally have multiple levels and surfaces.

As there are many shelter types available, the first decision is whether to deploy a reusable or single-use shelter. A reusable shelter has the advantage of not needing frequent replacement, and can be tinted to allow viewing without disturbing the animals. Shelters manufactured specifically for the research environment are typically autoclavable, while shelters for the pet market are typically only dishwasher-safe. PVC tubing can be used, although it is best to consider the potential effects of PVC on a research project before using them.

Shreddable and consumable houses, shelters, or tubes have the basic advantage of not needing to be washed. The shelter is only handled once and can provide additional stress relief to hamsters simply by being chewable. Multiple single-use shelters are available from research vendors, including shacks, homes, domes, refuges, mazes, huts, and tubes made of cardboard with certificates of analysis. They vary in size, shape and number of openings, and allow the hamster to control multiple facets of the environment.

The shelters are light enough to be moved around by the cage occupants. They can use it to block light, block drafts or wind, or as a step, if food or water is high up.

A key feature of single-use shelters is that the hamster is able to customize them. Some hamsters will chew additional openings into the shelter, while others may tip them over completely and build a nest on top of the shelter. Some may completely chew the shelter and create a new nest. In this respect, the single-use shelter can become an important part of any enrichment protocol.

Nesting material

Hamsters are prodigious nest builders. As such, the appropriate volume and type of material is a very important factor in the housing of hamsters. Within many research facilities, compressed pulp squares (1 in^2) are commonly used for mouse nesting material. These squares can be used for hamsters; however, at least three to four squares are necessary for each hamster. Also, the compressed nature of the squares may confuse some hamsters, so the square may need to be manually pulled apart by the caretaker.

Loose strand paper fibers offer a more dynamic nesting opportunity. This paper is more biologically relevant in its shape and characteristics. It facilitates complex nest building as it consists of long strands of loose material. Hamsters tend to carry it around in their cheek pouches and make the nest in a preferred location. They will transport it into shelters. Its crinkly characteristics make nests with walls and roofs that can easily cover the entire animal. While these materials may be preferred by the hamster, the loose nature can present a challenge when using automated bedding dispensers. Many vendors offer options to dispense these products in quantifiable amounts.

Another option is a rodent nesting sheet, which resembles tissues made of wood pulp. These sheets are 4 x 8 in., making it possible to create nests from large pieces. The sheets can also be used in most automated bedding dispensers. Other options include fleece strips, cloth squares, and paper pads. Although these are typically much sturdier than the other products, they may not be suitable for proper nest creation and burrowing. As such, they should be avoided.

In the noncertified market there are many common items that can be used. Paper towels allow for large pieces and customization. The common lab wipe makes a lightweight nesting material, again offering a large size. Both of these can be autoclaved if required. Bench paper comes in large rolls, allowing for customized strips; however, the plastic layer needs to be peeled off the back. Shredded newsprint is a viable option, but there may be unexpected effects associated with the ink.

Common sense is probably the best guide for providing nesting materials to hamsters. The nesting material should be clean, sturdy, and of sufficient volume to allow the animals to cover themselves to modify light levels, control ambient temperatures, and provide security. In group housing, the provision of enough material, presumably, will also decrease aggression by enabling more than one nest to exist.

Bedding

As hamsters are burrowing animals, the depth and type of bedding is an important consideration for their well-being. One study (Hauzenberger et al., 2006) examined stress responses in hamsters using three different bedding depths (10 cm, 40 cm, and 80 cm). The results suggested that shallower bedding levels are associated with indicators of increased

stress, such as more frequent wire-gnawing and lower body condition scores. Given that the typical bedding depth in a research facility is less than 1 cm, these findings suggest that lack of burrowing capability may be a stressor for most hamsters in research. Further, the bedding types commonly used in research (shavings, corncob, and compressed paper bits) are not conducive to creation of a burrow.

Since it usually is not practical to consider bedding depths of even 10 cm, the goal should be to provide enough bedding to enable the hamsters to "bulldoze" through the bedding (approximately 3–5 cm). Concurrent to the increased amount of bedding should be the provision of adequate nesting material, as described above.

Gnawing

The provision of hard, chewable objects provides an excellent option for hamsters to display species-typical behavior (beyond nesting and hoarding). Like most rodents, hamster teeth continually grow and must be worn down by regularly gnawing on hard items in the environment. Failure to provide opportunities for wearing down teeth can result in overgrown teeth or malocclusion, both significant health and welfare issues.

Chances to gnaw can be loosely divided into non-nutritive and nutritive opportunities. Non-nutritive options are typically favored by researchers, as these options are less likely to cause shifts in body weight, body fat, or biochemistry. Hard nylon chewing products (bones, chews, and pucks) are likely the best option for gnawing. They can be purchased with certified components, if required by the study, and come in flavored and unflavored varieties. Rat-sized products are an appropriate size for hamsters. Softer polyurethane bones and chewing toys are also available with certified components, but will not provide the same teeth-wearing

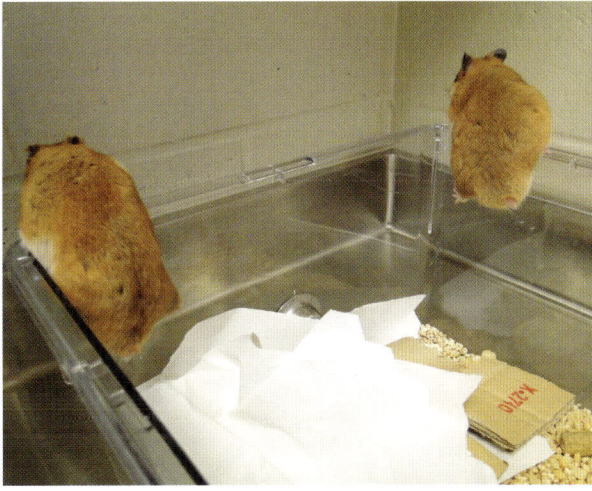

capabilities as the harder varieties. Wood chews (i.e., pre-cut blocks, manzanita sticks, etc.) are an excellent option and can also provide additional climbing opportunities. Care should be taken to ensure the wood will not interfere with the research objectives. Also, with larger branches, it is important to ensure that the sticks don't inadvertently provide a means of escape from the cage. Myriad other options are available through pet supply stores. All new toys should be carefully tested with just one or two hamsters before purchasing large quantities.

Since most non-nutritive gnawing options will last for many days (possibly over a cage change period) protocols should be developed, in conjunction with the veterinarian and IACUC, to determine when to discard the item and how to clean and sanitize it.

Virtually an endless variety of options exist to provide nutrition and satisfy the hamster's need to gnaw. Historically, these options have been avoided by researchers, due to concerns over additional calories, unaccounted trace minerals, and nonstandard components. One strategy is to use the standard diet. Scattering a few pellets on the cage floor when the cage is changed can offer hamsters a viable enrichment option. A similar strategy can be employed by putting a few pieces of dog or monkey chow in the cage. While the ingredients of these items will typically not affect the nutritional status of the hamsters, the veterinarian, IACUC, and researcher should be consulted before trying food items not specifically designed for rodents. In all cases, food pellets offer an opportunity to both hoard and gnaw food.

High-fiber food items (i.e., high-fiber rabbit food, alfalfa, timothy cubes, etc.) can offer an enrichment option, but should not be considered a gnawing alternative. The same can be said for seeds, granolas, dehydrated fruits and vegetables, and other treats. All are excellent food enrichment items and should have minimal effects on the animals or studies, when provided in moderation, but, again, should not be considered a gnawing alternative.

Running wheels

The hamster's natural foraging behaviors cause them to have a strong urge to travel. In captivity, this urge is best addressed with a running wheel.

When choosing a running wheel, it is important to use the correct size and adjust the caging to accommodate both the height and footprint of the wheel. The running wheel should be large enough for the hamster to stand on the bottom with just a small elevation in the back feet. If the wheel is too small, it will cause the hamster's back to be unnaturally flexed.

Comfortable quarters for the hamster

What does a hamster want? As discussed above, there are many ways to easily enrich a hamster's environment. Some items do appear to be more valuable than others. Hamsters prize and will work for

nesting material (Jansen et al., 1969). The nesting material that produces the best nest-building potential is more prized than low-potential material. As such, paper strips, large shavings, or sheet material that offer increased size options or structure potential are the best choices. Shelters are always used when large enough. Shelters that allow the hamster to modify them (i.e., cardboard or shaped pulp) are preferable. While hamsters appear to favor very deep bedding, this is not a realistic option in most facilities. Even so, the bedding provided should at least be sufficient to enable the animals to demonstrate burrowing behaviors. Food should be provided in such a way to facilitate hoarding, carrying, and gnawing. The hamster's desire to gnaw should be addressed with items other than food, such as hard nylon chewing products. All of these options can be accomplished within any research setting.

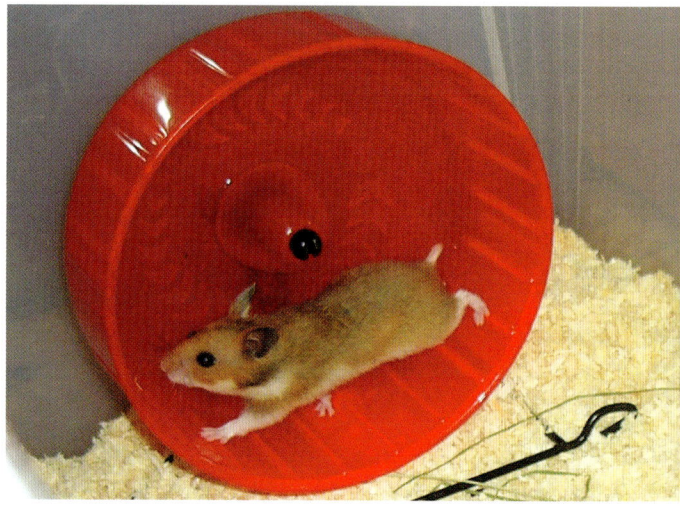

Hamsters want to move, they are agile climbers and wheel runners. As such, a running wheel should be provided whenever possible. If there is concern about the use of the wheel, the cages can be monitored with an infrared camera at night to determine how much the wheel is used.

It should be the goal of all facilities with hamsters to provide them with the best possible environment. Determining that environment can be difficult. One approach is to use a scoring system to benchmark progress toward the ideal environment. Shown below is one example of a scoring system. For each possible area; caging, provision of food, burrowing, bedding type, nesting material, forage/food scattering, running wheel, manipulanda, and mobility options, the current situation can be ranked to get a baseline and help determine opportunities for improvement. The system allows for some fixed items, such as cage size, while still leaving room to improve overall housing by focusing on other items, such as shelter, nesting, and food.

Caging

Cage size	Number animals/cage	Score
Cages 6–7 in. tall, floor area under 80 in.^{2}	2	1
Cages 6–7 in. tall, floor area 140–243 in.^{2}	5-10	1
Cages 8–9 in. tall, floor area 140–257 in.^{2}	8-10 3-7	2 2.5
Cages 9–9.5 in. tall, floor area 200 in.^{2}	6-9 3-5	2 2.5
Cages $\geq$ 9.5 in. tall, floor area $\geq$ 576 in.^{2†}	7-20 1-6	3 4

Provision of standard diet

Type	Score
Wire-bar food hopper, with no food scattered	0
Wire-bar food hopper, food scattered at cage change	2
Food hopper IVC type, low hanging, $\geq$ 7/16 in. spacing	1.5
Food hopper IVC type, low hanging, $\geq$ 7/16 in. spacing, plus scatter food on floor at cage change	2.5
J-type feeder inside cage or bowl in cage	3

Burrowing potential

Depth	Score
Bedding depth < 0.5 in.	0
Bedding depth 0.5–1 in.	1
Bedding depth 1–3 in.	2
Bedding depth > 3 in.	3

Bedding

Type	Score
Paper bedding pellet type	1
Wood pellets	1
Corn cob 1/8 in.	1
Corn cob 1/4 in.	1.5
Wood or paper chip	1.5
Mixed bedding and nesting material	2
Shavings, large	3

Shelter

Type	Score
No shelter	0
Reusable, plastic shelter	2
Disposable/chewable cardboard shelter	3
Wood, grass, or hay shelter	4
More than one shelter provided	1 point plus shelter-type points

Forage/food scattering

Type/frequency	Score
No food scattered	0
Food from standard diet scattered at cage change	1
Single nonstandard food type scattered at cage change	2
Rotating two or more nonstandard food types scattered at cage change	2.5
Scatter feeding occurs more often than just at cage changing	3

Nesting material

Type/amount	Score
Nesting material added is one nest square or less than 15 g	0
Nesting material is more than 15 g, but in chip form	1
Nesting material is long grained, or uses big sheets, or large shaving	3
Nesting material is long-strand hay	4
More than one type of nesting material is offered	5

Running wheel

Presence of Wheel	Score
Running wheel not present	0
Running wheel present	3

Manipulada/gnawing

Type	Score
Soft nylon bones, gummy bones for gnawing	1
Hard wood, manzanita, fruit wood gnaw blocks or sticks; hard nylon bones	2

Mobility

Device added	Score
No climbing or two-story options	0
Branches larger than 3 in. or tunnels flat in cage	1
Ladders, parrot toys hanging in cage	2
Branches, tunnels, other furniture, allowing two-story activity	3

A score of 9 should be considered the minimum score for an adequate environment. A score of 15 or higher would be an example of a best practice for housing hamsters.

This cage is scoring 16.5

Caging	2
Provision of standard diet	2
Burrowing potential	0
Bedding	1.5
Shelter	4
Forage/food scattering	2
Nesting material	3
Running wheel	0
Manipulanda/gnawing	2
Mobility	0
Total Score	16.5

REFERENCES

Anderson MC and Shettleworth ST 1977 Behavior adaptation to fixed-interval and fixed-time food delivery in golden hamster. *Journal of the Experimental Analysis of Behavior 25*: 33–49

Beaulieu A and Reebs SG 2009 Effects of bedding material and running wheel surface on paw wounds in male and female Syrian hamsters. *Laboratory Animals 43*: 85–90

Day D, Mintz E and Bartness T 1999 Diet self-selection and food hoarding after food deprivation by Siberian hamsters. *Physiology & Behavior 68*: 187–194

EU Wildlife and Sustainable Farming Project 2009 *Hamster, Cricetus cricetus factsheet* [pdf document]. Retrieved from http://ec.europa.eu/environment/nature/natura2000/management/docs/Cricetus%20cricetus%20factsheet%20-%20SWIFI.pdf

Gad SC (ed) 2007 *Animal Models in Toxicology, Second Edition*. CRC Press: Boca Raton, FL

Garretson JT and Bartness TJ 2014 Dynamic modification of hoarding in response to hoard size manipulation. *Physiology & Behavior 127*: 8–12

Gattermann R, Johnston J, Mcphee ME, Colak E, Song Z and Neumann K 2008 Golden hamsters are nocturnal in captivity but diurnal in nature. *Biology Letters 4*: 253–255

Harkness JE, Turner PV, VandeWoude S and Wheler CL (eds) 2010 *Harkness and Wagner's Biology and Medicine of Rabbits and Rodents, Fifth Edition*. Wiley-Blackwell: Ames, IA

Hauzenberger AR, Gebhardt-Henrich SG and Steiger A 2006 The influence of bedding depth on behaviour in golden hamsters (*Mesocricetus auratus*). *Applied Animal Behaviour Science 100*(3-4): 280–294

Jansen PE, Goodman ED, Jowaisas D and Bunnell BN 1969 Paper as a positive reinforcer for acquisition of a barpress response by the golden hamster. *Psychonomic Science 16*: 113–114

Kuhnen G 1999 The effect of cage size and enrichment on core temperature and febrile response of the golden hamster. *Laboratory Animals 33*: 221–227

Laber-Laird, K, Swindle MM and Flecknell, P (eds) 1996 *Handbook of Rodent and Rabbit Medicine*. Pergamon Press: Oxford, UK

National Research Council 2011 *Guide for the Care and Use of Laboratory Animals, Eighth Edition*. National Academies Press: Washington, DC

Phillips J, Robinson A and Davey G 1969 Food hoarding behaviour in the golden hamster (*Mesocricetus auratus*): Effects of body weight loss and hoard-size discrimination. *Quarterly Journal of Experimental Psychology 41B*: 33–47

Reebs SG and St-Onge P 2005 Running wheel choice by Syrian hamsters. *Laboratory Animals 4*: 442–451

Richards M 1969 Effects of oestrogen and progesterone on nest building in the golden hamster. *Animal Behaviour 17*: 356–361

van Hoosier GL and McPherson CW 1987 *Laboratory Hamsters*. Academic Press: Orlando, FL

Veillette M and Reebs SG 2010 Preference of Syrian hamsters to nest in old versus new bedding. *Applied Animal Behaviour Science 125*: 189-194

Willows Veterinary Centre and Referral Service [n.d.] Looking after your hamster. Retrieved from http://www.willows.uk.net/en-GB/general-practice-service/looking-after-your-pet/looking-after-your-hamster

Winnicker C 2012 *A Guide to the Behavior and Enrichment of Laboratory Rodents*. Charles River Laboratories: Wilmington, MA

Rabbits

Rabbits

Jennifer Lofgren, DVM, MS, DACLAM

UNIVERSITY OF MICHIGAN

Despite the overall decline in the number of rabbits used for research over the last several decades, the number of publications citing rabbit models is steadily increasing. While rabbits are commonly used for polyclonal antibody production, they are also widely used in other research disciplines, including infectious disease, cancer, and cardiovascular disease. Domestic breeds of rabbits (including the New Zealand White (NZW), the most frequently used in research) were selectively bred from the European rabbit (*Oryctolagus cuniculus*). With the advent of the eighth edition of the *Guide for the Care and Use of Laboratory Animals* (*Guide*), many institutions are revisiting how they house and care for rabbits (National Research Council, 2011).

As institutions engage in this process it will be important to understand the rabbit's natural behavior, and design enrichment and housing to foster their optimal welfare.

Species-typical characteristics of rabbits

The catalog of normal behaviors for rabbits, both domestic and wild, includes foraging, chewing, playing, interacting with other rabbits, rearing, hiding, and resting (Myers, 1961; Vastrade, 1986; Gunn and Morton, 1995; Morton & Jennings, 2003; Held et al., 2001; Hawkins et al., 2008). Wild rabbits are most active during the dawn and dusk hours (i.e., crepuscular time frame), rest during the day, and spend most of the night foraging and grooming (Villafuerte et al., 1993; Bakker et al., 2005; Diez et al., 2005). Domesticated rabbits have retained many social behaviors from the wild (Stodardt, 1964). In North America, the most common rabbit species, cottontails (*Sylvilagus spp.*), do not build large social warrens (Crowell-Davis, 2007). In contrast, the wild European rabbit, from which NZW rabbits are descended, is a social species; these rabbits live in large warrens in groups of up to four males and up to six females (von Holst et al., 1999) with an average 89% of males and 96% of females living in groups containing at least one other rabbit of the same sex (Cowan, 1987).

Addressing the species-typical characteristic of the rabbit in the research laboratory

Rearing up: As rabbits will stand on their hind legs to survey their surroundings, cages must be tall enough to easily accommodate this behavior. This behavior can be encouraged by hanging treats or toys from the top of the cage. While providing boxes and elevated platforms to climb on increases cage complexity and addresses the need to view the environment, it does not necessarily allow the rabbit to stand fully upright.

Foraging: Food treats, specifically hay, are preferred and hold the attention of laboratory rabbits for a greater period of time than nonfood enrichment items (Lidfors, 1997; Harris et al., 2001). Providing hay is the easiest way to encourage foraging behavior. Hay, principally timothy hay, is an important component of the rabbit diet, providing the fiber necessary for normal digestive motility. Eating grasses and other greens is normally a major component of a wild rabbit's activity and the motivation to engage in this behavior in a laboratory setting is not diminished. Hay can easily serve as forage if it is spread across the cage or pen; however, it can also be suspended from hay balls or wire whisks to encourage rabbits to assume rearing postures. Providing plenty of free choice hay can also decrease barbering behavior in socially housed rabbits (Mulder et al., 1992; Bays, 2006).

Small amounts of alfalfa hay and leafy greens, such as spinach or kale, can also be great treats to encourage grazing behavior. However, one needs to be careful not to offer too many of these type of greens as they can be high in calcium, potentially predisposing the rabbit to bladder stones. Greens lower in calcium, including most lettuces, dandelion greens, leafy herbs, and stems from broccoli or celery, can be fed more frequently without this concern (Bays, 2006). Foraging for freeze-dried fruits, vegetables, or cereals is a favorite activity of rabbits in a laboratory setting (Brown, 2009). This enrichment practice can be built into daily husbandry procedures and serves the dual purpose of engaging a natural behavior and providing an early warning system for unhealthy rabbits. Rabbits who ignores their treats can be easily identified and referred for veterinary attention before the development of more severe health problems. Foods high in carbohydrates and sugar should be used sparingly as a foraging treat, as they can result in gastrointestinal distress.

Hiding: As a prey species, wild rabbits create and then live in burrows; similarly, domestic rabbits will readily utilize hiding places (Bays, 2006; Hansen & Berthelsen, 2000). These can be complex structures built into cages or simple cardboard boxes or paper bags. Being able to hide or perch also provides a mechanism for coping when scared or stressed (Buijs et al., 2011a). Perching and hiding places are particularly important for socially housed rabbits, as they provide a means to create microenvironments and escape dominant animals (Buijs et al., 2011b). One important consideration for socially housed rabbits is the provision of multiple hiding spaces so as to avoid creating a resource that can be guarded.

Chewing and gnawing: Chewing and gnawing behaviors represent up to 20% of wild rabbit activity. For domesticated rabbits, chew toys that are not easily swallowed or cannot entrap the rabbit's head or appendages can encourage this natural behavior (Lidfors, 1997; Poggiagliolmi et al., 2011). These can include untreated cardboard tubes or wood blocks; hard plastic dumbbells, kong toys, balls, or rings; or stainless steel rattles, oversized chains, or rings. Some plastic baby toys, like oversized keys, can be repurposed for rabbit enrichment. Wooden blocks or sticks, in particular, significantly increase locomotion and intake behavior (Gunn & Morton, 1995; Maertens et al., 2012).

Contrary to their often-quiet demeanor, rabbits love to make noise by flinging toys through the air. So providing toys they can pick up with their mouths and toss can be of great value

Comfortable Quarters for Laboratory Animals

(Bradley, 2000). Toys with bells in them are also well suited for this type of noise-making.

As with any species, it is important to choose manipulanda that will not entrap the head, limbs, or teeth, and chains must be short enough that they cannot wrap around the neck or limbs of a rabbit (Shomer et al., 2001). Sharp edges and materials that can be chewed apart should also be avoided.

Social interactions: The *Guide* states that social housing should be provided for all social species unless scientific justification, veterinary concern, or incompatibility precludes it (National Research Council, 2011). The literature describing both a preference for and the benefits from social housing of rabbits is extensive (Huls et al., 1991; Brooks et al., 1993; Whary et al., 1993; Seaman, 2002; Chu et al., 2004; Nevalainen et al., 2007). Rabbits will spend a significant percentage of their resting time lying in close physical contact with another rabbit. The benefits of social housing include increased ability to cope with new stressors, increased physical fitness, decreased gastrointestinal stasis due to increased activity, and normalized physiological parameters.

Refining husbandry and research procedures for rabbits

Improving the housing environment: For rabbits up to 5.4 kg, the *Guide* specifies 4 square feet per rabbit (National Research Council, 2011). However, providing 6 square feet or more per rabbit may increase the chance of successful, stable, social housing (Wyatt, 2013). Providing hiding/escape places and visual barriers is imperative, as it allows animals to choose whether to be in visual or physical contact with conspecifics (Bauman, 2005). A hiding place should be provided for each animal in the social housing to

prevent aggression stemming from defense of the hiding place. When floor pens are not possible, commercially available double-wide cages can be used to encourage natural behaviors and facilitate the formation and maintenance of stable social pairs (Lofgren et al., 2010; Lofgren, 2014).

Rotating enrichment items every 2 weeks can help maintain novelty (Harris et al., 2001; Johnson et al., 2003). The exception to this guideline is the provision of hay, which is a preferred food treat that should be provided daily—both to engage foraging behavior and support overall digestive health. Finally, lowering maximum light levels to 60 lux and adding more natural dawn and dusk transitions to the light cycle significantly contribute to a normalized circadian rhythm of body temperature (Verwer et al., 2009).

Litter boxes have been successfully used with rabbits, reducing the need to clean the entire enclosure daily. They can easily be lifted out of the enclosure and replaced. As with food, water, and shelter, multiple litter boxes are recommended to reduce resource guarding.

Acclimating to handling procedures: Regular gentle handling of rabbits can significantly reduce the stress associated with research procedures (Swennes et al., 2011). It can also help prevent injury due to handling, such as back and leg injuries. Some studies have suggested that handling early in life can make the adult rabbits more amenable to handling (Verga et al., 2007). In a recent study, rabbits were gently handled and restrained daily for approximately 5 minutes, over a 3-week period, before being returned to their home cage and provided a food treat (Swennes et al., 2011). In the subsequent 3 weeks, rabbits who had been habituated to handling were easier to catch and restrain by unfamiliar staff persons than their unhandled counterparts. This study demonstrated that acclimation to handling reduces rabbit stress during capture and restraint. The benefits of habituation were further demonstrated in a vaccine study in

which rabbits were held in a person's lap for 5 minutes, every other day. These rabbits had an improved immunological response to the vaccine, lower physiological arousal, and were protected from the post-vaccine weight loss experienced by unhandled rabbits. In addition, the rabbits were easier to catch and engaged in more exploratory behavior (Verwer et al., 2009).

Verwer et al. (2009) also evaluated several additional changes to handling and husbandry that may be useful for reducing rabbit stress:

> » compartmentalizing the enclosure when catching rabbits, i.e., ushering them into

a smaller pen and then catching them rather than chasing the rabbits around the full enclosure;
» knocking before entering the rabbit room to prepare the rabbits for such entrances;
» limiting performance of husbandry and research procedures to a time of day when the animals are already active, near dawn and dusk, rather than in the middle of a rest period (Jilge, 1991);
» spacing out invasive procedures to no more than once a day, e.g., performing baseline testing the day before, rather than simply earlier on the same day as a vaccination.

Providing social housing: The presence of a conspecific is the preferred environmental enrichment for social species, as it offers ever-changing, interactive stimuli, unlike inanimate toys or static cage furniture that provide only temporary novelties (Stauffacher et al., 2001; Nevalainen et al., 2007). Additionally, companionship provides an element of control over the cage environment, which can improve an animal's ability to cope with stressors (Newberry, 1995; Garner, 2005). During preference testing, female NZW rabbits worked just as hard for limited social contact as they did for food, indicating the relative importance of social access to rabbits (Seaman, 2002). Female pair-housed NZW rabbits spend up to 88% of their time in close proximity (Huls et al., 1991; Brooks et al., 1993; Whary et al., 1993). Adult male NZW rabbits housed in side-by-side cages separated by a perforated social-access panel show a significant preference for the quarter of the cage that provides visual, olfactory, and protected tactile contact with the neighbor (Lofgren et al., 2010). Female and male NZW rabbits housed in same-sex pairs in double-wide cages beginning at weaning can remain compatible past sexual maturity (Lofgren, 2014). Occasional barbering and

aggressive behavior can be successfully interrupted with additional hay and enrichment items (Lofgren, 2014). Furthermore, males and females can be group housed in same-sex groups of 4–10 rabbits for up to several months, although males occasionally inflicted significant fight wounds upon each other after reaching sexual maturity (Wyatt, 2013).

Ideally, rabbits should be paired or grouped from weaning age. This is most easily achieved when breeding rabbits in your own facility. However, investigators often need to order adult animals from vendors. A recent pilot study with a major vendor demonstrated the feasibility and benefits of ordering pre-paired rabbits (Lofgren, 2014). Requests for pre-pairing or grouping should be made as early as possible in the ordering process. As social housing of rabbits becomes more common practice, it is likely that vendors will increasingly provide pre-paired or pre-grouped animals.

If a facility receives rabbits who have been singly housed prior to or during shipping, attempts at pairing or grouping should be made as close to weaning age as possible. Once territories have been claimed (sometimes evident through urine spraying) or patterns of aggression (such as repeated bouts of chasing) have developed, chances of successful pairing or grouping decrease (Morton et al., 1993; Lofgren, 2014). Territories have likely been staked if each rabbit has been individually housed in a given space for more than a few hours; consequently, simply combining these spaces will often lead to immediate aggression (Lofgren, 2014); therefore, introductions should only occur in neutral spaces. Unless breeding is desired, pairs or groups should be comprised of a single gender. After sexual maturity, the risk of aggressive encounters increases dramatically, making it much more difficult to safely pair- or group-house intact males (Bays, 2006). Ideally, rabbits should be spayed or neutered to both reduce the risk of common health or behavioral issues, such as uterine tumors, urine spraying, and fight wounds, and to increase the chances of maintaining a long-term, stable social group (Bays, 2006). With careful introductions and oversight, unfamiliar intact adult females may be socially housed. While intact adult males likely benefit from having some protected social interaction (for example, through a visual divider with perforations to allow nose-to-nose contact), full physical contact can result in fight behavior and significant injuries (Lofgren, et al., 2010; Wyatt, 2013). However, several academic and commercial institutions

earlier rabbit(s) to view the later rabbit(s) as intruders. Scatter foraging enrichment on the floor of the cage, so that the animals will sniff and explore the environment.

A staff member should monitor all introductions and animals should not be left alone during this process. The observing staff member should watch for interactions, including nipping, chasing and mounting. These behaviors are often part of establishing a dominant/submissive relationship and can be a necessary step for the creation of a stable pair or group (Harriman, 2008). As long as one rabbit is performing dominant behaviors and the other is running away or passively receiving the mounting behavior, intervention is not necessary. If both/multiple rabbits are engaging in aggressive behavior, usually obvious in the first 5 minutes, intervention may be necessary.

Aggression may stop when the observer enters the pen or opens the cage door. If this is the case, offer both/all animals preferred treats and gentle interaction. Soft strokes over the head and ears can mimic the nuzzling and grooming rabbits exchange when bonded, and may assist in the introduction process (Harriman, 2008).

If aggressive behavior does not stop when the observer enters the pen or opens the door, use a spray bottle to lightly spray the rabbits with water. Once sprayed, rabbits usually will discontinue the aggressive encounter and begin to groom themselves; this provides a time-out of sorts (Harriman, 2008). Eventually, just the presence of the spray bottle may remind rabbits to remain calm (Harriman, 2008). If the spray bottle is not effective at interrupting aggressive behavior, use a visual barrier, such as a clean dustpan, to separate the rabbits. Do not reach in with bare hands, as they can become the target of aggression. Protect arms and hands with thick gloves (Bays, 2006).

have successfully paired or grouped adult males, particularly if paired at weaning, group composition is not changed, and abundant hide and escape opportunities are available (Wyatt, 2013; Lofgren, 2014). It is important to remember that social cohesion can break down unexpectedly, even in long-term stable groups or pairs. Thus, caretakers must be vigilant for any signs of aggression between rabbits. Decisions to separate should be made with input from research, veterinary, and animal care staff.

As stated above, it is imperative to use a neutral location for introductions when pairing or grouping; this can often be achieved with a fenced-off play area on the floor or a fresh, clean, double-wide cage (Bays, 2006; Harriman, 2008; Lofgren, 2014). Multiple hide and escape opportunities (boxes, shelves, huts, hay bales, crates or tunnels) should be available.

Aggression can increase at feeding times, so avoid introducing rabbits until after they have been fed. Additionally, provide several sources of food, particularly hay, and water (Harriman, 2008). Rabbits should be added to the pen at the same time, as staggering their entry into the pen may allow the

Daily introductions of increasing duration (often 15–30 minutes or more) for up to 2 weeks may be necessary to assure the animals can be safely left together unattended (Harriman, 2008; Wyatt, 2013; Lofgren, 2014). Rabbits who are successfully grouped will groom one another, usually over the head and ears, and will sit and lie stretched out near one another. If these behaviors are consistently observed during introductions and are stable, overnight co-housing can be attempted.

Cage dividers that allow protected physical contact, namely sniffing, are a good option for that first night together. While often a predictor of positive bonding, it is important to note that calm behavior of animals through a divider or fence does not guarantee the animals will not be aggressive toward one another once in full physical contact (Harriman, 2008; Lofgren, 2014). If animals bite each other through the divider, as evidence of trauma to the lips or nose, they likely will not pair well together (Harriman, 2008).

Social compatibility of newly formed pairs or groups can be tracked in several ways. Using colored markers on the backs of the rabbits to track interactions from a distance for the first few days will make it easier to identify individual animals and evaluate their relationships (Whary et al., 1993). Video cameras can be useful for monitoring the animals during the first few days they are socially housed (Lofgren et al., 2010). The day after the animals have had unsupervised overnight social housing, rabbits should be weighed and examined individually for bite wounds. Some fight wounds are not readily visible, so a hands-on exam is recommended. Each facility should develop their own guidelines regarding how many or what size small wounds will be tolerated while the dominance hierarchy is determined. In many cases, there will be some small nibbles. Remember to document the findings in the individual rabbit record, both for USDA compliance as well as establishing a history if an exception to social housing amendment is ultimately necessary. Once the pair or group appears stable, the typical daily health checks should uncover any alarming changes in the social rank relationships between cage companions. Social cohesion may suddenly change and it can be difficult to identify an instigating factor. Re-pairing or re-grouping may be attempted, but should follow the same steps as an initial introduction.

When group- or pair-housing is not appropriate: Single-housing of rabbits may be necessary for veterinary or officially approved scientific reasons. While individual housing does not allow for unobstructed touch, a number of cage manufactures now make cages with perforated Plexiglas dividers to allow olfactory and limited physical contact, which may be more valuable than visual contact for rabbits (Lofgren et al., 2010).

This type of housing can be paired with a playpen area to allow for animals to receive rotating opportunities for greater enrichment, movement, and social interaction. As with primates, when using dividers that allow for visual contact, keep in mind that animals need to have the ability to "escape" from their neighbor. This can be achieved by providing a visual barrier such as a shelter or keeping half the divider opaque so the animals retain the ability to choose whether to be in visual contact with their neighbor.

Being creative: In some facilities, budgeting and allocation of money for the purchase of new cages can be a multi-year process. Some lower-cost options for providing social housing include using a floor pen or an empty child's swimming pool to facilitate exercise and socialization in the rabbit room (J. Lanzim, personal communication, July 2, 2013). These areas can be enriched with foraging treats and hide/escape items, and used for exercise and socialization of compatible rabbits. Ideally, this type of pen should be used daily for 30–60 minutes, but a less time-intensive approach is to integrate this practice on cage-change day. Another budget-friendly option for improving social contact between cage neighbors is to modify existing caging with a polycarbonate vision and olfactory panel. Alternatively, cubicles, unused primate caging, cat caging, or dog kennels can be repurposed as enrichment caging or full-time housing.

Acknowledgements

Special thanks to Dr. Michael Esmail and Ms. Michele Cunneen for their contributions to the content of this manuscript.

REFERENCES

Bakker ES, Reiffers RC, Olff H and Gleichman JM 2005 Experimental manipulation of predation risk and food quality: effect on grazing behaviour in a central-place foraging herbivore. *Oecologia 146*: 57–167

Baumans V 2005 Environmental enrichment for laboratory rodents and rabbits. *ILAR Journal 46*(2): 162–170. http://dels-old.nas.edu/ilar_n/ilarjournal/46_2/pdfs/v4602baumans.pdf

Bays TB 2006 Rabbit behavior. In: Bays TB, Lightfoot TL and Mayer J (eds) *Exotic Pet Behavior: Birds, Reptiles, and Small Mammals*. Saunders: St. Louis, MO

Bradley TA 2112 2000 Rabbits: Understanding normal behavior. *Exotic DVM Magazine 2*(1): 19–24

Brooks DL, Huls W, Leamon C, Thomson J, Parker J and Twomey S 1993 Cage enrichment for female New Zealand White rabbits. *Lab Animal 22*(5): 30–38

Brown C 2009 Novel food items as environmental enrichment for rodents and rabbits. *Lab Animal 38*(4): 119–120

Brown C 2010 Organic wheatgrass as environmental enrichment. *Lab Animal 39*(3): 74–75

Buijs S, Keeling LJ, Rettenbacher S, Maertens L and Tuyttens FA 2011a Glucocorticoid metabolites in rabbit faeces—influence of environmental enrichment and cage size. *Physiology and Behavior 104*: 469–473

Buijs S, Keeling LJ and Tuyttens FAM 2011b Behaviour and use of space in fattening rabbits as influenced by cage size and enrichment. *Applied Animal Behaviour Science 134*: 229–238

Chu L, Garner JP and Mench JA 2004 A behavioral comparison of New Zealand White rabbits (*Oryctolagus cuniculus*) housed individually or in pairs in conventional laboratory cages. *Applied Animal Behaviour Science 85*: 121–139

Cowan D1 1987 Group living in the European rabbit (*Oryctolagus cuniculus*): Mutual benefit or resource localization. *Journal of Animal Ecology 56*: 779–795

Crowell-Davis SL 2007 Topics in medicine and surgery: Behavior problems in pet rabbits. *Journal of Exotic Pet Medicine 16*: 38–44

Diez CPJ, Prieto R, Alonzo M and Oimedo J 2005 Activity patterns of wild rabbit (*Oryctolagus cuniculus*, L.1758), under semi-freedom conditions, during autumn and winter. *Wildlife Biology in Practice 1*: 41–46

Hansen LT and Berthelsen H 2000 The effect of environmental enrichment on the behaviour of caged rabbits (*Oryctolagus cuniculus*). *Applied Animal Behaviour Science 68*: 163–178

Harriman, M 2008 *Introducing Rabbits: Bonding Techniques for Matchmakers* [film]. Drollery Press: Alameda, CA

Harris LD, Custer LB, Soranaka ET, Burge R and Ruble GR 2001 Evaluation of objects and food for environmental enrichment of NZW rabbits. *Contemporary Topics in Laboratory Animal Science 40*(1): 27–30

Hawkins P, Hubrecht R, Buckwell A, Cubitt S, Howard B, Jackson A and Poirier GM 2008 *Refining Rabbit Care: A Resource for Those Working with Rabbits in Research*. Royal Society for the Prevention of Cruelty to Animals: Horsham, UK. http://www.rspca.org.uk/sciencegroup/researchanimals/implementing3rs/refiningrabbitcare

Held SDE, Turner RJ and Wootton RJ 2001 The behavioural repertoire of non-breeding group-housed female laboratory rabbits (*Oryctolagus cuniculus*). *Animal Welfare 10*: 437–443

Huls WL, Brooks DL and Bean-Knudsen D 1991 Response of adult New Zealand White rabbits to enrichment objects and pair-housing. *Laboratory Animal Science 41*: 609–612

Garner JP 2005 Stereotypies and other abnormal repetitive behaviors: Potential impact on validity, reliability, and replicability of scientific outcomes. *ILAR Journal 46*(2): 106–117. http://dels-old.nas.edu/ilar_n/ilarjournal/46_2/html/v4602garner.shtml

Gunn D and Morton DB 1995 Inventory of the behaviour of New Zealand White rabbits in laboratory cages. *Applied Animal Behaviour Science 45*: 277–292

Jilge B 1991 The rabbit: a diurnal or a nocturnal animal? *Journal of Experimental Animal Science 34*: 170–183

Johnson CA, Pallozzi WA, Geiger L, Szumiloski JL, Castiglia L, Dahl NP, ... Klein HJ 2003 The effect of an environmental enrichment device on individually caged rabbits in a safety assessment facility. *Contemporary Topics in Laboratory Animal Science 42*(5): 27–30

Lidfors L 1997 Behavioural effects of environmental enrichment for individually caged rabbits. *Applied Animal Behaviour Science 52*: 157–169

Lofgren JLS 2014 unpublished data

Lofgren JLS, Wrong C, Hayward A, Karas AZ, Morales S, Quintana P, ... and Fox JG 2010 Innovative social rabbit housing. *American Association for Laboratory Animal Science Meeting Official Program*: 131

Maertens L, Buijs S and Davoust C 2012 Gnawing blocks as cage enrichment and dietary supplement for does and fatteners: intake, performance and behaviour. *World Rabbit Science 21*(S1): 185–192

Morton DB, Jennings M, Batchelor GR, Bell D, Birke L, Davies K, ... Turner RJ 1993 Refinements in rabbit husbandry: Second report of the BVAAWF/FRAME/RSPCA/UFAW Joint Working Group on Refinement. *Laboratory Animals 27*: 301-329. http://la.rsmjournals.com/content/27/4/301.full.pdf

Morton DB and Jennings M 2003 Refinements in rabbit husbandry. *Laboratory Animals 27*: 301–329

Mulder A, Nieuwenkamp AE and van der Palen JG 1992 Supplementary hay reduces fur chewing in rabbits. *Tijdschrift Voor Diergeneeskunde 117*: 655–658

Myers K and Poole WE 1961 A study of the biology of the wild rabbit, *Oryctolagus cuniculus* (L.), in confined populations. II. The effects of season and population increase on behaviour. *CSIRO Wildlife Research 6*: 1–41

National Research Council 2011 *Guide for the Care and Use of Laboratory Animals, Eighth Edition*. National Academies Press: Washington, DC

Newberry RC 1995 Environmental enrichment: increasing the biological relevance of captive environments. *Applied Animal Behaviour Science 44*: 229–243

Nevalainen TO, Nevalainen JI, Guhad FA and Lang CM 2007 Pair housing of rabbits reduces variances in growth rates and serum alkaline phosphatase levels. *Laboratory Animals 41*(4): 432–440

Poggiagliolmi S, Crowell-Davisa SL, Alworthb LC and Harveyb SB 2011 Environmental enrichment of New Zealand White rabbits living in laboratory cages. *Journal of Veterinary Behavior: Clinical Applications and Research 6*: 343–350

Seaman SC 2002 *Laboratory Rabbit Housing: An investigation of the social and physical environment. A summary of the report to the UFAW/Pharmaceutical Housing and Husbandry Steering Committee (PHHSC), based on a PH.D. thesis (Seaman, 2002)* pp 2–8. http://www.ufaw.org.uk/pdf/phhsc-schol1-summary.pdf

Shomer NH, Peikert S and Terwilliger G 2001 Enrichment-toy trauma in a New Zealand White Rabbit. *Contemporary Topics in Laboratory Animal Science 40* (1): 31–32

Stauffacher M, Peters A, Jennings M, Hubrecht RC, Holgate B, Francis R, ... Hansen AK 2002 *Future Principles for Housing and Care of Laboratory Rodents and Rabbits: Report for the Revision of the Council of Europe Convention ETS 123 Appendix A for Rodents and Rabbits: Part B* pp 36–37. Council of Europe: Strasbourg, France

Stodart EM 1964 A comparison of behavior, reproduction, and mortality of wild and domestic rabbits in confined populations. *CSIRO Wildlife Research 9*: 144–159

Swennes AG, Alworth LC, Harvey SB, Jones CA, King CS and Crowell-Davis SL 2011 Human handling promotes compliant behavior in adult laboratory rabbits. *Journal of the American Association for Laboratory Animal Science 50*(1): 41–45. http://www.ncbi.nlm.nih.gov/pmc/articles/PMC3035402/

Vastrade F 1986 The social behaviour of free ranging domestic rabbits (*Oryctolagus cuniculus* L.) *Applied Animal Behaviour Science 16*: 165–177

Verga M, Luzi F and Carenzi C 2007 Effects of husbandry and management systems on physiology and behaviour of farmed and laboratory rabbits. *Hormones and Behavior 52*: 122–129

Verwer, CM, van der Ark A, van Amerongen G, van den Bos R and Hendriksen CF 2009 Reducing variation in a rabbit vaccine safety study with particular emphasis on housing conditions and handling *Laboratory Animals 43*:155–164

Villafuerte R, Delibes M and Moreno S 1993 Environmental factors influencing the seasonal daily activity of the European rabbit (*Oryctolagus cuniculus*) in a Mediterranean area. *Mammalia 57*: 341–347

von Holst DH, Kaetzke P, Khaschei M and Schönheiter R 1999 Social rank, stress, fitness, and life expectancy in wild rabbits. *Naturwissenschaften 86*: 388–393

Whary M, Peper R, Borkowski G, Lawrence W and Ferguson F 1993 The effects of group housing on the research use of the laboratory rabbit. *Laboratory Animals 27*: 330–341

Wyatt, J 2013 *Strategies for Socially Housing Rabbits.* American College of Laboratory Animal Medicine Forum: Williamsburg, VA

Ferrets

Ferrets

Alvaro Duque, PhD
DEPARTMENT OF NEUROBIOLOGY, YALE SCHOOL OF MEDICINE

Jodi Scholz, DVM, DACLAM
DEPARTMENT OF COMPARATIVE MEDICINE, MAYO CLINIC

The ferret species commonly encountered in research institutions is the domestic ferret (*Mustela putoris furo*). A related species, the black-footed ferret (*Mustela nigripes*), is native to North America and is endangered. They may share a common ancestor, the European polecat (*Mustela putorius eversmanni*) (Poole, 1972), although the exact ancestry of the domestic ferret is still unclear (Boyce et al., 2001). The ferret was domesticated at least 2,000–3,000 years ago and has become a common household pet in the last few decades because of its social and playful behavior (Boyce et al., 2001; Plant and Lloyd, 2010; Thompson, 1951). Unlike their wild polecat relatives, which are nocturnal solitary hunters (Poole, 1972; Einon, 1995), domestic ferrets demonstrate an affinity toward humans, are social, and are quite active during the day (Boyce et al., 2001; Poole, 1972).

There are numerous sources of information regarding general aspects of ferret housing and management, veterinary care, nutrition, behavior, and physiology (e.g., Boyce et al., 2001; Brown, 2003; Marini et al., 2002; Plant & Lloyd, 2010; Quesenberry & Orcutt, 2003; Fox & Bell 1998; Fox 1998). In contrast, relatively few studies or reviews describing environmental enrichment for ferrets have been published (Baumans et al., 2007; Einon, 1995). Controlled studies have primarily focused on social isolation during early life and the effects of an enriched environment on learning behaviors (Chivers & Einon, 1982; Weiss-Buerger, 1981). A larger number of studies have been performed evaluating the effects of environmental conditions such as floor space on pelt quality in mink, which have been extensively raised for fur. As mink are closely related to ferrets, some extrapolation of the findings in these studies to ferrets may be appropriate. Specific aspects of domestic ferret behavior are described in further detail below, with recommendations for supporting species-appropriate behaviors in the laboratory.

Assessment of well-being

Environmental enrichment can be defined as the provision of an environment and incorporation of practices that provide animals with sensory and motor stimulation to facilitate the expression of species-specific behavior, reduce or prevent maladaptive behavior, and promote the psychological well-being of animals (National Research Council, 2011). Physical well-being may be easier to support and measure than psychological well-being, although they go hand in hand. Altered psychological states often result in physical manifestations that

may be hard to detect. Normal behavior of an animal in a laboratory environment can best be determined by comparing it to that exhibited by (1) other individuals of the same species under the same conditions, (2) similar species also living in a laboratory, or (3) closely related species in the wild. For example, despite some behavioral differences between wild and domestic ferrets, domestic ferrets maintain instinctive behaviors, including social play, aggression, hunting, and territorial marking, as well as maternal and sexual behaviors.

In general, proper assessment of ferret well-being requires close observation of their appearance and behavior. Body posture and activity level are essential aspects of interpreting well-being in ferrets (Boyce et al., 2001). Common postures and active behaviors include sideway approaches, rolling, chasing, wrestling, and tail wagging; often, excited ferrets demonstrate behavior that resembles dance movements (Boyce et al., 2001; Schilling, 2000). In contrast, a ferret in distress is inactive, does not explore his/her environment, has an unkempt coat that appears scruffy, may have eyes that appear puffy and half closed, and if approached, may react with aggression (Wolfensohn & Lloyd, 2003). An unusual amount of time in a "slumped" posture may also be noted (Boyce et al., 2001). Other general signs of distress are similar to those expressed by other species and include restlessness, altered eating, drinking and sleeping habits and, potentially, gastrointestinal upset.

Vocalization may indicate either excitement or distress in ferrets. Sounds that indicate excitement are described as "buck-a-buck" (also referred to as "the dook" or "clucking") and a closely related vocalization called a "chuckle" (Boyce et al., 2001; Brown, 2003). A high pitch "scream" or "screech" is a clear sign of pain, anger or terror. The bark and the hiss are other sounds that may be associated with distress, fear, impatience or warning (Boyce et al., 2001; Brown, 2003).

Play and social behavior

Play is an important part of normal development and appears to contribute significantly toward the animal acquiring normal social skills. For example, play is believed to be a means for young ferrets to learn dominance behavior (Boyce et al., 2001; Gupta, 1988; Lazar & Beckhorn, 1974). Play is also related to sexual behavior; neck biting is a common play behavior in pre-pubertal ferrets, and adult male ferrets bite at the female's neck prior to mating (Fox & Bell, 1998). Young ferrets spend a significant amount of time playing with each other, and the intensity and aggressiveness of the play increases during adolescence (Baum, 1998). Play behavior continues into adulthood, although by about 4–5 years of age, playing occurs primarily in short, occasional bouts (Boyce et al., 2001). Because ferret play is quite exuberant and can be intense, it may take a few moments of observation to distinguish social play from serious aggression (Boyce et al., 2001). General play behavior, different aspects of aggressive behavior, and analysis of social play of the ferrets and related species have been described elsewhere in detail (Bunnell, 1981; Chivers & Einon, 1982; Jeppesen & Falkenberg, 1990; Mankovich, 1982; Poole, 1966; Poole, 1972; Poole, 1978). Sex differences in play have also been studied in ferrets; for example, play behavior may be more aggressive in males as compared to females. Exposure to androgens during the postnatal period influences these differences (Biben, 1982; Mankovich, 1982; Stockman et al., 1986).

As with other species, social isolation during development has negative consequences in ferrets. For example, lack of conspecific interactions was found to cause hyperactivity that persisted into adulthood (Chivers & Einon, 1982). In mink, social impoverishment has been found to induce stereotyped behavior (Bildsøe et al., 1990)—a finding

that would be expected in ferrets, as well. In contrast, socially housed ferrets raised in an enriched environment were found to be superior in maze learning and reversal (Weiss-Buerger, 1981).

In addition to interactions with conspecifics, ferrets are very interactive with humans, which is one reason their popularity as companion animals has increased greatly over the past few decades (Brown, 2003). Human interaction can itself be enriching for ferrets, and regular handling in the laboratory can prevent fearful behavior and help buffer stress during research procedures (Ball, 2006).

Recommendations: Ferrets should be socially housed in an enriched environment (as described in this chapter) that encourages play behavior. This is especially important for young animals. When they are housed together, estrus females may become pseudopregnant, which does not affect their health and general well-being. Intact males should not be co-housed in the presence of estrus females due to high potential for fighting. Gentle human interaction is strongly recommended.

Burrowing, hunting, and swimming behaviors

The ancestors of the domestic ferrets probably lived in burrows, where they slept and stored prey (Lode, 1989; Thompson, 1951), and the skills and the desire to dig and store items appears to have been retained. Pet ferrets are notorious for digging into soft materials available in homes and bringing small objects to their dens (Brown, 2003; Schilling, 2000). Similarly, domestic ferrets have retained hunting instincts (Apfelbach & Wester, 1977). For example, a study by Russell (1990) demonstrated that isolated ferrets otherwise raised in enriched conditions (with a daily change of play objects) would choose the arm of a maze leading to the more prey-like play objects, were superior in capturing crickets and moving prey models, and demonstrated more elaborate prey-catching responses than those raised in an impoverished environment.

Ferrets have good binocular vision and are skilled in tracking objects that move at 25–45 cm/sec, the escape speed of a mouse (Brown, 2003). Indeed, they seem to very much enjoy the thrill of a chase. Hunting behavior can be elicited by quickly moving small objects across the floor and allowing the ferret to chase after them (Schilling, 2000).

Most authors agree that ferrets are good swimmers; however, not all ferrets seem to enjoy water to the same degree. Some publications tailored to owners of pet ferrets recommend bathing ferrets at intervals that differ according to the season, generally around every 2 weeks (Ovechka, 2002; Shefferman, 2001). Bathing too frequently may deplete the ferret pelt of essential oils and dry their skin (Schilling, 2000; Shefferman, 2001).

At the authors' institution,[1] a ferret "play cage" has been developed by repurposing a nonhuman primate cage. Minor modifications to the cage were needed in order to prevent escape (e.g., through the food hopper). The cage is supplied with a wide variety of enrichment objects such as PVC tubes, digging basins, and glove boxes, and ferrets are placed in the cage for several hours, at least once weekly. In our experience, play behavior begins immediately after the ferrets are placed in the cage and continues throughout the day, especially if the ferrets are young.

Recommendations: Enrichment devices that encourage the natural behaviors of burrowing and hunting should be provided. PVC pipes, commercially available tunnels marketed for ferrets, used (clean) glove boxes, paper bags, and cardboard tubes are all readily available materials that allow ferrets to express burrowing behaviors. Similarly, a box (such as a rodent cage) with bedding substrate or shredded paper encourages digging behavior. Hunting

[1] At the time this chapter was written, both authors were at the Yale School of Medicine.

behavior can be elicited by providing small plastic balls or other objects that can be chased, and human caretakers can also encourage hunting behavior by using mobile toys that the ferrets chase. As ferrets—like all other animals kept in research labs—tend to lose interest rather quickly in enrichment gadgets, frequent rotation of different enrichment objects is highly recommended. A swimming basin can be provided for ferrets who enjoy water, though it is recommended that ferrets be supervised while playing with water and dried when the water is removed. Keeping ferrets in play cages is the optimal housing for these very inquisitive and playful animals. If this cannot be arranged, regular rotation in a generously furnished play cage is recommended, particularly for young ferrets and adults housed long-term.

Sleeping and resting behaviors

Despite the above-mentioned activities, ferrets spend 50–75% of their day sleeping (Boyce et al., 2001; Plant & Lloyd, 2010). As ferrets age, play behavior becomes less frequent and the animals sleep more; they also tend to sleep more deeply as they get older (Boyce et al., 2001). Specific sleep patterns vary by individual and the environment; for example, sleeping habits of pet ferrets depend on the schedule of their owners (Boyce et al., 2001). Ferrets prefer to sleep in dark, enclosed areas, generally as a group (Ball, 2006; Brown, 2003; Plant & Lloyd, 2010).

Recommendations: Ferret cages should be supplied with hammocks and enclosed areas for sleeping and rest, keeping in mind that they often prefer to sleep in groups.

Comfortable Quarters for Laboratory Animals

Maternal and reproductive behavior

Ferrets become sexually mature at about 9–12 months of age or the spring of the year after they are born (Fox & Bell, 1998; Wolfensohn & Lloyd, 2003). Females are induced ovulators and are seasonally polyestrous. Photoperiod plays an important role in ferret reproduction as estrus is triggered in females by increasing daylight. Detailed information on breeding and reproduction is available elsewhere (e.g., Bell, 2003; Fox & Bell, 1998; Marini et al., 2002).

Ferrets are altricial animals; newborn ferrets are blind and deaf at birth with eyes and ears opening on postnatal days 32–34 (Fox, 1998; Marini et al., 2002; Plant & Lloyd, 2010). Ferret mothers express typical maternal behaviors observed in other mammals (Lazar & Beckhorn, 1974). When provided with bedding material and a nest box within 8 hours of delivery, ferret mothers collect their young into the nest and provide warmth and care for them (Baum, 1998). Ferret mothers should be allowed to wean their offspring naturally. If artificial weaning is required, the kits should be at least 6 weeks old (Brown, 2003; Plant & Lloyd, 2010; Wolfensohn & Lloyd, 2003). During the pre-weaning period, interactions between the mother and kits are essential for normal social development; if kits are weaned early, they can demonstrate rough or aggressive behaviors that will likely persist unless behavioral modifications are made by human handlers (Boyce et al., 2001).

Recommendations: Periparturient females should be kept in a quiet area and provided a nest box with nesting material. The walls of the nest box should be high enough to prevent kits from climbing over it, yet low enough to allow the mother to easily enter and exit the nesting box; a height of 6 inches (15 cm) is recommended (Bell, 2003; Marini et al., 2002). Newborn ferrets should be kept with their mother for at least 6–8 weeks. If wire-bottom caging or caging not specifically designed for ferrets is used for housing, particular attention should be paid to the flooring; large grid size or large separations (i.e., larger than 1.0 x 0.5 in.) pose a risk of injury to the kits as they emerge from the nesting box (Ball, 2006; Marini et al., 2002).

Feeding behaviors and nutrition

Ferrets are obligate carnivores and thus have short gastrointestinal tracts and minimal gut flora. Because of these characteristics, they do not digest fiber well or utilize carbohydrates efficiently (Brown, 2003). Ideally, ferrets should be fed a diet specifically formulated for ferrets, although some facilities successfully maintain ferrets on feline or mink diets (Ball, 2006; Brown, 2003; Marini et al., 2002). Ferrets will eat 9–10 small meals each day if provided with free access to food (Fox, 1998). Dietary preferences are formed by 4 months of age, so introduction of new foods later in life may be difficult (Ball, 2006; Brown, 2003). Even though ferrets generally enjoy cereals and fruits, treats high in carbohydrates should be avoided or kept to a minimum. Rather, treats or nutritional supplements that are specifically formulated for ferrets or high in protein (e.g., Nutri-Cal nutritional supplement or meat-based baby food) are preferred (Ball, 2006; Brown, 2003). Such foods may be useful for distracting ferrets during medical procedures, thus decreasing stress (Quesenberry & Orcutt, 2003).

Recommendations: A diet appropriate for ferrets should be provided. Treats can be offered in moderation for enrichment and to encourage human interaction. Foods high in carbohydrates should be avoided.

Ferret-adequate housing

Ferret-specific cages are currently available and are generally very similar to rabbit cages. Either molded plastic cages or

stainless steel cages have been successfully used; galvanized metal should be avoided, as zinc toxicosis secondary to licking galvanized bars has been reported in ferrets (Straube & Walden, 1981). Both suspended and solid-floor cages are suitable, although at least one source recommends solid flooring whenever possible to allow provision of bedding material (Ball, 2006; Plant & Lloyd, 2010). Because of the active and inquisitive nature of ferrets, any housing that is not specifically designed for ferrets should be carefully evaluated for safety. Specific risks include sharp edges and potential escape routes. Ferrets are notorious for escaping; any space large enough for their heads will inevitably result in escape (Moody et al., 1985; Plant & Lloyd, 2010; Wolfensohn & Lloyd, 2003). Recommended grid size for flooring is 1.0 x 0.5 inches, or 0.25 inches if wire mesh or slatted flooring is used (Fox, 1998). A useful feature of some available housing systems is flexibility to join cages together, providing increased floor space when populations are small or when social groups are formed.

Detailed studies investigating optimal floor space for ferrets are lacking. Certainly, severe restrictions in mobility and cage space have been shown to be detrimental in ferrets and related species, resulting in skeletal changes and reduction in pelt quality in mink (e.g., Bildsøe et al., 1991; Einon, 1995). No specific guidelines or requirements for floor space are provided in the Animal Welfare Regulations or the *Guide for the Care and Use of Laboratory Animals*, but a commonly used rule of thumb is a cage size of 4 square feet (3,721 cm^2) for two adult ferrets. European guidelines describing minimum floor space for ferrets are more specific, ranging from 1,500 cm^2 for animals ≤ 600 g to 5,400 cm^2 for a female with a litter (Plant & Lloyd,

2010). Recommended temperature and humidity ranges for housing ferrets are 40–65°F and 40–65%, respectively (Fox, 1998; Marini et al., 2002).

Various enrichment items, as described above, when added to the primary or exercise enclosure, will encourage natural behaviors and provide additional resting places (beyond the necessary bedding area). A wide variety of ferret-appropriate enrichment items are commercially available from vendors of products for animals in the laboratory and pet toy suppliers, although items not specifically designed for the laboratory may need to be evaluated for their ability to withstand sanitation procedures (see below). Improvised devices such as PVC tubes for tunnels and surgical drapes for hammocks (Baumans et al., 2007) may be more affordable than commercially available devices. However, any improvised items must be carefully evaluated for safety prior to and during use, as ferrets are very inquisitive and have a tendency to chew and ingest foreign objects; foam, rubber, and other items of similar texture, in particular, must be avoided (Ball, 2006; Brown, 2003).

Sanitation of caging and enrichment objects

At the authors' institution, ferret cages are spot cleaned daily to remove fecal waste and sanitized weekly. Prior to sanitation, enrichment devices are removed from the cages and the cages are washed in a rack washer using a pre-wash, 10-minute detergent wash, acid wash, and two 10-minute rinses. Hammocks and beds are washed in a washing machine for two cycles. Enrichment devices and toys are sanitized in a tunnel washer (heavily soiled items are pre-treated prior to washing). Nonsanitizable enrichment items such as glove boxes are primarily used in the enrichment cages and are disposed of daily.

REFERENCES

Baumans V, Coke C, Green J, Moreau E, Patterson-Kane E, Reinhardt A, ... Van Loo P (eds) 2007 *Making Lives Easier for Animals in Research Labs: Discussions by the Laboratory Animal Refinement & Enrichment Forum* p 87. The Animal Welfare Institute: Washington, DC

Apfelbach R and Wester U 1977 The quantitative effect of visual and tactile stimuli on the prey-catching behaviour of ferrets (*Putorius furo* L.). *Behavioural Processes 2*(2): 187-200

Ball RS 2006 Issues to consider for preparing ferrets as research subjects in the laboratory. *ILAR Journal 47*: 348-357

Baum MJ 1998 Use of the ferret in biomedical research. In: Fox JG (ed) *Biology and Diseases of the Ferret* pp 511–520. Lippincott Williams & Wilkins: Philadelphia, PA

Bell JA 2003 Periparturient and neonatal diseases. In: Quesenberry KE & Carpenter JW (eds) *Ferrets, Rabbits and Rodents: Clinical Medicine and Surgery* pp 50-57. WB Saunders: Saint Louis, MO

Biben M 1982 Sex differences in the play of young ferrets. *Biology of Behaviour 7*: 303–308

Bildsøe M, Heller KE and Jeppesen LL 1990 Stereotypies in adult ranch mink. *Scientifur 14*: 169–177

Bildsøe M, Heller KE and Jeppesen LL 1991 Effects of immobility stress and food restriction on stereotypies in low and high stereotyping female ranch mink. *Behavioural Processes 25*: 179–189

Boyce SW, Zingg BM and Lightfoot TL 2001 Behavior of *mustela putorius furo* (the domestic ferret). *Veterinary Clinics of North America: Exotic Animal Practice 4*: 697–712

Brown SA 2003 Basic anatomy, physiology and husbandry. In: Quesenberry KE & Carpenter JW (eds) *Ferrets, Rabbits and Rodents: Clinical Medicine and Surgery* pp 2–12. WB Saunders: Saint Louis, MO

Bunnell T 1981. *Playful behaviour in relation to family group activity in polecats (*Mustelidae*)* [dissertation]. The Open University: Milton Keynes, UK

Chivers SM and Einon DF 1982 Effects of early social experience on activity and object investigation in the ferret. *Developmental Psychobiology 15*: 75–80

Einon D 1995 The effects of environmental enrichment in ferrets. In: Smith CP and Taylor V (eds) *Environmental Enrichment Information Resources for Laboratory Animals: 1965–1995: Birds, Cats, Dogs, Farm Animals, Ferrets, Rabbits, and Rodents* pp 113–126. Animal Welfare Information Center: Beltsville, MD and Universities Federation for Animal Welfare: Potters Bar, UK

Fox JG 1998 Housing and management. In: Fox JG (ed) *Biology and Diseases of the Ferret* pp 173–182. Williams & Wilkins: Baltimore, MD

Fox JG and Bell JA 1998 Growth, reproduction and breeding. In: Fox JG (ed) *Biology and Diseases of the Ferret* pp 211–227. Williams & Wilkins: Baltimore, MD

Gupta AS 1988 The structure and development of play in ferrets and dogs. *Index to Theses Accepted for Higher Degrees in the Universities of Great Britain and Ireland 37*: 1608–1609

Jeppesen LL and Falkenberg H 1990 Effects of play balls on peltbiting, behaviour, and level of stress in ranch mink. *Scientifur 14*: 179–186

Lazar JW and Beckhorn GD 1974 Social play or the development of social behavior in ferrets (*Mustela putorius*). *American Zoologist 14*: 405–414

Lode T 1989 Le comportement de mise en reserve alimentaire des proies chez le putois (*Mustela putorius*) [Prey storing in the polecat (*Mustela putorius*)]. *Cahiers Ethologie Appliquee 9*(1): 19–30

Mankovich NJ 1982 Sex differences in play-wrestling in the ferret (*Mustela putorius*). *Dissertation Abstracts International (B) 43*: 651

Marini RP, Otto G, Erdman S, Palley LS and Fox JG 2002 Biology and diseases of ferrets. In: Fox JG, Anderson LC, Loew FM and Quimby FW (eds) *Laboratory Animal Medicine* pp 483–517. Academic Press: New York, NY

Moody KD, Bowman TA and Lang CM 1985 Laboratory management of the ferret for biomedical research. *Laboratory Animal Science 35*(3): 272–279

National Research Council 2011 *Guide for the Care and Use of Laboratory Animals, Eighth Edition*. National Academy Press: Washington, DC

Ovechka G 2002 *Ferrets as a new pet*. TFH Publications: Neptune City, NJ

Plant M and Lloyd M 2010 The ferret. In: Hubrecht R & Kirkwood J (eds) *The UFAW Handbook on the Care and Management of Laboratory Animals* pp 418–431. Wiley-Blackwell: Chichester, UK

Poole TB 1966 Aggressive play in polecats. *Symposia of the Zoological Society of London 18*: 23–44

Poole TB 1972 Some behavioural differences between the European polecat, *Mustela putorius*, the ferret, *M. furo*, and their hybrids. *Journal of Zoology 166*: 25–35

Poole TB 1978 An analysis of social play in polecats (Mustelidae) with comments on the form and evolutionary history of the open-mouth play face. *Animal Behaviour 26*: 36–49

Quesenberry KE and Orcutt C 2003 Basic Approach to Veterinary Care. In Quesenberry KE & Carpenter JW (eds) *Ferrets, Rabbits and Rodents: Clinical Medicine and Surgery* pp 50–57. WB Saunders: Saint Louis, MO

Russell J 1990 *Predatory Object Play in the Ferret* [dissertation]. University of London: London, UK

Schilling K 2000 *Ferrets for Dummies*. Wiley Publishing: New York, NY

Shefferman MR 2001 *The Ferret*. Howell Book House: New York, NY

Stockman ER, Callaghan RS, Gallagher CA and Baum MJ 1986 Sexual differentiation of play behavior in the ferret. *Behavioral Neuroscience 100*: 563–568

Straube EF and Walden NB 1981 Zinc poisoning in ferrets (*Mustella putoris furo*). *Laboratory Animals 15*: 45–47

Thompson APD 1951 A history of the ferret. *Journal of the History of Medicine and Allied Sciences 6*: 471

Weiss-Buerger M 1981 An investigation on the influence of exploration and playing on learning by polecats (*Mustela putorius X Mustela furo*). *Zeitschrift für Tierpsychologie 55*(1): 33–62

Wolfensohn S and Lloyd M 2003 *Handbook of Laboratory Animal Management and Welfare*. Blackwell Publishing: Ames, IA

Zebrafish

Zebrafish

Christian Lawrence, MS

AQUATIC RESOURCES PROGRAM, BOSTON CHILDREN'S HOSPITAL

Over the past several decades, the zebrafish (*Danio rerio*) has become an important research animal model. Many of the same characteristics that make this diminutive tropical freshwater minnow a favorite of fish hobbyists have also contributed to its emergence as a model for human development and disease, toxicology, genetics, and behavior. An important factor in this emergence is the relative ease of keeping and raising zebrafish in captivity. Zebrafish are tolerant of a wide range of environmental conditions and will readily breed and produce large quantities of offspring, even under variable or suboptimal conditions. Ironically, the fact that zebrafish thrive so well in captivity has allowed researchers to overlook the importance of developing standards for care and management, based on the biology and behavior of these animals.

As the zebrafish has become more prevalent in animal research programs, there is a growing need to develop standards of care that are conducive to the well-being of these animals. Such standards will not only improve animal well-being, but will ultimately serve to improve the quality and efficacy of research in which the fish are involved.

The purpose of this chapter is to provide those charged with the care of the zebrafish in laboratory settings with recommendations for managing the fish in accordance with their species-specific requirements. These recommendations are based upon updated, biological and behavioral data in the scientific literature and the practical experience of the author.

Natural history

The ability of caregivers and managers to promote the well-being of the many captive or domesticated animals—including zebrafish—is dependent upon their understanding of the natural history of the species in question. The zebrafish is a member of the *Cyprinidae* (minnow) family of fishes, and occurs in nature across much of India, Bangladesh, and lowland Nepal (Spence et al., 2008). This geographic region is characterized by its broad diversity of habitat types and monsoonal climate with pronounced dry and rainy seasons. Zebrafish are most commonly encountered in floodplain habitats, occupying the upper to middle zone of the water column in standing or slow-moving bodies of water. They are often associated with abundant submerged aquatic vegetation, and are frequently found in rice paddies or farm ponds constructed for agriculture (Spence et al., 2007). Zebrafish are an active, shoaling species, most typically associating with each other

in small, mixed-sex schools of 5–20 individuals (Pritchard et al., 2001). The fish spawn primarily during the rainy season in shallow water, along the margins of water bodies. They are egg scatterers; the male fertilizes the eggs once the female releases them, and there is no parental care (adults will eat their own eggs if given the chance). Once fertilized, the eggs drop to the substrate where, depending on conditions, they will develop and hatch within 2–4 days. After hatching, larval fish inflate their gas bladders (an organ that controls buoyancy in fishes) by swimming up and gulping air at the surface of the water. After this, the animals begin actively seeking prey in shallow, weedy zones rich in zooplankton, their primary dietary item.

Zebrafish are typically an annual species, and reach sexual maturity within several months after hatching. Like other animals low on the food chain, their reproductive strategy is to grow up quickly and produce as many offspring as possible before they are eaten.

The above-mentioned biological factors must be taken into account when designing laboratory enclosures for zebrafish. For example, it should be possible for caregivers to exert control over water flow rates into tanks; low or no flow conditions are typically required for larval stages, whereas increased rates of flow are appropriate for adults in order to facilitate water exchange and remove solid wastes from enclosures. Enclosures should also be large enough to allow the animals to engage in normal swimming and schooling behavior. Additionally, larval and juvenile fish should not be housed with adults, to eliminate the possibility of the adults cannibalizing the younger individuals.

Species-typical behavior

Zebrafish display a rich repertoire of behaviors that are only marginally understood. Most of what is known about typical zebrafish behavior comes from observations made in the laboratory (Spence et al., 2008).

Olfaction governs many behaviors in zebrafish. They use their sense of smell to detect and discern between different dietary items (Lindsay & Vogt, 2004), distinguish kin from non-kin (Gerlach & Lysiak, 2006), and avoid predators (Speedie & Gerlai, 2008). Olfaction also plays a critical role in reproduction. It has been shown that pheromones released by the fish control mating behaviors and promote or suppress ovulation, probably depending on climatical factors (Chen & Martinich, 1975; Gerlach, 2006).

Although zebrafish are classified as a schooling species, they can be very territorial. Under certain circumstances—typically associated with competition for resources—they readily form dominance hierarchies, and aggressive interactions occur between and within sexes (Larson et al., 2005; Paull et al., 2010). Fish will compete for food or access to it (Hamilton & Dill, 2002) and spawning sites (Spence & Smith, 2005). In captive situations, these interactions can be mediated by increased fish density (within reason). In general, aggressive interactions and territorial behaviors are highest at low holding densities, and decrease as the number of animals occupying a given space is increased (Spence et al., 2008).

Water quality

Although zebrafish are exceptionally tolerant of a wide range of environmental conditions, the operational goal of those charged with their care must be the maintenance of a stable, clean, and favorable environment. Water quality and nutrition are the most important determinants of fish health and productivity. Therefore, the water in which zebrafish are housed has to be managed in such a way that it remains consistently within a specified range of chemical and physical parameters that are known to be most favorable to the species.

Recommended water quality parameters for zebrafish in the laboratory (Harper & Lawrence, 2011)

Parameter	Target Range
pH	Stable, within 6.8–8.5
Salinity	Stable up to 0–5 g/L
Alkalinity	Stable 50–150 mg/L
Hardness (g/L)	Stable, 75–200 mg/L
Total Ammonia Nitrogen (mg/L)	Zero
Nitrite (NO_2)	Zero
Nitrate (NO_3)	Up to 200 mg/L
Dissolved Oxygen (DO_2)	No less than 4 mg/L
Carbon Dioxide (CO_2)	No more than 20 mg/L
Temperature	Stable within 24–30°C

Comfortable Quarters for Laboratory Animals

The first step in this process is to ensure that the source water being used in fish housing is suitable. Contaminants, such as chemicals and heavy metals, are typically removed by running source water through deionizing resins and/or reverse osmosis filters. Once impurities are removed from the water, it can be treated with synthetic sea salts and/or buffers to create water of the appropriate salinity and alkalinity.

Once the source water is prepared, the most critical challenge to maintaining the environmental quality in a fish housing system revolves around the fact that fish excrete wastes directly into the water. The primary component of this waste, ammonia-nitrogen, is toxic to the fish, and needs to be removed. Flow-through systems remove ammonia by flushing; clean water is pumped into tanks, fish excrete wastes into the water, and the effluent is flushed out. The flow is unidirectional—clean water in, effluent water out—and may be continuous or periodic. In a recirculating system, clean water is pumped into tanks, fish excrete wastes into the water, and the effluent water is pumped into a "treatment" zone where wastes are removed before the water is returned, clean, to the fish.

Regardless of which aquaculture system is employed, the goal is always the same: maintenance of the optimal living environment for the fish. While an in-depth discussion of how this is achieved is beyond the scope of this chapter, extensive reviews of the subject are available elsewhere (Harper & Lawrence, 2010; Lawrence, 2007). Many research applications and procedures require fish to be kept in static water for varying periods of time. It is critical to remember that water quality deteriorates with time, at a rate dependent on fish density—the more fish in the water, the more quickly the water quality deteriorates in static situations. Therefore, appropriate

measures have to be taken to ensure that water quality is maintained for as long as fish must remain there. These include limiting densities of fish in the water, reductions in feeding, and manual water changes.

Tank materials and design

Tanks are the primary enclosure used to house zebrafish. They may be freestanding, but are more typically designed to be supported, along with many other tanks, on a rack in an application reminiscent of books on shelves. Tanks are commonly made from glass, acrylic, fiberglass, polycarbonate, or

polysulfone. Polycarbonate tanks are most commonly used, as they are both durable and withstand repeated sterilization in autoclaves. However, it is important to consider that both polycarbonate and polysulfone may leach bisphenol A (BPA), a synthetic estrogen mimic that has been shown to cause reproductive problems in various animals, including fish (Howdeshell et al., 2003; Segner et al., 2003). While it is unclear if polycarbonate tanks utilized in most commercially available zebrafish housing systems leach BPA in significant enough amounts to harm zebrafish, it is an issue that managers, caregivers, and scientists should be aware of.

Removal of solid wastes is a very important factor in tank design, as bacterial breakdown of solids (feces and uneaten food) can produce significant amounts of ammonia, which is toxic to the fish and can interfere with the previously mentioned "treatment zones" of aquaculture systems. Solid wastes are ideally removed by water flow from the tank to the filter system, although manual removal is sometimes necessary. Tank shape can facilitate waste removal. More often than not, the tanks are square or rectangular with sloped or V-shaped bottoms. Solids collect in the low part of the slope and are moved out of the tank by flowing water, where they can be flushed or siphoned out. Tanks that have "dead zones" where solids collect and remain must be avoided.

Tanks are typically transparent to facilitate unobstructed visualization of the fish by caregivers, but lids used to cover the tanks and prevent fish escape are often tinted blue or green to reduce light waves most conducive to algal growth. Algae and cyanobacteria are natural denizens of any aquaculture system, and colonize the surfaces of tanks, gutters, and piping. While these organisms are generally not harmful to the fish, they must not be allowed to grow to the extent that they prevent caregivers from being able to observe the animals.

Tank size
The size of tanks used to house zebrafish varies considerably, depending on the experimental or breeding application. Most commercial system vendors offer several different tank sizes ranging from less than 1 liter up to 10 liters. Tanks of 1 to 3 liters are normally used to house

larval stages of fish, or adult individuals, pairs, or small groups. Larger tank sizes are typically used to house multiple adults. For larval housing, tanks should be outfitted with screens to keep young fish in the tanks while allowing water and solids to flow out.

Housing densities

The welfare of zebrafish in laboratory settings is significantly affected by housing densities (i.e., the number of fish kept in a given amount of space). Recommendations for housing density are based on current understanding of the fish's behavior, particularly concerning the relationship between housing density and stress. Stress can be inferred from various data, most notably by measuring production of the stress hormone cortisol. The fish's behavior and reproductive performance are also very useful indicators of stress.

Typically, adult zebrafish show the highest levels of cortisol production in very low- or very high-density conditions. At low (< 1 fish/L) densities, adult zebrafish will spend nearly all of their time establishing and defending territories. While this behavior is natural, the intensity and frequency of these interactions are increased within the confines of a holding tank. Constant engagement in these aggressive activities is stressful for dominant and especially for subordinate individuals (Filby et al., 2010).

Aggressive interactions between fish decrease considerably as the number of individuals in a tank increase, as it becomes progressively more difficult for individuals to defend territories. Once densities reach a certain point, it is no longer "economically feasible" for fish to establish and maintain territories and so they stop doing it. Cortisol levels are lowest when fish are held under these conditions.

At high densities (> 40 fish/L), cortisol levels tend to increase as fish become crowded (Ramsay et al., 2006). This effect is more pronounced in situations where the fish are underfed or during experimental fasting, but does not appear to result in increased aggression. Generally, it should also be noted that growth rates are depressed at densities above 20 fish per liter. Thus, recommended housing densities for mature fish is within a range of 5–20 fish/liter—not too many and not too few fish per liter.

Space requirements also vary with life stage. The aggressive behaviors described above are driven by reproductive urges (competition for mates and spawning sites) and therefore only start to occur once the fish approach sexual maturity. Immature fish do not exhibit these behaviors and may be kept at densities of up to 50 individuals per liter without any negative effect on growth and survival rates.

Recommended housing densities for zebrafish in the laboratory

Tank Type	Age of Fish	Density (fish/liter)
Nursery	Up to 45 days	Up to 50
Adult community	45 days and beyond	Between 5 and 10

Sexually segregated housing

While fish kept primarily in sexually segregated groups exhibit improved reproductive performance when compared to fish maintained in mixed gender arrangements (Kurtzman et al., 2010), this strategy should be employed with caution. In favorable environments, adult female zebrafish will constantly produce eggs that may only be released during spawning. If females are not exposed to males, mature eggs are not released and must be resorbed. However, under typical laboratory conditions, the rate of ovulation in adult female zebrafish exceeds that of resorption. In some cases, this imbalance may cause the oviduct to become plugged or clogged with degenerating eggs. This condition, referred to as "egg-binding," results in chronic inflammation of the abdomen (Kent et al., 2012) and impairs the animal's well-being.

Egg-binding is more likely to occur when fish are housed in sexually segregated groups over long periods of time. Fish kept in mixed gender groups show lower rates of egg retention because females shed eggs during spawning that naturally takes place in tanks. For these reasons it is advisable to keep zebrafish in mixed gender groups for maintenance purposes, and house them in sexually segregated groups only if they are allowed to spawn on a weekly or biweekly basis.

Housing of individuals or single pairs

Many experimental conditions require that zebrafish be individually housed, or kept in pairs. Although they are a schooling species, zebrafish appear to tolerate long-term isolation, at least when measured by body condition and reproductive performance. However, without data on the effects of isolation on zebrafish stress and behavior, it is uncertain how such isolation might affect fish well-being and

study outcomes. Regardless, females who are isolated should be allowed to spawn with males at least once every 2 weeks to prevent egg binding and/or reproductive senescence. The fish should not be kept in pairs for extended durations, as a dominant-subordinate relationship may be established, leading to constant and intense aggression. Subordinate partners are at risk of being subjected to chronic stress, manifesting in reduced growth, impaired reproductive function, and increased susceptibility to disease, or they may even be killed by the dominant partner. Thus, zebrafish should only be kept in pairs when absolutely necessary. In experimental circumstances that do require pair-housing, the animals should be kept in this situation for no more than 7 days, and shelter (see below) should be placed in tanks to provide subordinate animals with refuge.

Well-being and environmental enrichment

There are three basic approaches that can be employed to assess the well-being of zebrafish: behavioral cues, performance, and physiology. Careful observation and understanding of fish behavior is a simple and straightforward way to assess welfare. Like all animals, zebrafish will display species-typical behaviors that are indicative of their well-being. Certain behaviors can be considered "normal"—that is, animals displaying them are unlikely to be distressed or experiencing adverse conditions; other "maladaptive" behaviors show that the fish are not able to adapt to a given situation, hence are distressed.

A list of normal and maladaptive zebrafish behaviors is shown on the opposite page. Caregivers should be trained to recognize and distinguish between these behaviors so that they can quickly react to adverse conditions that jeopardize the animals' well-being.

Typical behaviors of zebrafish in laboratory settings

Behavior type	Normal behavior	Abnormal or maladaptive behavior
Swimming	Moderately paced, constant	Darting sharply, erratic, freezing
Position	Generally parallel to surface	Head up, tail down (perpendicular to surface)
Distribution in water column	Throughout	Concentrated along bottom or top
Schooling	Loose	Tight
Aggression	Occasional chasing, displaying, biting	Constant chasing, displaying, biting
Feeding	Active, consuming all available feed presented at each feeding	Limited or no response to feed when presented
Spawning	High activity clusters of males and females in corners of tanks near surface, particularly early in the morning	No apparent clustering or grouping of males and females in tanks, increased aggression during morning hours
Ventilation rates	No discernible or only occasional movement of operculum (gill covering)	Rapid, constant movement of operculum

Performance indices (i.e., growth, survival and reproductive rate) are commonly used as surrogate measures of well-being in zebrafish (e.g., Castranova et al., 2011). In general, one can infer that the welfare of fish is good under conditions that also support normal growth and survival rates, and normal reproduction. Conversely, it is reasonable to conclude that welfare may be poor when performance is depressed in one or more of these areas.

A more objective way of assessing welfare in zebrafish is to monitor physiological indicators of stress, reproduction, and health in individuals or groups of animals. Elevated production of stress hormones and decreases in sex steroid production and immune response are negative indicators of welfare, while normal levels of stress hormones and basal or increased levels of steroid production and immune response metrics are considered neutral or positive. The context of each particular analysis has to be taken into account when interpreting these kinds of results.

The most comprehensive way to measure the well-being of zebrafish is to combine behavioral observations, performance, and physiological assessment into a single analysis. Perhaps the best-known example of this in the literature is a study by Filby et al. (2010) that characterized the consequences of social status in zebrafish. The social status of individual animals in groups of zebrafish was defined first by behavioral observations (dominant vs. subordinate), and then correlated with performance (growth, reproductive output) and physiological indicators (cortisol, sex steroid production and immune function).

While there are no established standards for providing environmental enrichment for laboratory zebrafish, there are a number of simple options described below. Even though they have not yet been tested, it is reasonable to assume that they do foster the well-being of zebrafish in the research laboratory.

Diet: While captive or domesticated zebrafish will accept a wide variety of different feed types, in the wild, they are primarily zooplanktivores (Spence et al., 2007). This becomes evident in the laboratory, as the fish show superior growth and survival when their diets are comprised of live zooplankton such as *Artemia*, rotifers,

or *Paramecium* (Harper & Lawrence, 2010). Furthermore, because the provision of these organisms in the diet allows for the fish to engage in species-typical foraging behavior, their inclusion can be considered a form of environmental enrichment.

It should be noted that live feeds present a potential biosecurity risk, and may be considered a source of nonprotocol induced variation in specific research protocols.

Lighting/Photoperiod: Zebrafish are considered diurnal, and are primarily active during the daylight hours. Some behavior, particularly spawning, is most intense at dawn and to a lesser extent in the evening (Harper & Lawrence, 2010). Zebrafish do sleep primarily at night. This circadian pattern of activity determines many biochemical, physiological, and behavioral processes in the animal, and therefore must be maintained for captive fish.

The fish should be housed in rooms with a controlled photoperiod (usually 12–14 hours light: 10–12 hours dark). While the provision of regular light/dark periods in and of itself cannot be considered a form of environmental enrichment, the manner in which it is administered can have an effect on fish well-being. Special light controls have been implemented at the author's facility to slowly ramp lights up to full intensity (54–354 lux) in the morning and slowly ramp them down to dark in the evening. This simulation of dawn and dusk can be considered a form of enrichment and should be employed when possible.

Plastic plants: In nature, zebrafish are associated with abundant aquatic vegetation (Engeszer et al., 2007; Spence et al., 2006). The fish like to utilize plants as cover and protection from predation, and during spawning and oviposition (Spence et al., 2007). In the laboratory, zebrafish display a preference for structured environments

(containing vegetation) when given a choice (Kistler et al., 2011). While the provision of live plants in zebrafish holding tanks would probably not be practical due to perceived maintenance and biosecurity issues, plastic plants should be available for fish as an essential form of environmental enrichment. This enrichment approach serves several purposes. First, the inclusion of artificial plants in housing tanks provides subordinate fish with a refuge from aggressive or dominant fish. This is particularly important when fish are kept in pairs, or at low densities. The provision of plastic plants in housing enclosures also stimulates natural spawning and enhances cycling of eggs by females when males and females are housed together in groups. As discussed earlier, this is a simple maintenance strategy that helps prevent egg retention and binding in females that may sometimes lead to health problems. Plants may also be added to breeding tanks to enhance egg production.

There are many types of plastic plants that can be readily purchased from aquarium suppliers. The best designs are those that float and extend well beneath the surface of the water so that the fish can swim through and easily maintain position within them. Preference should be given to plastic plants that are easy to remove and can be sanitized.

Further, plants with loose parts should be avoided to prevent clogging of filters.

Handling

Many experimental applications in which zebrafish are used imply frequent handling and manipulation of the animals. While most domesticated strains of fish used in the laboratory tolerate these disturbances, investigators and caregivers must be aware that these activities are always stressful for the animals. Therefore, handling should be minimized as much as possible and it should always be performed efficiently and expediently.

The skin of all fish is coated with a protective layer of mucus (i.e., "slime coat" or "slime layer") that acts as a barrier against infection and helps the animal to maintain blood salt balance. Therefore, protective measures must be taken during handling to prevent damage to the mucus, including the use of soft nylon nets, and keeping the skin of the fish moist when they are removed from the water. Oils, soaps, and lotions damage the slime coat, and so people should wash and rinse their hands thoroughly prior to handling the animals. Zebrafish also possess specialized cells in the skin that release an alarm pheromone into the water when the protective layer of mucus is damaged during injury. When other fish sense this pheromone, it elicits a strong escape response that includes rapid darting, usually at or along the bottom of the water column. Caregivers should be aware of this, as its occurrence after procedures is a sign that the animals were improperly handled.

REFERENCES

Castranova D, Lawton A, Lawrence C, Baumann DP, Best J, Coscolla J, ... Weinstein BM 2011 The effect of stocking densities on reproductive performance in laboratory zebrafish (*Danio rerio*). *Zebrafish* 8(3): 141–146

Chen LC and Martinich RL 1975 Pheromonal stimulation and metabolite inhibition of ovulation in the zebrafish, *Brachydanio rerio*. *Fish Bulletin 73*: 889–894

Engeszer R, Patterson L, Rao A and Parichy D 2007 Zebrafish in the wild: A review of natural history and new notes from the field. *Zebrafish 4*(1): 21–40

Filby AL, Paull GC, Bartlett EJ, Van Look KJ and Tyler CR 2010 Physiological and health consequences of social status on zebrafish (*Danio rerio*). *Physiology and Behavior 101*: 576–587

Gerlach G 2006 Pheromonal regulation of reproductive success in female zebrafish: female suppression and male enhancement. *Animal Behaviour 72*: 1119–1124

Gerlach G and Lysiak N 2006 Kin recognition and inbreeding avoidance in zebrafish, *Danio rerio*, is based on phenotype matching. *Animal Behaviour 71*: 1371–1377

Hamilton IM and Dill LM 2002 Monopolization of food by zebrafish (*Danio rerio*) increases in risky habitats. *Canadian Journal of Zoology 80*: 2164–2169

Harper C and Lawrence C 2010 *The Laboratory Zebrafish*. CRC Press: Boca Raton, FL

Howdeshell KL, Peterman PH, Judy BM, Taylor JA, Orazio CE, Ruhlen R and Welshon WV 2003 Bisphenol A is released from used polycarbonate animal cages into water at room temperature. *Environmental Health Perspectives 111*: 1180–1187

Kent ML, Spitsbergen JM, Matthews JM, Fournie JW and Westerfield M 2012 *Diseases of Zebrafish in Research Facilities: ZIRC Health Services Zebrafish Disease Manual sec. 6(c)*. Zebrafish International Resource Center: published online at http://zebrafish.org/zirc/health/diseaseManual.php

Kistler C, Hegglin D, Würbel H and König B 2011 Preference for structured environment in zebrafish (*Danio rerio*) and checkered barbs (*Puntius oligolepis*). *Applied Animal Behaviour Science 135*: 318–327

Kurtzman M, Craig M, Grizzle B and Hove J 2010 Sexually segregated housing results in improved early larval survival in zebrafish. *Lab Animal 6*: 183–189

Larson ET, O'Malley DM and Melloni RH 2005 Aggression and vasotocin are associated with social rank in zebrafish. *Behavioral Brain Research 167*: 94–102

Lawrence C 2007 The husbandry of zebrasfish (*Danio rerio*): A review. *Aquaculture 269*: 1–20

Lindsay SM and Vogt RG 2004 Behavioral responses of newly hatched zebrafish (*Danio rerio*) to amino acid chemostimulants. *Chemical Senses 29*(2): 93–100

Paull G, Filby A, Giddings H, Coe T, Hamilton P and Tyler C 2010 Dominance hierarchies in zebrafish (*Danio rerio*) and their relationship with reproductive success. *Zebrafish 7*(1): 109–117

Pritchard VL, Lawrence J, Butlin RK and Krause J 2001 Shoal choice in zebrafish, *Danio rerio*: the influence of shoal size and activity. *Animal Behaviour 62*: 1085–1088

Ramsay JM, Feist GW, Varga ZM, Westerfield M, Kent ML and Schreck CB 2006 Whole-body cortisol is an indicator of crowding stress in adult zebrafish, *Danio rerio*. *Aquaculture 258*: 565–574

Segner H, Navas J, Schäfers C and Wenzel A 2003 Potencies of estrogenic compounds in in vitro screening assays and in life cycle tests with zebrafish in vivio. *Ecotoxicology and Environmental Safety 54*: 315–322

Speedie N and Gerlai R 2008 Alarm substance induced behavioral responses in zebrafish (*Danio rerio*). *Behavioural Brain Research 188*(1): 168–177

Spence R and Smith C 2005 Male territoriality mediates density and sex ratio effects on oviposition in the zebrafish, *Danio rerio*. *Animal Behaviour 69*: 1317–1323

Spence R, Ashton R and Smith C 2007 Oviposition decisions are mediated by spawning site quality in wild and domesticated zebrafish, *Danio rerio*. *Behaviour 144*: 953–966

Spence R, Fatema MK, Ellis S, Ahmed ZF and Smith C 2007 Diet, growth and recruitment of wild zebrafish in Bangladesh. *Journal of Fish Biology 71*: 304–309

Spence R, Gerlach G, Lawrence C and Smith C 2008 The behaviour and ecology of the zebrafish, *Danio rerio*. *Biological Reviews 83*: 13–34

Frogs

Frogs

Russell Yothers, LATg, CLABP, CATEP

DIVISION OF LABORATORY ANIMAL RESOURCES, UNIVERSITY OF PITTSBURGH

Although there has been a recent surge in biomedical research with frogs, they have a long history as research subjects, across many scientific disciplines. The most common frogs in research are those of the genera *Bufo, Hyla, Rana* (*Rana pipiens, Rana catesbeiana*, etc.), and *Xenopus* (*Xenopus laevis* and *Xenopus tropicalis*). Clawed frogs (*Xenopus* spp.) were used in a method of early pregnancy detection (Bellerby, 1934). Northern leopard frogs (*Rana pipiens*) have been involved in neurology studies since the 1950s (Fatt, 1952), and American bullfrogs (*Rana catesbeiana*) have been subjects of physiology research and teaching. Although frogs are less frequently found in classrooms today due to availability of computer models, many people can still recall frog dissection as a standard teaching tool within basic biology curricula. Northern leopard frogs (*R. pipiens*) enabled Briggs et al. (1952) to conduct their early work of cloning via somatic cell nuclear transfer, which was later used in the better-known cloning of Dolly the sheep (Campbell et al., 1996).

More recently, frogs have been widely used in genetic and developmental research. This is in large part due to their fecundity. For example, due to the large number of eggs produced over a lifetime (about 5,000) and the large egg size (about 4,000 times the size of a mouse egg), each *Xenopus laevis* female can produce approximately the same amount of embryologic material as 10^6 mice (Gurdon, 2002). Klein et al. (2002) described a National Institutes of Health initiative to create catalogs of genetic and genomic information for both *X. tropicalis* and *X. laevis* to aide in further research using these models. Other research with frogs includes the use of bullfrogs in ecological studies (Halverson et al., 2006; Laurila et al., 2006), and examination of various compounds produced in skin secretions for their antiviral, antifungal, or antibacterial properties (Mangoni, 2006). Until recently, most laboratory frogs were obtained from the wild. While these frogs were less expensive than captive-raised, there were many issues of local population impacts and unknown health status. Therefore, if frogs are to be used in research or in teaching, care should be taken to acquire only healthy, captive-bred animals from reputable breeders and suppliers.

Species-typical characteristics of frogs

Frogs, of which 6,800 species have been identified, are ectothermic (cold-blooded) tetrapod amphibians with large eyes on short heads attached to compact bodies by little or no neck. Their specialized hind legs end

in five toes that may or may not be webbed, and are adapted for hopping, jumping, running, climbing, swimming, or burrowing (Dodd, 2013). Since a discussion of all frog species used in biomedical research is beyond the scope of this chapter, the focus will be on the most common laboratory frogs.

Among the more commonly used laboratory species, the American bullfrog is semi-aquatic, with legs built for jumping great distances. Adult size is reported to vary greatly in different geographies, with a minimum snout-ischium length of about 95 mm in males and 108 mm in females (Howard, 1981). Clawed frogs are completely aquatic, with webbed toes containing claws that are well adapted for swimming. Tropical clawed frogs (*X. tropicalis*) are smaller than the African clawed species (*X. laevis*), with a typical male/female length of 36/50 mm for the former, compared to 82/110 mm for the latter.

Bullfrogs are typically found in warmer, stagnant grassland ponds, marshes, lakes, and streams with dense vegetation. This is somewhat different from ponds used by African clawed frogs, which are usually devoid of any higher plant vegetation, and covered in green algae. As all frogs are ectotherms, their activity is more reliant on ambient environmental temperature and humidity than the time of day.

Breeding and feeding activities tend to take place at night, when the air temperature is cooler and the humidity is higher. During the warmer daytime temperatures, activity tends to revolve around maintaining body temperature, either by basking in the sun or cooling under rocks or in the shade. Further, the activity of an individual species can vary widely, based on the available microclimate. For all extra-tropical species, a common seasonal theme is to maintain activity throughout warmer months and to then enter into torpor

In some species, toxic or noxious secretions for defense against predators will also be secreted by granular glands and exist in the mucus layer on the skin surface.

Of the species commonly found in laboratories, tadpoles are herbivorous and tend to eat algae and aquatic vegetation, while adults are carnivorous and will readily eat most invertebrates and smaller vertebrates. American bullfrogs will attack any animal smaller than themselves, and will readily eat mice, snakes, or other frogs. "In *Xenopus*, prey capture employs a combination of toothed jaws that improve the grip on the prey, forelimbs that are used to fork the prey into the mouth, and the strong hindlimbs that can be used to rake the prey with the sharp claws. This shredding action enables *Xenopus* to tackle larger food items than could otherwise be ingested whole; indeed, groups of *Xenopus* may attack the same prey and can tear the body into fragments that can then be ingested. This method of feeding is particularly useful for scavenging" (Tinsley et al, 1996).

Frogs are polygamous, with males calling to attract females to a favored mating site. American bullfrogs become extremely aggressive during the mating season, attacking all other males until a female follows to the

during the colder months. This is achieved by self-burial into mud, pond caves, or shallow hiding places.

A common characteristic shared by most frogs is the existence of a fully aquatic larval stage, often referred to as a tadpole. Tadpoles will typically lack legs, have tails for swimming, gills for respiration, and smooth, moist skin that allows water to freely enter and waste to be expelled.

This larval stage varies greatly in length of time, both within and between species. The metamorphosis to adult in the American bullfrog can be as early as a few months to upwards of 3 years. *X. laevis* have been shown to develop in the laboratory setting at 10–12 weeks (Bles, 1905).

As is typical across most of the species, all North American frogs possess lungs, but also exchange gasses across their thin skin membrane. This skin is protected from trauma and pathogens by a layer of slimy mucus, though since their skin is only semi-impermeable to water loss, they must remain in moist conditions to prevent desiccation.

egg-laying site. Ryan (1980) describes this as a resource defense polygyny, indirectly gaining access to females through defense of a critical resource, while Emlen (1976) describes it as a lek, or defending communal display grounds. Male clawed frogs produce mating calls through rapid contractions of intrinsic laryngeal muscles, as they lack the vocal sac found in most frog species. When the female hears these calls, she will respond with a "rapping" acceptance call or a "ticking" rejection. Regardless of attraction technique, acceptance typically leads to the male grasping the female in amplexus, the female releasing the eggs, and external fertilization occurring in the water.

Addressing the species-typical characteristic of frogs in the research laboratory

As virtually all frog species used in research are aquatic or semi-aquatic, one of the most important considerations in their laboratory care is the quality of their water. It is important to provide water that is clean and of the correct temperature, pH, alkalinity, oxygen level, and other parameters. "Standards for acceptable water quality, appropriate parameters to test, and testing frequency should be identified at the institutional level" (National Research Council, 2011, p 78), by those knowledgeable of the natural habitat and history of the specific species. Green (2010, p 36) provides a table of values for *Xenopus laevis*; she recommends 17–24°C, 6.5–8.5 pH, 500–3000 μS conductivity, 50–200 mg/L ($CaCO_3$) alkalinity, and >7 mg/L dissolved oxygen; she also stresses the need for all chlorine or chloramines to be removed. These parameters can also be followed in a typical laboratory if setting general levels across multiple species, though further investigation may be warranted for housing more exotic frogs. Currently, requirements for space and tank densities for frog species are not well

established, but should account for the behavioral needs of the species. Semi-aquatic species, such as bullfrogs, must be provided with both a terrestrial space to bask and feed outside of the water, as well as sufficient water to allow them to submerge and hydrate. Tank height must also account for jumping, and not allow for escape, or injury on lids. *Xenopus* spp. are completely aquatic and have typically been housed at 2 liters of water per frog (National Research Council, 2011, p 83), though different systems allow variable housing densities. In their natural habitat, the common laboratory species interact exclusively for mating; however, most frogs are typically tolerant of each other and can be housed in groups, given sufficiently sized tanks—the exception being male bullfrogs during the mating season.

Frogs will spend a vast majority of their time remaining still and under cover to keep alert for potential predators, as well as potential prey. As such, their tanks should be provisioned with ways to remain hidden.

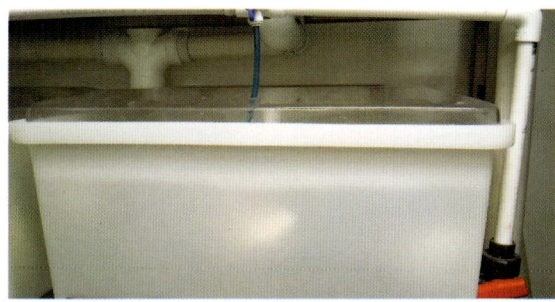

Opaque or translucent tanks are recommended, as well as PVC tubes or other shelters. Faux foliage can be used to float on top of the water to provide a barrier at that level, which is also easily moved for observation. If faux foliage is used, it should be cleaned when the tank is sanitized, in a disinfecting level (per label) bleach solution, followed by a very thorough rinse in clean, unchlorinated water.

Most of the frogs species used in research will readily consume a pelleted diet designed for them. Crickets and mealworms are also easily obtained and maintained sources of live feed, allowing for typical hunting behavior. For many aquatic species, it is critical to schedule cleaning soon after feeding, as residual food very quickly produces molds and an uninhabitable environment.

Handling of frogs should be carefully considered, based on the species, and kept to the minimum required. Gloves are suggested, as a means of minimizing the opportunity for contact with toxins and poisons in the protective mucus layer and to prevent damage to the layer. Also, gloves prevent the spread of *Salmonella* infections, as the *Salmonella* bacteria are frequently harbored by frogs. Importantly, Gutleb et al. (2001) suggests that latex may be toxic to some species, and any gloves used must certainly be of the powder-free variety. The careful and proper use of a suitably sized net can also reduce handling concerns. Nets should be deep enough to encase the entire frog, and have an opening sufficient to allow entry without causing trauma. A slow scooping motion should be used, taking care not to startle any frogs, as they may injure themselves or others while trying to escape.

Refining husbandry and research procedures for frogs

As little is known regarding appropriate tank densities and the effect of group housing on an otherwise solitary species, more research in this area must be done. This is also true regarding what may be the minimum space requirements that would allow adequate space to swim for all species and adequate space to jump for some semi-terrestrial species. A reliable method for identifying individual frogs also requires further investigation. Due to their delicate skin membrane, most marking systems commonly used in other species, such as tattoos or dyes, are largely ineffective, temporary, or unsafe. While toe clipping is becoming less acceptable in rodent species, it is, unfortunately, still fairly common in frogs. Other methods, such as transponders, are being explored, though a primary "best practice" recommendation has yet to be identified.

Furthermore, it is not a common practice in the laboratory to allow frogs an opportunity to experience the biologically typical state of torpor, or hibernation, by means of lowering temperatures to the point that this is induced. This may be standard practice in some systems, but it is far from universal. The effect of allowing or preventing this is not known, and warrants further study.

REFERENCES

Bellerby CW 1934 A rapid test for the diagnosis of pregnancy. *Nature 133*: 494–495

Bles EJ 1905 The life-history of *Xenopus laevis*, Daud. *Transactions of the Royal Society of Edinburgh 41*: 789–821

Briggs R and King TJ 1952 Transplantation of living nuclei from blastula cells into enucleated frogs' eggs. *Proceedings of the National Academy of Sciences 38*: 455–463

Campbell K, McWhir J, Ritchie W and Wilmut I 1996 Sheep cloned by nuclear transfer from a cultured cell line. *Nature 380*: 64–66

Dodd C 2013 *Frogs of the United States and Canada, Volume 1*. Johns Hopkins: Baltimore, MD

Emlen S 1976 Lek organization and mating strategies in the bullfrog. *Behavioral Ecology and Sociobiology 1*: 283–313

Fatt P and Katz B 1952 Spontaneous subthreshold activity at motor nerve endings. *The Journal of Physiology 117*: 109–28

Green SL 2010 *The Laboratory* Xenopus *sp: A Volume in the Laboratory Animal Pocket Reference Series*. CRC Press: Boca Raton, FL

Gurdon J 2002 Perspective on the Xenopus Field. *Developmental Dynamics 225*: 379

Gutleb A, Bronkhorst M, Berg J and Murk A 2001 Latex laboratory-gloves: An unexpected pitfall in amphibian toxicity assays with tadpoles. *Environmental Toxicology and Pharmacology 10*: 119–121

Halverson MA, Skelly DK and Caccone A 2006 Kin distribution of amphibian larvae in the wild. *Molecular Ecology 15*: 1139–1145

Howard RD 1981 Sexual dimorphism in bullfrogs. *Ecology 62*: 303–310

Klein SL, Strausberg RL, Wagner L, Pointius J, Clifton SW and Richardson P 2002 Genetic and genomic tools for Xenopus research: The NIH Xenopus Initiative. *Developmental Dynamics 225*: 384–391

Laurila A, Pakkasmaa S and Merila J 2006 Population divergence in growth rate and antipredator defences in *Rana arvalis*. *Oecologia 147*: 585–595

Mangoni ML 2006 Temporins, anti-infective peptides with expanding properties. *Cellular and Molecular Life Sciences 63*: 1060–1069

Mazzoni R, Cunningham AA, Daszak P, Apolo A, Perdomo E and Speranza G 2003 Emerging pathogen of wild amphibians in frogs (*Rana catesbeiana*) farmed for international trade. *Emerging Infectious Diseases 9*: 995–998

National Research Council 2011 *Guide for the Care and Use of Laboratory Animals, Eighth Edition*. National Academy Press: Washington, DC

O'Rourke D 2007 Amphibians used in research and teaching. *ILAR Journal 48*: 183–187.

Ryan Michael J 1980 The reproductive behavior of the bullfrog (*Rana catesbiana*). *Copeia 1*: 108–114.

Tinsley RC and Kobel HR (eds) 1996 *The Biology of Xenopus*. Clarendon Press: Oxford, UK

Zayas JG, O'Brien DW, Tai S, Ding J, Lim L and King M 2004 Adaptation of an amphibian mucociliary clearance model to evaluate early effects of tobacco smoke exposure. *Respiratory Research 5*: 9

Cattle

Cattle

David W. Cawston, BS, MHA, RLAT

Cattle in the research environment can be easily cared for and managed with attention to a main principle of cattle behavior: herd-based living. Providing cattle with access to their herd-mates should be the main focus in providing them with a positive, cattle-appropriate environment not only on the farm but also in the research setting. Through an understanding of the animals' natural social system and its details, species-adequate management of cattle can be achieved. In addition to an understanding of the social behavior of cattle, both humane stockmanship and thoughtful facility design contribute greatly to their welfare. This chapter identifies the species-typical behaviors of cattle and recommend ways to house the animals in the best possible environment.

Cattle-typical behavior

Cattle live in herds as a primary means of protection from predators. The size of the herd is dependent upon the natural resources available. The social structure of a herd is based upon social rank relationships. Shifts in rank position are inherent in the dynamic interdependency between age, body weight and current social rank. Up to the age of about 9 years, cattle tend to gain in dominance until they occupy the best positions in the herd's social hierarchy. This

is usually a short-lived dominance. In general, after the age of about 10 years, cattle show a progressive decline in status, which is paralleled by a gradual loss of body weight and body strength. Tension observed in cattle occurs more often among animals who are of similar age and size. Younger/smaller animals do not typically create problems as they are naturally submissive towards older and larger animals and avoid social conflicts with adults in the herd (Wagnon et al., 1966; Reinhardt & Reinhardt, 1975).

Cattle are typically diplomatic when they demonstrate or assert their social positions, as they use facial gestures and various body positions before resorting to overt aggression. Overt, possibly injurious, aggression is very rare in free-ranging cattle (Schloeth, 1961; Reinhardt et al., 1986). Dominant animals have privileges that are respected by their subordinate rank partners. Subordinate animals will move out of the way when a dominant partner shows the intention to get access to a resource such as water, food, comfortable resting place, or shade.

This behavior can become an issue when groups of cattle are kept in confined housing conditions, as low-ranking animals may be deprived of essential resources. Subordinate

animals may also have problems keeping an acceptable distance from their dominant partners because of spatial restrictions. This in turn can trigger threat displays or overt aggression in dominant animals; the lower ranking partners will try to yield—often without success. Thus, the lack of space can cause high social tensions and social distress, particularly in low-ranking members of the herd (Keeling, 2001; Huzzey et al., 2006; de Vries et al., 2013).

In a typical cattle herd, mother cows wean their calves when they have reached the age of about 10 months, although there is a marked gender difference. Female calves are prevented by their mothers from suckling when they are about 9 months old, but male calves are weaned by their mothers at approximately 11 months of age (Reinhardt & Reinhardt, 1981). Calving intervals are significantly shorter when cows are allowed to wean their calves naturally than when their calves are prematurely weaned by forcefully removing them from their mothers, possibly due to the intense stress caused by the separation (Reinhardt, 1982). Young bulls can start copulating with cows several months after they have been weaned naturally, although adult bulls tend to prevent them from mating before about 16 months of age. In an undisturbed setting, heifers conceive at the average age of 25 months.

Cattle develop strong cohesive relationships with other herd members, primarily within a matrilineal lineage. Mother cows prefer their progeny over unrelated calves as grooming and grazing partners. These affiliations can last for many years beyond the time when the calves have become fully mature sexually. Comparable attachments also exist between siblings. This social cohesiveness is a key component of a successful herd and is reflected in their many synchronized behaviors (i.e., grazing, resting, movement).

Being removed from familiar conspecifics is very distressing for cattle. This is particularly true when the strong mother-calf bond is forcefully broken in the process of premature weaning (von Keyserlingk, 2007); but adult cattle also show intense distress responses—such as behavioral and vocal agitation, increased heart rate, and increased cortisol output—when they are removed from other conspecifics and kept alone (Hopster & Blokhuis, 1993).

Cattle spend most of the 24-hour day at rest, either sleeping or ruminating. When they have a choice, cattle prefer lying on a relatively soft and dry surface rather than on a hard and wet one (Von Wander, 1976; Irps, 1983; Jensen et al., 1988). As such, cattle used in biomedical research should be provided with bedding to create a comfortable resting surface. Acceptable options for bedding can include straw, corn stover, old hay, soybean residue, oat hay, wheat straw, and wood chips/shavings or paper. Mixing bedding types is recommended. Sawdust or paper can be used to soak up the copious urine and fecal production, while straw or shavings provide comfortable bedding. When cleaning or rousing resting cattle, it is important to remember that typical cattle behavior involves urinating and defecating upon standing. Urination and defecation while lying down can be a sign of health issues. Availability, cost, and concern about contamination potential for experimental design are factors that may be considered when determining the bedding options.

A strong social disposition, along with general amiability, are the key characteristics that led to the domestication of cattle more than 10,000 years ago. When they feel threatened, cattle will not defend themselves, but rather run away as a group. They are usually not aggressive animals and do not shun contact with humans who are trustworthy. In fact, they are curious and will approach a friendly human.

Comfortable Quarters for Laboratory Animals

They not only love to be groomed by other conspecifics but they equally enjoy it when a person rubs and scratches them firmly but also gently. When no grooming partner is available, cattle will scratch themselves on low branches, tree trunks, or other suitable objects to relieve itching caused by flies, parasites, or dirt.

Cattle are remarkably sensitive to prolonged exposure to direct sunlight and are prone to suffer from heat stress when they have no access to shade (Kidd, 1993). Heat stress causes cattle to exhibit changes in their behavior, physiology, and immune function; it not only creates a welfare concern but also affects reproduction (Silanikove, 2000).

Addressing cattle-typical behavior in the research lab

Cattle should always be housed in a social setting. It has been shown in heifers that they need the presence of at least one other conspecific to cope with distressing circumstances such as being moved to an unfamiliar environment (Veissier, 1992). The presence of a familiar person also buffers behavioral but not physiological stress responses to an unfamiliar environment (Rushen et al., 2001).

If a cow, heifer, or calf is to be housed individually for officially approved research-related reasons, prior housing arrangements must be made to assure that the research subject can maintain uninterrupted visual and acoustical contact with at least one nearby, familiar cow, heifer, or calf. A big mirror can distract a socially isolated heifer for a short while (Piller et al., 1999), but it is not an acceptable substitute for having a companion nearby.

Transferring cows or heifers into different groups is a source of stress and must be avoided in the research setting (Schein et al., 1955; Porzig, 1969; Arave & Albright, 1976; Dobson et al., 2001; Rousing, 2006). When cattle are removed from their familiar environment and introduced into a different one, numerous stressful and even distressful situations are created that are bound to destabilize the animals' physiological equilibrium. All of the following may occur: (a) The animal who is removed from the herd will be distressed; (b) with one herd member being removed, the old herd will have to reorganize its hierarchical structure in order to restabilize the social system—this process is bound to temporarily increase social tensions between all herd members and increase stress or distress in some individuals who lose high rank positions; (c) the introduction of the new animal into the already established social rank system of the new group will necessitate a shifting of rank positions so that the newcomer gets integrated; this process

will also lead to heightened social tension within this group and will be associated with high stress levels for the newcomer.

Cattle need sufficient space to avoid social conflicts arising from subordinate animals being unable to maintain an appropriate distance from dominant herd members, especially in situations of competition over access to food, water and comfortable resting sites. Animals of lower social rank are displaced during feeding, drinking and resting more often than higher-ranking animals (Huzzey et al., 2006). The design of a research facility that works with cattle must take the social hierarchy of a cattle herd into account. There must be enough feeding spaces, enough drinking spaces, and a comfortable resting area that is large enough so that even the lowest ranking member of the herd can access these basic resources without fear of being pushed away by higher ranking herd members. Visual barriers have a protective effect for subordinate animals during feeding time (Bouissou, 1978; Huzzey et al., 2006).

Cattle prefer somewhat cooler temperatures than most research species. Cattle housed in a laboratory setting should be kept at a thermo-neutral temperature range of about 18°C (65°F; Keown & Grant, n.d.).

Unlike at production facilities, cattle in research settings do not need extra grain and silage. Diet should be primarily a high-quality hay, provided by a reputable source and kept in a clean and dry location. Alfalfa cubes and grain-based diets should be avoided in the laboratory setting, as they do not have sufficient fibrous content to maintain proper gut health. However, it is acceptable to provide them in small quantities as dietary treats.

Prematurely removing calves from their mothers creates a distressing situation, both

for the calves and the mothers. Calves who have been naturally weaned by their mothers will show significantly lower physiological stress responses to disturbing situations than calves who have been prematurely removed from their mother in order to be artificially weaned (Lay et al., 1992).

Research facilities must take care to avoid common maladies such as lameness and leg injuries. Solid, slightly rough flooring is optimal to give cattle the best footing and sense of security. Raised flooring, while attractive because it allows for dispersal of urine and feces, should be avoided, unless it is does not move and is not slippery when wet. Adequate, deep bedding and/or the use of stall mattresses are recommended. Frequent addition of new bedding material, as well as providing regular upkeep to that bedding material, are important, as cattle produce large quantities of urine and feces (Chapinal, 2013). Care should be taken to ensure that dirty bedding does not become caked on the skin or in the hooves. Routine examination of hooves, followed by appropriate trimming and/or cleaning, will minimize hoof-related health problems.

When moving cattle, it is important to provide them with solid footing to prevent slips and reduce anxiety from walking. This can be difficult in many research facilities, as floors are designed to be easy to clean (i.e., smooth). Concurrent with problems from walking on smooth surfaces is the issue of flooring changes. Cattle can become unnerved when they encounter a change in flooring and may stop walking. Pulling them is not recommended, as this is highly stressful to the animals and can result in slips and injuries. Placing mats on the floor can be helpful. Gentle pushing from alongside the animal can also be used.

Tie stalls—where the animals are tethered to the front of the stalls by neck collars, keeping them in the stalls and preventing them from roaming the facility—do not provide proper housing conditions for cattle and must be avoided in the research setting; it hinders normal lying-down behavior and can frequently result in knee and hock inflammations (Krohn & Munksgaard, 1993). If the research objectives require tethering, then the tether must be long enough to allow the animal to stand and lie down in a species-typical manner, defecate away from bedding, and attend to other physiological needs.

Stall design must allow cattle to exhibit species-typical body movement and postures. Anderson (2001) states that cows should have a resting area that provides them the freedom to—

» stretch the legs forward;
» lie on one side, with unobstructed space for neck and head;
» rest the head against one side without hindrance;
» rest the whole body, including the tail, on the platform;
» stand or lie without fear or pain from neck rails, partitions, or supports; and
» rest on a clean, dry and soft bedding.

Comfortable Quarters for Laboratory Animals

Cattle (particularly calves) will take advantage of many inanimate objects as enrichment, engaging in play behavior with them and using them to groom. Recently, commercially produced automatic articulating brushes have been introduced to provide cattle with a way to engage in this behavior. The brush remains idle and starts moving upon contact with the animal. The author has witnessed several of these brushes and they are widely used by the animals; the brush allows them to be groomed over the entire top half of their body and face. The animals wait in line for their turn to use such a brush.

Individuals working with cattle in the research setting must have a good understanding of typical cattle behavior and how to work with the animals to accomplish their goals. The fact that cattle are relatively docile animals who show little or no self-defense behaviors makes them susceptible to callous treatment in stockyards (e.g., yelling or shouting, slapping, punching, hitting with an object, tail twisting, or use of an electrical prod). Cattle are sensitive animals, who respond positively to kind and gentle interactions (Kidd, 1994).

REFERENCES

Anderson N 2001 Time-lapse video opens our eyes to cow behavior and comfort. *American Association of Bovine Practitioners 31*: 35-42

Arave CW and Albright JL 1976 Social rank and physiological traits of dairy cows as influenced by changing group membership. *Journal of Dairy Science 59*: 974-981

Bouissou MF 1978 Relations sociales chez les bovins domestiques dans les conditions d'élevage moderne. *Proceedings of the First World Congress on Ethology Applied to Zootechnics*: 267-274

Chapinal N 2013 Herd-level risk factors for lameness in free stall farms in northeaster United States and California. *Journal of Dairy Science 96*: 318-328

de Vries M, Bokkers EAM, van Schaik G, Botreau, R, Engel B, Dijkstra T and de Boer IJM 2013 Evaluating results of the Welfare Quality multi-criteria evaluation model for the classification of dairy cattle welfare at the herd level. *Journal of Dairy Science 96*: 6264-6273

Dobson H, Tebble JE, Smith RF and Ward WR 2001 Is stress really all that important? *Theriogenology 55*: 65-73

Hopster H and Blokhuis HJ 1993 Consistent stress response of individual dairy cows to social isolation. *Proceedings of the International Congress on Applied Ethology*: 123-126

Huzzey JM, DeVries TH, Valois P and Von Keyserlingk MAG 2006 Stocking density and the feed barrier design affect the feeding and social behavior of dairy cattle. *Journal of Dairy Science 89*: 126-133

Irps H 1983 Results of reseach projects into flooring preferences of cattle. In: Baxter SH, Baxter MR and MacCormack JAC (eds) *Farm Animal Housing and Welfare* pp 200-215. Marinus Nijhoff: The Hague, Netherlands

Jensen R, Recèn B and Ekesbo I 1988 Preference of loose house dairy cows for two different cubicle floor coverings. *Swedish Journal of Agricultural Research 18*: 141-146

Keeling JL 2001 *Social Behavior in Farm Animals*. CABI Publishing: New York, NY

Keown JF and Richard JG [n.d.] How to Reduce Heat Stress in Dairy Cattle. University of Missouri Extension. http://extension.missouri.edu/p/G3620

Kidd R 1993 Help livestock keep their cool: Water and shade are keys to comfort. *The New Farm 15*(5): 8-12

Kidd. R 1994 Put away your prod: Herd stock with less stress by understanding how they think. *The New Farm 16*(5): 6-10 & 44

Krohn CC and Munksgaard L 1993 Behaviour of dairy cows kept in extensive (loose housing/pasture) or intensive (tie stall) environments II. Lying and lying-down behaviour. *Applied Animal Behaviour Science 37*: 1-16

Lay DC, Friend TH, Randel RD, Bowers CL, Neuendorff DA, Grissom KK and Jenkins OC 1992 Does maternal deprivation affect a calf's physiological and behavioral reactions to later stress? *Journal of Animal Science 70*(Supplement 1): 162

Piller CAK, Stookey JM and Watts JM 1999 Effects of mirror-image exposure on heart rate and movement of isolated heifers. *Applied Animal Behaviour Science 63*: 93-102

Porzig E and Wenzel G 1969 Verhalten der Milchkühe nach der Umstellung aus dem Abkalbestall in den Boxenlaufstall. *Tierzucht 23*: 535-537

Reinhardt C, Reinhardt A and Reinhardt V 1986 Social behaviour and reproductive performance in semi-wild Scottish Highland cattle. *Applied Animal Behaviour Science 15*: 125-136

Reinhardt V 1982 Reproductive performance in a semi-wild cattle herd (*Bos indicus*). *Journal of Agricultural Science 98*: 567-569

Reinhardt V and Reinhardt A 1981 Natural sucking performance and age of weaning in zebu cattle (*Bos indicus*). *Journal of Agricultural Science 96*: 309-312

Rousing T and Wemelsfelder F 2006 Qualitative assessment of social behavior of dairy cows housed in loose housing system. *Applied Animal Behaviour Science 10*: 40-53

Rushen J, Munksgaard L, Marnet PG and DePassillé AM 2001 Human contact and the effects of acute stress on cows at milking. *Applied Animal Behaviour Science 73*: 1-14

Schloeth R 1961 Das Sozialleben des Camargue Rindes. *Zeitschrift für Tierpsychologie 18*, 574–627.

Schein MW, Hyde CE and Fohrman MH 1955 The effect of psychological disturbance on milk production in dairy cattle. *Proceedings of the Association of Southern Agricultural Workers 52nd Convention*: 79-88

Silanikove N 2000 Effects of heat stress on the welfare of extensively managed domestic ruminants. *Livestock Production Science 67*: 1-18

Veissier I and Le Neindre P 1992 Reactivity of Aubrac heifers exposed to a novel environment alone or in groups of four. *Applied Animal Behaviour Science 33*: 11-15

Von Keyserlingk MAG 2007 Maternal Behavior in cattle. *Hormones and Behavior 52*: 106-113

Von Wander JF 1976 Haltungs- und verfahrenstechnisch orientierte Verhaltensforschung. *Züchtungskunde 48*: 447-459

Pigs

Pigs

Evelyn K Skoumbourdis, MS, LATg

Swine and human beings have had an important relationship since the swine's domestication, about 9,000 years ago in central Asia. The domesticated swine has been a secure food resource for many human societies. The utilization of swine in medical research dates as far back as the 16th century; however, there has been a significant increase in their use as research models over the past 50 years (Bollen et al., 2000). Swine have been used for a diverse range of studies, including cardiovascular, integumentary, and behavioral. Swine have also become a primary surgical training model for organ transplantation and other complex surgical procedures. (Smith & Swindle, 2006). They are hearty, highly intelligent animals who adapt relatively well

to the laboratory environment. However, those in charge of their care must take into consideration their natural behaviors and dispositions so the animals are not stressed by species-inadequate housing and handling conditions. Stressed swine will be difficult to manage and will not provide reliable research data.

Natural environment and species-typical behaviors of pigs

Wild pigs (*Sus scrofa*) are highly adaptable and are able to live in almost any environment, but tend to be found mostly in river valleys and wooded areas. Their home range may be anywhere from 1 square mile (1.6 km^2) to up to 10 square miles (16.1 km^2), depending upon food availability, with a smaller range normally preferred (Arey & Brooke, 2006).

Wild pigs reside in small, strongly bonded social groups of three to five animals, with social rank determined by sows and their young. Younger males will often form bachelor groups, but become solitary once they reach adulthood and sexual maturity. Social rank begins very early in swine, with piglets assuming a teat order just a few days after birth, with the more dominant piglets consistently using the anterior teats for the superior milk supply. Once established, social

Comfortable Quarters for Laboratory Animals

rank will remain stable unless new animals are introduced to the group, at which time some fighting may ensue to reestablish the hierarchy (Arey & Brooke, 2006; Bollen et. al., 2000).

Pigs are opportunistic omnivores and will spend a majority of their day rooting for food and eating many small meals throughout the day (Bollen et al., 2000). Rooting behavior is extremely important (sometimes seen as more important than the food consumption itself) and pigs will often spend time simply exploring their environment. They tend to be most active in the early morning and then again in the evening. The rest of the day is spent resting, rubbing against trees/rocks/shrubs, and wallowing in water and/or mud. These activities allow for the removal of parasites and also help the pig to keep cool and free from itchy, dry skin (Arey & Brooke, 2006).

Housing

When housing swine in the lab, it is important to remember the natural environment to which they would gravitate. The *Guide for the Care and Use of Laboratory Animals* provides size guidelines for the housing of one to several animals (National Research Council, 2011). Common sense should be taken into account, regardless of pen size guidelines, in order to ensure the pen is large enough for the animal(s) to fully turn around without bumping into a pen side, and that there is enough room to accommodate a defecation area separate from the feeding area. Solid, rough surface flooring with a substrate such as straw works well. Straw bedding reduces physical discomfort in pigs at all stages of life. Additionally, it provides a suitable opportunity for pigs to forage, promotes activity, and reduces abnormal behaviors such as apathy, stereotypies and anti-social behavior (Arey, 1993; Burbidge et al., 1994; Bolhuis et al., 2005). If however, straw is not ideal due to floor drains, a drain trap may be placed (Batchelor, 1991) or coated grated floors may be utilized. Uncoated slatted flooring is not recommended, as it does not provide good footing and pigs tend to slip. If housed on raised floors, swine should be provided with a large strong mat for resting and their hooves should be trimmed often, as they will not wear down naturally on the grates.

If possible, providing indoor/outdoor housing is an optimal way to enrich the pigs' environment, although occasional access to the outdoors is a reasonable compromise. If provided, outdoor flooring should be rough concrete and easily sanitized. Also, if swine are allowed free access to the outdoors, precautions should be taken to keep the animals from overheating or getting sunburned. Water baths, sprinklers, and sunscreens are highly recommended.

The housing area should not be barren. Swine require an enriched environment, including toys, novel objects, and foods, in order to stave off boredom. If no opportunities for rooting are provided, pigs will quickly turn to substitutes such as flooring, pen mates, and/or human caretakers. Additionally, the housing area should contain a properly placed and fixed object for the animal(s) to rub against. The author has found round, heavy-duty

brushes from floor buffers to be ideal. The brushes are easily hung onto pens via heavy chain and brackets, and are strong enough to withstand larger swine heavily rubbing against them and/or rooting against the bristles.

Socialization

With the exception of intact boars, swine should be housed in pairs or groups, unless there is a scientifically supported justification for single housing approved by the veterinarian or Institutional Animal Care and Use Committee. If pigs must be housed alone, they should have access to visual, auditory, olfactory, and somatic contact with another pig to obviate physiological stress responses to social isolation (Herskin & Jensen, 2000). Side-by-side pens with open bars allow for snout touching, active play, and sleeping next to each other. Also, unless prohibited due to health or research status, when their pens are cleaned, pigs should be free to run and play together in the room. This allows not only for socialization, but also for the pigs to stretch and expend pent-up energy.

Domestic swine have a social structure similar to that of their wild counterparts and will develop strong social bonds. However, fighting may occur if new animals

Comfortable Quarters for Laboratory Animals

are introduced or brought into an already established group (Rushen, 1990; Barnett et al., 1994; Arey & Franklin, 1995; D'Earth, 2004). Thus, it is recommended that new animals be introduced in a neutral environment in order to lessen the likelihood of fighting. If it is necessary to pair or re-pair swine often, it is best to have a separate room for the introductions. The author, along with colleagues, found that a playroom where swine were allowed to root for treats in pine shavings, as a pair or group, decreased the likelihood of aggression and facilitated relatively smooth introductions. This led to the formation of rank relationships that remained stable when the animals were moved as a group to a new stall (Casey et al., 2007).

Training

Swine are intelligent and are easily trained to cooperate with husbandry and research procedures, such as standing still for biophysical measurements (Chilcott et al., 2001), permitting physical examinations, electrocardiography, dermal dosing (Blye et al., 2006), and even nasal dosing (Brodersen et al., 2010). Regardless of the procedure, force should not be used at any time. Swine are strongly averse to force and will not only become uncooperative, but will develop fear and distrust of the handler, which may result in the animal resorting to defensive, potentially injurious aggression toward all staff. Additionally, rough handling has profound effects on the normal physiology and behavior of swine (Hemsworth & Barnett, 1991). Swine who are fearful of humans show a marked increase in corticosteroid levels both in the presence and in the absence of people (Barnett & Hemsworth, 1986). Thus, rather than force, positive reinforcement and gentle handling methods should be utilized. Techniques such as target training and clicker training can allow a handler to lead a swine into a holding cage or

onto a weight scale (Pell et al., 2010; Neubauer et al., 2011), or to train a swine to cooperate for blood collection. Squeeze bottles containing juice are very useful, as they allow reward to be squeezed directly into the swine's mouth during training. Other small food items such as jelly beans, pieces of cookie, mini marshmallows, or manufactured treats also serve well as reward items. However, one must be careful not to use too many treats, as the animal may start to refuse to eat the standard diet.

Should restraint be necessary for a project, swine should never be tethered, as they become highly stressed (Barnett et al., 1985; Schouten et al., 1991) and will often vocalize from fear. Pig boards are useful for short-term restraint of larger swine, but for longer periods of time, a sling is recommended. Swine often learn quickly to walk into the sling, but may also be trained to be placed by lifting (Williams & Watson, 2003). Over just a few sessions, swine become quite comfortable and will remain relaxed—or even fall asleep—for several hours (Panepinto et al., 1983; Grandin et al., 1989).

Enrichment

Due to their high level of intelligence and behavioral needs, swine require an enriched environment. If left in a barren environment, swine may resort to stereotypical behaviors such as pacing, or may begin to chew on enclosures, feed/water bowls, or even pen mates. Thus, it is very important to provide environmental enrichment for swine, with rooting activities being the most important of all, as rooting is a behavioral need that is performed regardless of feeding level or nutritive feedback (Beattie & O'Connell, 2002).

Swine have an innate need to root and will do so on any and every surface within their reach. Straw/hay with hidden treats such as jelly beans, pieces/chunks of apple, whole apples (pigs LOVE apples!), manufactured treats, etc. allow swine to express this important species-typical behavior. Swine will also root through straw/hay when no treats are present (Fraser et al., 1991), making this activity an option even in those cases where novel food items may be restricted due to research protocols. In such cases, scattering food on the floor, or on the straw bedding, is an excellent alternative to bowls, as this allows the pigs to perform their natural rooting behavior without added confounds (Beale et al., 2007). In cases of raised floors and/or drains where straw/hay may be problematic for sanitation, it is recommended to provide opportunities for the expression of this activity in another separate area specifically dedicated for the purpose (Casey et al., 2007). Other enrichment devices such as balls, Kongs, and thick-walled cardboard tubes are recommended for indoor pens with drains.

Filling toys such as Kongs with pieces of fruit or other treats adds more incentive for the pig to use them. Boomer Balls (of appropriate size) filled partially with juice and then frozen can provide a challenge and a great deal of fun for all swine.

Pigs also enjoy playing with hanging items (Young et al., 1994). Tug toys, pieces of fire hose, Jolly Apples and other hanging items—such as forage balls baited with treats (Huntsberry et al., 2008)—are well received by swine of all sizes. The author has found that short sections of hanging chain seem to always be of interest to swine, although they prefer to play with soft pliable objects when given a choice (Grandin, 1988). Pieces of thin cloth such as bed sheets can make great tug toys for swine (Grandin & Curtis, 1984). They are readily tied to pens and are easy to change/replace when they become soiled—a necessary practice since swine will ignore or avoid any enrichment item that

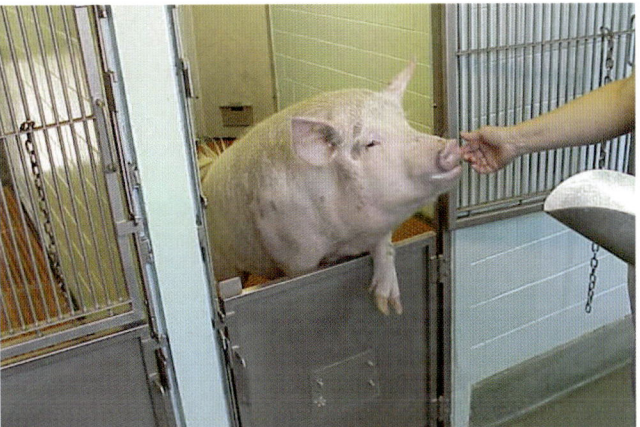

becomes soiled with fecal material (Grandin, 1988). Any enrichment item that becomes soiled must be cleaned or removed from the pen in a timely fashion.

Human interaction

Swine are very social and affectionate animals, making it easy for their caregivers to forge a positive relationship with them. Regular, positive human interaction not only helps pigs become more comfortable in their surroundings (Geers et al., 1995) and overcome fear and stress responses to people (Gonyou et al., 1986), but may also serve as an appropriate enrichment for the singly housed swine. The simple act of squatting down, speaking softly, and providing a good snout rub can allow for a bond to form between the pig and the person almost immediately. Additionally, interactive enrichment activities help to make the swine more comfortable in their surroundings. Playing "sprinkler" with a garden hose (being careful not to get the pig overly wet!) or tossing cut-up apples into a water bowl for bobbing are great activities for all staff in the institution to share with the swine. The more positive interaction pigs receive, the more likely they will be calm and cooperative during husbandry and research procedures.

Final thoughts

The late Maurice Sendak centered one of his final books upon an orphaned piglet: Bumble-Ardy, who's only wish is to have a birthday party. In an interview with Avi Steinberg of the *Paris Review*, in 2011, when asked why he chose a pig, he said, "I've always loved pigs: the shape of them, the look of them, and the fact that they are so intelligent. ... The prospect of drawing pigs was something I could look forward to, and I needed something to look forward to."

Working with pigs is indeed something to look forward to. It is highly rewarding, and, at times, is a great deal of fun. A naturally cheerful species, pigs are capable of generating great joy, and often give more than they receive. The author hopes the tips provided in this chapter will bring many pigs, and many people, more fulfilling days in the laboratory.

REFERENCES

Arey DS 1993 The effect of bedding on the behaviour and welfare of pigs. *Animal Welfare 2*: 235–246

Arey DS and Brooke P 2006 *Animal Welfare Aspects of Good Agricultural Practice: Pig Production.* Compassion in World Farming Trust: Hampshire, UK. http://www.fao.org/fileadmin/user_upload/animalwelfare/gap_book_pig%20production.pdf

Arey DS and Franklin MF 1995 Effects of straw and unfamiliarity on fighting between newly mixed growing pigs. *Applied Animal Behaviour Science 45*: 23–30

Barnett JL, Cronin GM, McCallum TH and Newman EA 1994 Effects of food and time of day on aggression when grouping unfamiliar adult pigs. *Applied Animal Behaviour Science 39*: 339–347

Barnett JL and Hemsworth PH 1986 The impact of handling and environmental factors on the stress response and its consequences in swine. *Laboratory Animal Science 36*: 366–369

Barnett JL, Winfield CG, Cronin GM, Hemsworth PH and Dewar AM 1985 The effect of individual and group housing on behavioural and physiological responses related to the welfare of pregnant pigs. *Applied Animal Behaviour Science 14*: 149–161

Batchelor GR 1991 Environment enrichment for the laboratory pig. *Animal Technology 42*: 185–189

Beale C, King L and Young B 2007 Investigation of different methods of food delivery as enrichment for singly housed male Göttinger Minipigs. *Animal Technology and Welfare 6*(1): 33–36

Beattie VE and O'Connell NEO 2002 Relationship between rooting behaviour and foraging in growing pigs. *Animal Welfare 11*: 295–303

Blye R, Burke R, James C, Fitzgerald AL and Cox ML 2006 The use of operant conditioning of Göttingen minipigs for topical safety studies. *American Association for Laboratory Animal Science Meeting Official Program*: 153–154

Bolhuis EJ, Schouten WGP, Schrama JW and Wiegant VM 2005 Behavioural development of pigs with different coping characteristics in barren and substrate-enriched housing conditions. *Applied Animal Behaviour Science 93*: 213–228

Bollen, PJA, Hansen, AK and Rasmusen, HJ 2000 *The Laboratory Swine.* CRC Press: Boca Raton, FL

Brodersen T, Glerup P, Molgaard S, Andersen L and Sorensen DB 2010 The use of positive reinforcement with Göttinger minipigs. *American Association for Laboratory Animal Science Meeting Official Program*: 167–168

Burbidge JA, Spoolder HAM, Lawrence AB, Simmins PH and Edwards SA 1994 The effect of feeding regime and the provision of a foraging substrate on the development of behaviours in group-housed sows. *Applied Animal Behaviour Science 40*: 72

Casey B, Abney D and Skoumbourdis E 2007 A playroom as novel swine enrichment. *Lab Animal 36*(2): 32–34

Chilcott RP, Stubbs B and Ashley Z 2001 Habituating pigs for in-pen, non-invasive biophysical skin analysis. *Laboratory Animals 35*: 30–35

D'Earth RB 2004 Consistency of aggressive temperament in domestic pigs: Effects of social experience and social disruption. *Aggressive Behavior 30*: 435–448

Fraser D, Phillips PA, Thompson BK and Tennessen T 1991 Effect of straw on the behaviour of growing pigs. *Applied Animal Behaviour Science 30*: 307–318

Geers R, Janssens G, Villé H, Bleus E, Gerard H, Janssens S and Jourquin J 1995 Effect of human contact on heart rate of pigs. *Animal Welfare 4*: 315–359

Gonyou HW, Hemsworth PH and Barnett JL 1986 Effects of frequent interactions with humans on growing pigs. *Applied Animal Behaviour Science 16*: 269–278

Grandin T 1988 Environmental enrichment for confinement pigs. *Livestock Handling Committee Proceedings of the 1988 Annual Meeting, Kansas City, Missouri.* http://grandin.com/references/LClhand.html

Grandin T and Curtis SE 1984 Material affected cloth-toy touching and biting by pigs. *Journal of Animal Science 59* (Supplement 1): 85

Grandin T, Dodman N and Shuster L 1989 Effect of naltrexone on relaxation influenced by flank pressure in pigs. *Pharmacology, Biochemistry and Behavior 33*: 839–842

Hemsworth PH and Barnett JL 1991 The effects of aversively handling pigs, either individually or in groups, on their behaviour, growth and corticosteroids. *Applied Animal Behaviour Science 30*: 61–72

Herskin MS and Jensen KH 2000 Effects of different degrees of social isolation on the behaviour of weaned piglets kept for experimental purposes. *Animal Welfare 9*: 237–249

Huntsberry ME, Charles D, Adams KM and Weed JL 2008 The foraging ball as a quick and easy enrichment device for pigs (*Sus scrofa*). *Lab Animal 37*: 411–414

National Research Council 2011 *Guide for the Care and Use of Laboratory Animals, Eighth Edition.* National Academy Press: Washington, DC

Neubauer T, Betts T and Evans C 2011 The use of enrichment to facilitate data collection in a pig study. *American Association for Laboratory Animal Science Meeting: Abstracts of Poster Sessions*: 21

Panepinto LM, Phillips RW, Norden S, Pryor PC and Cox R 1983 A comfortable minimum stress method of restraint for Yucatan miniature swine. *Laboratory Animal Science 33*: 95–97

Pell C, Armellino K, Williams A, Farthing J and Hickman D 2010 Swine enrichment techniques that increase efficiency and promote animal welfare in the laboratory environment. *American Association for Laboratory Animal Science Meeting Official Program*: 164

Rushen J 1990 Social recognition, social dominance and the motivation of fighting by pigs. *Current Topics in Veterinary Medicine and Animal Science 53*: 135–143

Schouten WGP, Rushen J and de Passillé AM 1991 Heart rate changes in loose and tethered sows around feeding. *Applied Animal Behaviour Science 30*: 173–196

Smith AC and Swindle MM 2006 Preparation of swine for the laboratory. *ILAR Journal 47*: 358–363

Williams N and Watson J 2003 Use of behavior modification (clicker training) to facilitate handling and restraint and provide environmental enrichment in Göttinger minipigs. *American Association for Laboratory Animal Science Meeting Official Program*: 97 (Abstract)

Young RJ, Carruthers J and Lawrence AB 1994 The effect of a foraging device (The 'Edinburgh foodball') on the behaviour of pigs. *Applied Animal Behaviour Science 39*: 237–247

Sheep

Sheep

Louis DiVincenti, Jr, DVM, MS, DACLAM, DACAW

CHIEF OF LARGE ANIMAL MEDICINE & RESEARCH, UNIVERSITY OF ROCHESTER

Sheep (*Ovis aries*) have been domesticated for thousands of years. They are used in biomedical research in the United States in studies ranging from investigations of asthma and respiratory disease to development of novel cardiac interventions (Scheerlinck et al., 2008). This chapter will discuss some characteristics of sheep that are important to consider in the research laboratory, and will provide recommendations for refining research protocols involving sheep.

Species-typical characteristics of sheep

Kilgour (1976), the father of modern sheep ethology, described the sheep as a "defenseless, vigilant, tight-flocking, visual, wool-covered ruminant ... displaying a follower-type dam-offspring relationship, with strong imitation between young and old."

These characteristics are fundamental to describing and understanding the normal behavior of sheep in research facilities. The social nature of sheep is their most notable characteristic in terms of understanding and considering their behavior, and the instinct of sheep to flock tightly and to maintain a visual link with other sheep is perhaps the most important consideration for housing sheep in the laboratory.

Sheep establish well-defined relationships and strong bonds within the group; they can remember the faces of conspecifics for up to 2 years (Kendrick et al., 1996). Individual sheep tend to synchronize their activities with other sheep in the flock, as they walk and run together, follow one another, graze together, and lie down or ruminate together (Hutson, 2007). When separated from the group, individuals run towards the other sheep even when the path is obstructed by a handler or dog (Kilgour, 1977). Lambs instinctively follow their mothers. This following behavior remains in adulthood and can be utilized to handle sheep efficiently without undue stress.

Sheep are primarily grazers with absent maxillary incisors, allowing grazing close to the ground; however, sheep will also browse when they have access to low branches or shrubs. Although flocks move together in a leader-follower pattern, high behavioral synchronization for other activities, like grazing, is not as apparent as in cattle. Sheep are not territorial, but do utilize a home range, which may be shared with other groups. When a home range is shared, sheep recognize members of their own group and avoid animals from other groups (Dwyer, 2008). Ewes and juvenile animals form matrilineal groups, and rams form small bachelor groups. Dominance between rams

may be established by physical contact in the form of nudging and head butting, especially in the mating season. Submissive individuals lower and twist their head sideways, and avoid the dominant animal in the future (Ekesbo, 2011). In contrast, dominance is not apparent among similar-age ewes. Outside of competitive situations, agonistic behaviors have not been reported among ewes, even in studies in which flocks from different origins were mixed (Dwyer, 2008).

As a "defenseless" prey species, the sheep's main anti-predator strategies are flocking and flight (Dwyer, 2008). Sheep are disturbed easily, and can be frightened by sudden sounds or fast movements, especially in unknown or confined areas, an important consideration for handling sheep in the laboratory. If one animal adopts an "alarm posture"—with head raised rigidly, while using slow, tense steps—this rapidly alerts others in the group (Geist, 1971). With a visual field of 280°, sheep are able to maintain awareness of potential predators and spatial relationships with others in the flock (Hutson, 2007). In response to perceived danger, sheep will readily flee. In the research setting, all handlers should be aware of this instinct, as the wild flight of panic can cause harm to the sheep and/or the handlers (Ekesbo, 2011).

Ensuring the welfare of sheep

Since the Brambell Report was first published in 1965 in response to concern for farm animal welfare in the United Kingdom, the "Five Freedoms" in that report have become the minimum standards of care for farm animals (Brambell Committee, 1965). The enumerated freedoms are as follows:

1. Freedom from thirst, hunger and malnutrition—by ready access to fresh water and a diet to maintain full health and vigor.
2. Freedom from thermal or physical distress—by providing an appropriate environment, including shelter and a comfortable resting area.

3. Freedom from pain, injury and disease—by prevention or by rapid diagnosis and treatment.
4. Freedom to display most normal patterns of behavior—by providing sufficient space, proper facilities, and company of the animals' own kind.
5. Freedom from fear and distress—by ensuring conditions and treatment to avoid mental suffering.

Although the initial intent of the Brambell Report was to improve the welfare of animals on the farm, these principles apply to sheep in a research setting, as well.

Freedom from thirst, hunger and malnutrition: An adult sheep may drink up to 6 liters per day, so continuous access to fresh, drinkable water is essential. Roughage is particularly important in ruminants such as sheep. The major component of the diet for sheep should be bulky food with a high fiber content. Diets lacking in fiber are associated with increases in abnormal oral behaviors in sheep such as mouthing bars, chewing chains, mandibulation (licking lips and mouthing air), and repetitive licking (Done-Currie et al., 1984; Mardsen & Wood-Gush, 1986; Cooper et al., 1994; Yurtman et al., 2002). Hay or an increased fiber diet reduces these oral stereotypies (Done-Currie et al., 1984), and adequate roughage reduces the incidence of maladaptive behaviors such as wool-biting (Vasseur et al., 2006).

Sheep on pasture may graze up to 11 hours and ruminate up to 8 hours per day, and even animals whose nutritional needs are met by a concentrated feed ration still exhibit food-searching behaviors (Ekesbo, 2011). Hanging hay in suspended baskets or nets or providing fresh browse are relatively inexpensive and easy ways to provide opportunities for indoor-housed animals to engage in feeding behavior for longer periods, to help satisfy the biological need for grazing.

Freedom from thermal or physical distress: Sheep in the wild are one of the most successful animal species, with a nearly global distribution over a wide range of terrains and climates (Dwyer, 2008). Outdoor housing on pasture provides sheep with the most opportunity to express species-typical behaviors. Animals housed exclusively outdoors must be provided with shelters to withstand weather extremes (National Research Council, 2011), and sheep will use such shelters particularly as protection against strong wind. Sheep are extremely well-adapted to cold, with a lower critical temperature as low as 0°C in fully fleeced adult animals (Terrill & Slee, 1991). Protection from predation, particularly domestic dogs, is also important in outdoor housing situations (Dwyer, 2008).

When an institution has not been able to provide housing for the duration of a research study, the author has utilized a local farm in the area to house animals who needed to be followed for an extended period but required only minimal experimental sampling. In the author's experience, transporting the animals to the farm resulted in minimal stress and improved their well-being for the 6–12 months they were able to be outdoors, compared to the indoor pens available in the medical center setting. Sheep may be housed in outdoor groups even during studies. The sheep in the photo to the right underwent cardiac surgery and were housed in an indoor pen for recovery. Afterwards, the animals were moved to a large, outdoor pen where they could engage in species-typical behaviors for the remainder of the study.

In contrast to outdoor housing, the indoor environment in research is typically well controlled within a narrow range of environmental parameters. For indoor

Comfortable Quarters for Laboratory Animals

housing, the *Guide for the Care and Use of Laboratory Animals* states that sheep should be maintained at a dry-bulb temperature of 16–27°C (61–81°F; National Research Council, 2011). Animals kept indoors should be sheared to prevent heat stress. Flooring should be solid or slatted, with a slip-resistant surface. Routine foot checks should be part of the program of veterinary care, especially for animals housed indoors. The floor surface should be such that it provides some wear on the hoof to minimize the requirement for hoof trimming, but avoids excessive hoof wear. In indoor facilities where floors are slippery and replacement is not an option, rubber matting can be used to improve sheep comfort. In the author's experience, providing sheep with firm footing greatly increases their compliance with light restraint for minimally invasive procedures.

Hay and straw are highly recommended bedding options for sheep, not only on farms, but also in the research setting. Wood chips, corn cobs, and paper products have been used in indoor facilities, and these alternative bedding options may have advantages in terms of moisture absorption and cleaning needs. However,

straw and hay provide the animal with more opportunity for foraging, so should be used whenever possible.

Regulations regarding minimum cage sizes for sheep have been promulgated, but animals must also be provided with adequate space for normal ambulation. Pen sizes should be large enough, or cleaned frequently enough, so that all members of the group may simultaneously lie in clean, dry areas of the pen.

Freedom from pain, injury and disease: Health, i.e., normal biologic function, is one of the most basic aspects of welfare. Like other research animals, sheep must be provided with adequate veterinary care, including a routine preventive health program of vaccinations, anthelmintics, and ectoparasite control; frequent observations by trained personnel to monitor for health problems (including the need for hoof trimming); and access to an experienced veterinarian for prompt diagnosis and treatment of health problems.

Prey animals, in general, instinctively hide signs of pain or disease, and sheep in particular have been described as "stoical" or "physically tough" (Webster, 1994). With the exception of lambs separated from their mothers, sheep are rarely vocal in response to stressors (Dwyer & Lawrence, 2008). However, sheep do feel pain and are subject to the same behavioral and physiologic consequences of pain as other research animals who may display signs of pain more readily. Unlike cattle, sheep do not vocalize in response to painful procedures (Hutson, 2007), so absence of vocalization does not indicate the animal is not experiencing pain. In sheep, signs of pain include more subtle changes in appetite or facial expression, reduced cud chewing, adoption of a rigid stance with lowered head, bruxism (teeth grinding), and withdrawal from the group. In the absence of specific evidence to the contrary, any procedure expected to cause pain or distress in humans should be considered painful or distressing in sheep (Interagency Research Animal Committee, 2011), and appropriate analgesics must be given under the direction of a veterinarian.

Many sheep used in research are not purpose-bred and are acquired from farms where they may be subject to standard agricultural practices. In some cases, these practices include dehorning, castration, and/or tail-docking without anesthesia or analgesia (Federation of Animal Science Societies, 2010). Responsible research institutions should request that these procedures not be done, unless required by the research, and then, only with appropriate anesthesia and analgesia.

Freedom to display most normal patterns of behavior: As previously discussed, gregariousness is a key behavioral characteristic of sheep. Sheep also display leadership, a social behavior in which one animal initiates a movement and

is subsequently followed by the other members of the group (Nowak et al., 2008). Failure to provide appropriate social companionship to sheep is associated with a myriad of endocrine, hematological, biochemical, and behavioral alterations. Isolation reduces growth rate and decreases feeding time (Abdel-Rhaman, 2000). Isolation persistently elevates cortisol level and heart rate (Cockram et al., 1994), and actually activates the endocrine stress response (e.g., hypothalamic-pituitary-adrenal axis) to a greater degree than does handling or restraint (Baldock & Sibly, 1990).

Although social companionship is required for adequate welfare, aggression occurs when social groups are mixed or stocking densities become so high as to restrict access to resources (Arnold & Maller, 1974; Done-Currie et al., 1984). In stable groups, dominance hierarchies are well defined and maintained through nonaggressive behaviors (Guilhem et al., 2000), so the occurrence of aggressive behaviors can indicate some form of management and/or husbandry-related stress within the group.

In many species, demonstration of stereotypic behavior is a hallmark of poor welfare, but sheep may be less likely than other species to display maladaptive behaviors (Houpt, 1987; Lawrence & Rushen, 1993). However, individually housed sheep demonstrate stereotypical oral and locomotor behaviors (Done-Currie et al., 1984; Marsden & Wood-Gush, 1986; Yurtman et al., 2002). Improper housing may incite stereotypical behaviors. For example, wool-pulling only occurs in indoor-housed sheep, and the behavior is obviated when sheep are turned out in pastures (Dwyer, 2008) or the fiber content of the diet is increased (Vasseur et al., 2006). As stated above, sheep typically graze for up to 11 hours a day (Lynch et al., 1992), so providing grazing opportunities is important to foster the expression of species-typical behavior. In indoor facilities, equine hay nets provided as "puzzle feeders" can be a good substitute for grazing (Atkins et al., 2007). In the author's experience, these nets not only increase foraging time, but also reduce hay wasting by the animals.

Freedom from fear and distress: Sheep are particularly apprehensive when subject to restraint and handling. This fear induces physiological changes that can confound research data. For example, stressed sheep have a reduced lymphocyte blastogenic response when challenged with certain mitogens (Minton et al., 1992; Minton et al., 1995). Restraint, confinement, and transport can block or delay leutinizing hormone (LH) secretion, resulting in suppression of follicular growth and reduction in estradiol (Rasmussen & Malven, 1983; Dobson & Smith, 1995). Isolation is also a potent fear inducer in sheep, and animals do not adapt to isolation even when subjected to it repeatedly (Niezgoda et al., 1987).

Positive human contact can reduce fearfulness and subsequently the stress of handling procedures (Dwyer & Lawrence, 2008). Sheep may form a bond with a caretaker that allows the animal to develop coping strategies to handling and experimental manipulations (Wolfle, 1996).

Refining husbandry and research procedures for sheep

Although sheep have specific nutritional, physiologic, and social needs, their domestication has made them an adaptable, placid laboratory animal species. Catering to the sheep's unique needs will not only improve their welfare, but as previously discussed, will ease experimental manipulations and improve research outcomes.

Relatively little research on handling methods has been done with sheep, but the same broad principles that apply to proper cattle handling are likely applicable to handling sheep, as well. For example, handlers should avoid the use of fearful stimuli and punishment, and instead choose positive reinforcements. They should understand sheep behavior and take advantage of sheep-typical leading/following, and form positive relationships with the animals. In the United States, the Animal Welfare Act requires that research facilities provide adequate training (Animal Welfare Act, 2012). This training should include basic instructions in animal behavior.

Following arrival at the institution, sheep should be given an acclimation period of at least 1 week before any aversive or negative experiences. Importantly for research, this period allows the animal's immunologic and physiologic stress responses triggered by shipping and transport to return to baseline. During this time, sheep can adapt to the new housing situation and husbandry routine. This time period may be especially crucial for animals transferred from outdoor pastures to indoor housing, as this transition is particularly disorienting and distressing for sheep (Done-Currie et al., 1984; Fordham et al., 1991). Although a 1-week period is customary, it may take up to 4 weeks before cortisol levels return to baseline after sheep are transported from pasture to indoor housing (McNatty & Young, 1973). Transport has also been associated with unwillingness to eat new foods (Dwyer, 2008), so animals must be observed closely for adequate food and water consumption. Consistency in husbandry and handling routine and stable group composition are essential for adaptation in the sheep (Fraser, 1995), so staff should avoid changing the composition of an established group of sheep. The animals develop expectations based on their previous experiences, and deviation from the expected routine reportedly causes increased heart rate and agitation (Greiveldinger et al., 2007).

The acclimation period also provides time for staff to gain the trust of the animals, so that future manipulations may be less stressful. Since sheep can distinguish visually between familiar and unfamiliar humans (Kendrick et al., 2001), an acclimation period gives the animals an opportunity to become familiar with their caretakers. Gentling, or stroking the head of the sheep, talking quietly, and hand-feeding results in familiarization with staff and habituation with routine husbandry procedures. Well-familiarized sheep approach humans more readily, have shorter flight distances, and lower heart rates (Hargreaves & Hutson, 1990; Mateo et al., 1991). These sheep will typically accept a potentially stressful situation, such as blood collection, more readily (Kilgour, 1987). Over the course of the acclimation period, the caretakers will become familiar with the animals and sensitive to subtle behavioral changes that may indicate pain or distress during future research procedures.

The sheep's innate following behavior should be exploited to provide a more positive handling experience for both the animal and the researcher or animal care technician. Encouraging this leader-follower behavior can facilitate routine procedures such as weighing and veterinary examinations (Hutson, 2007). In most cases, the animals will proceed as a group to a target location away from the handler simply as a response to the handler encroaching on the leader's flight distance. The use of fear stimuli, such as loud noises or rapid movements, unnecessarily frightens the animals, and activating the animals' flight instinct will make them less compliant and less likely to do what is expected of them.

The single most important welfare aspect of sheep is the biologically inherent need for social companionship. It is imperative to maintain stable social groups in research facilities. Companions assist the individual sheep in coping with disturbing situations,

and they buffer the stress and fear response experienced in husbandry and research procedures (Federation of Animal Science Societies, 2010; Gonzáleza et al., 2013; Porter et al., 1995). Prolonged isolation may be associated with reductions in food and water intake and (as noted above) activation of the animal's endocrine stress response (Apple et al., 1993; Carbajal & Orihuela, 2001). When animals must be singly housed, slatted or chain-link fences should be used to allow for visual and protected physical contact (Baldock & Sibly, 1990). When sheep are isolated for use in metabolism studies, the stress effect of this isolation must be considered, as individual housing may increase the sheep's metabolism up to 15% (Van Adrichem & Vogt, 1993). The presence of a single companion is sufficient to mitigate the physiological and behavioral effects of isolation (Carbajal & Orihuela, 2001). Mirrored panels or familiar sheep-face pictures may be used when no conspecifics are within view to mitigate anxiety and avoid

Comfortable Quarters for Laboratory Animals

panic responses to isolation (da Costa et al., 2004; McLean, 2004; Parrott et al., 1988). In the author's experience, intermittent supervised periods of free contact with conspecifics serve as a social facilitator to improve appetite when animals must be singly housed for experimental purposes. Such periods of social relief foster the well-being of the lonely sheep.

When caretakers have formed a positive relationship with the individual sheep, the human-animal bond can serve as a substitute for conspecific social contact. For example, lambs vocalized and moved less when in the presence of a shepherd than when isolated (Boivin et al., 1997). However, research facilities should facilitate constant social companionship, with strategies such as basing per diems on pairs of animals or creating a pen charge for two or more animals. These strategies allow the housing of companion animals not being actively used for research at no additional cost to the researcher.

Stable social relationships are essential; changing the composition of established groups by introducing or removing sheep may be a significant source of stress resulting from disrupted social relationships and rank-determining aggressive interactions (Sevi et al., 2001). When neonates or pre-weanlings are needed for research, every effort should be made to keep the lambs with their mothers until the age of natural weaning. Separation of mothers from offspring is one of the few occasions that result in vocalization in sheep, and lambs prematurely separated exhibit behavioral and physiologic responses indicative of stress, including increased cortisol levels and impaired immune responses (Moberg & Wood, 1981; Napolitano et al., 1995; Price & Thos, 1980).

Staff should be trained in and actively employ positive reinforcement training when handling sheep. Food rewards like barley or grain are readily accepted by sheep, and although some consider sheep as incapable of learning, sheep can be easily conditioned (Hutson, 2007). For example, in the author's facility, sheep have been taught to drink from a syringe, stand still for venipuncture, and shift from one pen to another during cleaning. Sheep reportedly remember distressful experiences for at least 12 weeks (Rushen, 1986) and up to 1 year (Hutson, 1985), but also will return to places where they were manually restrained if the experience was positive (Grandin, 1989). Using food rewards not only reduces the stress response to handling, but also improves the speed of handling, and these positive effects on handling are maintained for at least a year (Hutson, 1985). Training sheep to participate in procedures provides the animal with an element of control, which has been shown to affect the sheep's emotional response to a disturbing situation (Greiveldinger et al., 2007).

Summary

Taking into account the unique behavioral needs of sheep will improve animal welfare and research outcomes. The management of sheep in research institutions must meet the following requirements:
  » Sheep are housed in stable social groups.
  » Sheep are provided appropriate housing with bedding and bulky food.
  » Staff members are trained in sheep behavior.
  » Gentle familiarization is employed for new animals and positive reinforcement training is used with all animals to promote cooperation during potentially aversive experimental procedures.

REFERENCES

Abdel-Rahman MA 2000 Behavioural and endocrinological changes of sheep subjected to isolation stress. *Assiut Veterinary Medicine Journal 42*: 1–12

Animal Welfare Act, 7 USC § 2131 *et seq.* 2012. http://www.gpo.gov/fdsys/pkg/USCODE-2012-title7/pdf/USCODE-2012-title7-chap54.pdf

Apple JK, Minton JE, Parson KM and Unruh JA 1993 Influence of repeated restraint and isolation stress and electrolyte administration on pituitary-adrenal secretions, electrolytes, and other blood constituents of sheep. *Journal of Animal Science 71*: 71–77

Arnold GW and Maller RA 1974 Some aspects of competition between sheep for supplementary feed. *Animal Production 19*: 309–319

Atkins PL, Millsap L, Thain D and Van Andel R 2007 An environmental enrichment tool used to prevent barbering in *Ovis aries. American Association for Laboratory Animal Science Meeting Official Program*: 52 (Abstract)

Baldock NM and Sibly RM 1990 Effects of handling and transportation on the heart rate and behaviour of sheep. *Applied Animal Behaviour Science 28*: 15–39

Boivin X, Nowak R, Despres G, Tournadre H and LeNeindre P 1997 Discrimination between shepherds by lambs reared under artificial conditions. *Journal of Animal Science 75*: 2892–2898

Brambell Commitee 1965 *Report of the Technical Committee to Enquire into the Welfare of Animals Kept under Intensive Livestock Husbandry Systems.* Her Majesty's Stationary Office: London, UK

Carbajal S and Orihuela A 2001 Minimal number of conspecifics needed to minimize the stress response of isolated mature ewes. *Journal of Applied Animal Welfare Science 4*: 249–255

Cockram MS, Ranson M, Imlah P, Goddard PJ, Burrells C and Harkiss GD 1994 The behavioural, endocrine and immune response of sheep to isolation. *Animal Production 58*: 389–400

Cooper JJ, Emmans GE and Friggens NC 1994 Effect of diet on behaviour of individually penned sheep. *Animal Science 58*: 44

da Costa AP, Leigh AE, Man M-S and Kendrick KM 2004 Face pictures reduce behavioural, automonic, endocrine and neural indices of stress and fear in sheep. *Proceedings of the Royal Society of London, Series B 271*: 2077–2084

Dobson H and Smith RF 1995 Stress and reproduction in farm animals. *Journal of Reproduction and Fertility 49* (Supplement): 451–461

Done-Currie JR, Hecker JF and Wodzicka-Tomaszewka M 1984 Behaviour of sheep transferred from pasture to an animal house. *Applied Animal Behaviour Science 12*: 121–130

Dwyer CM 2008 Environment and the sheep. In: Dwyer CM (ed) *The Welfare of Sheep* pp 1–40. Springer: New York, NY

Dwyer CM and Lawrence AB 2008 Introduction to animal welfare and the sheep. In: Dwyer CM (ed) *The Welfare of Sheep.* Springer: New York, NY

Ekesbo I 2011 Sheep. In: Fraser AF and Broom DM (eds) *Farm Animal Behavior: Characteristics for Assessment of Health and Welfare* pp 82–92. CABI: Oxfordshire, UK

Federation of Animal Science Societies 2010 *Guide for the Care and Use of Agricultural Animals in Research and Teaching, Third Edition.* Federation of Animal Science Societies: Champaign, IL

Fordham DP, Al-Gahtani S, Durotoye LA and Rodway RG 1991 Changes in plasma cortisol and ß-endorphin concentrations and behaviour in sheep subjected to a change in environment. *Animal Production 52*: 287–296

Fraser AF 1995 Sheep. In: Rollin BE and Kesel ML (eds) *The Experimental Animal in Biomedical Research, Volume II, Care, Husbandry, and Well-Being: An Overview by Species* pp 87–118. CRC Press: Boca Raton, FL

Gonzáleza M, Averósa X, de Herediaa IB, Ruiza R, Arranza J and Estevez I 2013 The effect of social buffering on fear responses in sheep (*Ovis aries*). *Applied Animal Behaviour Science 49*: 13–20

Grandin T 1989 Voluntary acceptance of restraint by sheep. *Applied Animal Behaviour Science 23*: 257–261

Geist V 1971 *Mountain Sheep: A Study in Behavior and Evolution.* University of Chicago Press: Chicago, IL

Greiveldinger L, Veissier I and Boissy A 2007 Emotional experience in sheep: predictability of a sudden event lowers subsequent emotional responses. *Physiology & Behavior 92*: 675–683

Guilhem C, Bideau E, Gerar JF and Maublanc ML 2000 Agonistic and proximity patterns in enclosed mouflon (*Ovies gmelini*) ewes in relation to age, reproductive status, and kinship. *Behavioural Processes 50*: 101–112

Hargreaves AL and Hutson GD 1990 The effect of gentling on heart rate, flight distance and aversion of sheep to a handling procedure. *Applied Animal Behaviour Science 26*: 243–252

Houpt K 1987 Abnormal behavior. In: Price EO (ed) *The Veterinary Clinics of North America: Farm Animal Behavior* pp 357–367. Saunders: Philadelphia, PA

Hutson GD 1985 The influence of barley food rewards on sheep movement through a handling system. *Applied Animal Ethology 14*: 263–273

Hutson GD 2007 Behavioural principles of sheep-handling. In: Grandin T (ed) *Livestock Handling and Transport, Third Edition* pp 127–148. CABI: Wallingford, UK

Interagency Research Animal Committee 2011 U.S. Government Principles for the Utilization and Care of Vertebrate Animals Used in Testing, Research, and Training. In: National Research Council *Guide for the Care and Use of Laboratory Animals, Eighth Edition* pp 199–200. National Academy Press: Washington, DC

Kendrick KM, Atkins KA, Hinton MR, Heavens P and Keverne B 1996 Are faces special for sheep? Evidence from facial and object discrimination learning tests showing effects of inversion and social familiarity. *Behavior Proceedings 38*: 19–35

Kendrick KM, da Costa AP, Leigh AE, Hinton MR and Pierce JW 2001 Sheep don't forget a face. *Nature 414*: 165–166

Kilgour R 1976 Sheep behavior: its importance in farming systems, handling, transport and pre-slaughter treatment. In: Truscott GMC and Troth FH (eds) *Sheep Assembly and Transport Workshop* pp 64–84. Western Australian Department of Agriculture: Perth, Australia

Kilgour R 1977 Design sheep yards to suit the whims of sheep. *NZ Farmer 98*(6): 29–31

Kilgour R 1987 Learning and the training of farm animals. In: Price EO (ed) *The Veterinary Clinics of North America: Food Animal Behavior* pp 269–284. Saunders: Philadelphia, PA

Lawrence AB and Rushen J 1993 Introduction. In: Lawrence AB and Rushen J (eds) *Stereotypic Animal Behaviour* CAB International: Oxon, UK

Lynch, JJ, Hinch, GN and Adams, DB 1992 *The Behavior of Sheep: Biological Principles and Implications for Production*. CAB International: Wallingford, UK and CSIRO Australia: East Melbourne, Australia

Mateo JM, Estep DQ and McCann JS 1991 Effects of differential handling on the behaviour of domestic ewes (*Ovis aries*). *Applied Animal Behaviour Science 32*: 45–54

Marsden MD and Wood-Gush DGM 1986 A note on the behaviour of individually-penned sheep regarding their use for research purposes. *Animal Production 42*: 157–159

McLean CB and Swanson LE 2004 Reducing stress in individually housed sheep. *American Association for Laboratory Animal Science Meeting Official Program*: 144 (Abstract)

McNatty KP and Young A 1973 Diurnal changes of plasma cortisol levels in sheep adapting to a new environment. *Journal of Endocrinology 56*: 329–330

Minton JE, Coppinger TR, Reddy PG, Davis WC and Blecha F 1992 Repeated restraint and isolation stress alters adrenal and lymphocyte functions and some leukocyte differentiation antigens in lambs. *Journal of Animal Science 70*: 1126–1132

Minton JE, Apple JK, Parsons KM and Blecha F 1995 Stress-associated concentrations of plasma cortisol cannot account for reduced lymphocyte function and changes in serum enzymes in lambs exposed to restraint and isolation stress. *Journal of Animal Science 73*: 812–817

Moberg GP and Wood VA 1981 Neonatal stress in lambs: behavioral and physiological responses. *Developmental Psychobiology 14*: 155–162

National Research Council 2011 *Guide for the Care and Use of Laboratory Animals, Eighth Edition*. National Academy Press: Washington, DC

Napolitano F, Marino V, de Rosa G, Capparelli R and Bordi A 1995 Influence of artificial rearing on behavioral and immune response of lambs. *Applied Animal Behaviour Science 45*: 245–253

Niezgoda J, Wronska D, Pierzchala K, Bobek S and Kahl S 1987 Lack of adaptation to repeated emotional stress evoked by isolation of sheep. *Zentralblatt für Veterinärmedizin 34*: 734–739

Nowak R, Porter RH, Blache D and Dwyer CM 2008 Behavior and welfare of the sheep. In: Dwyer CM (ed) *The Welfare of Sheep*. Springer: New York, NY

Parrott RF, Houpt KA and Misson BH 1988 Modification of the responses of sheep to isolation stress by the use of mirror panels. *Applied Animal Behaviour Science 19*: 331–338

Porter R, Nowak R and Orgeur P 1995 Influence of a conspecific agemate on distress bleating by lambs. *Applied Animal Behaviour Science 45*: 239–244

Price EO and Thos J 1980 Behavioral responses of short-term isolation in sheep and goats. *Applied Animal Ethology 6*: 331–339

Rasmussen DD and Malven PV 1983 Effects of confinement stress on episodic secretion of LH in overiectomized sheep. *Neuroendocrinology 36*: 392–396

Rushen J 1986 Aversion of sheep to electro-immobilization and physical restraint. *Applied Animal Behaviour Science 15*: 315–324

Scheerlinck JPY, Snibson KJ, Bowles VM and Sutton P 2008 Biomedical applications of sheep models: From asthma to vaccines. *Trends in Biotechnology 26*: 259–266

Sevi A, Taibi L, Albenzio M, Muscio A, DellAquila S and Napolitano F 2001 Behavioral, adrenal, immune, and productive responses of lactating ewes to regrouping and relocation. *Journal of Animal Science 79*: 1457–1465

Terrill CE and Slee J 1991 Breed differences in adaptation of sheep. In: Majala K (ed) *Genetic Resource of Pig, Sheep and Goat* pp 195–233. Elsevier: Amsterdam, Netherlands

Van Adrichem PWM and Vogt JE 1993 The effect of isolation and separation on the metabolism of sheep. *Livestock Production Science 33*: 151–159

Vasseur S, Paull DR, Atkinson SJ, Colditz IG and Fisher AD 2006 Effects of dietary fibre and feeding frequency on wool biting and aggressive behaviours in housed Merino sheep. *Australian Journal of Experimental Agriculture 46*: 777–782

Webster J 1994 *Animal Welfare: A Cool Eye Towards Eden*. Blackwell Science: Oxford, UK

Wolfle TL 1996 How different species affect the relationship. In: Krulisch L, Mayer S and Simmonds RC (eds) *The Human/Research Animal Relationship* pp 85–91. Scientists Center for Animal Welfare: Greenbelt, MD

Yurtman IY, Savas T, Karaagac F and Coskuntuna L 2002 Effects of daily protein intake levels on the oral stereotypic behaviours in energy restricted lambs. *Applied Animal Behaviour Science 77*: 77–88

Dogs

Dogs

Kaile Bennett, BS, LATg
UNIVERSITY OF MICHIGAN

Dogs have a special status in our culture, where they are known as "man's best friend" or "constant protector" and often viewed as members of the family. Americans spend billions of dollars every year on their companion dogs (Associated Press, 2014). This places raised expectations on the laboratory animal community to not only care for the basic necessities of laboratory dogs, but also address their social and emotional

well-being. Some of these expectations are described in the *Guide for the Care and Use of Laboratory Animals,* which addresses the issues of social housing and human interaction (National Research Council, 2011).

The importance of increased awareness of social and behavioral needs

Domestic dogs are highly social animals, relying on human or conspecific interactions to fulfill their social requirements. Multiple reports have noted that social isolation is a significant stressor for dogs (e.g., Wolfe, 1992). Within the laboratory setting, there are many opportunities to provide for their social needs, as will be discussed below.

It should be the norm to house dogs with conspecifics (Overall & Dyer, 2005) and any solitary housing should be considered the exception, subject to regular reevaluation by the attending veterinarian and/or the Institutional Animal Care and Use Committee. Behavioral and personality assessments should be done before pairs or groups are formed and care must be taken to house animals accordingly. While pair-housed dogs are more common in the laboratory, larger groups can be formed if sufficient space is available and study parameters permit it (Field & Jackson, 2007).

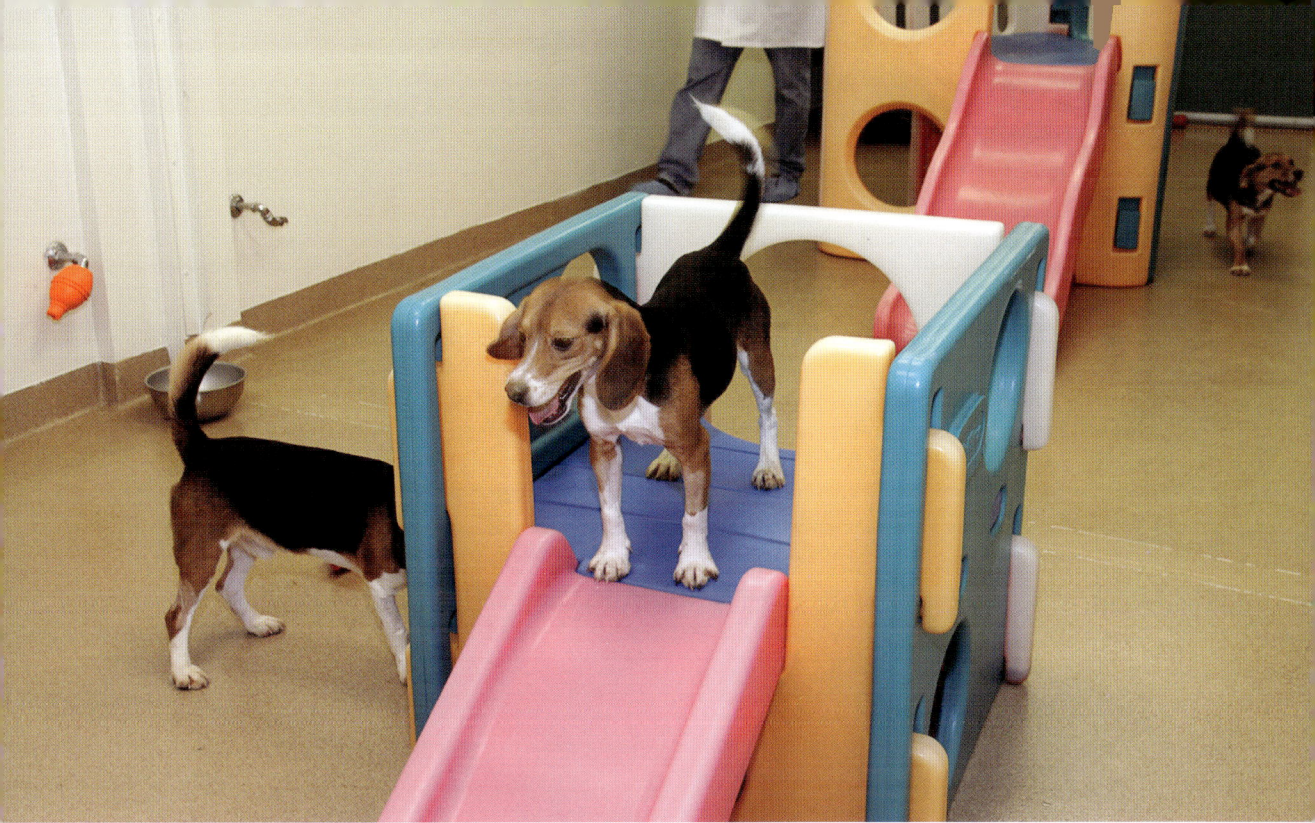

If social housing is not possible, housing should allow for visual access to other canines. Some facilities have created play rooms or play areas, where animals are allowed to exercise and socialize (Loveridge, 1994; Evans et al., 1999; Andrews-Kelly, 2010; Shulder & Ogbin, 2010). The play areas can include a number of enrichment devices such as platforms and ramps (reminiscent of canine agility courses) and, if access to outside is prohibited, turf flooring made to resemble grass (Hubrecht, 2002). Enrichment may also involve tapping into canines' innate abilities—exploring the environment through scent or puzzle solving are two activities that engage and promote exercise and use of space (Haug, 2006). In the author's experience, having a structured schedule for human interaction and positive reinforcement training can work to both improve welfare and socialization and improve efficiency in husbandry. Training a room full of long-term canines in basic commands such as sit, come and leash walking can speed up the performance of husbandry practices and make them less stressful for the animals. In such an environment, exercise time involves obedience training and removal/return to housing is voluntary.

Even when social housing and group play facilities are available, daily human interaction is an important part of the daily regimen for dogs in a laboratory. The USDA stresses access to at least one enrichment device, as well as human interaction and social housing, as the rule rather than the exception (Overall & Dyer, 2005). These guidelines should be considered as starting points to a more facility-based and inventive housing program.

Mimicking a natural setting for animals in laboratories is encouraged. In broad terms, for domestic dogs, the natural setting is social living with daily human interaction, training, and novel enrichment devices (Wolfle, 1992). This allows for flexibility, and institutions often take advantage of this flexibility to create a plan that fits within their own constraints and staffing allowances.

Because of their ability to be trained, it is possible to modify behaviors of dogs in the laboratory to minimize the stress of handling or of minor research procedures (Trussell et al., 1999; Hubrecht, 2002; Roddis, 2005; Hussain et al., 2006; Tabers et al., 2009; Savastano, 2013). Positive reinforcement training is the preferred method (Laule, 1999). This can lead to a trust bond between technician and canine that will lead to easier daily activities. Socialization and basic training at the supplier have a considerable impact on how laboratory canines will react to handling during research and husbandry practices (Fox, 1975; Freedman et al., 1991; Trussell et al., 1999). Institutions should work with suppliers to ensure any canines are socialized to both other canines and humans.

Enrichment devices must be of a quality and construction that make them safe to leave with unsupervised dogs, and must be cleaned/checked daily for wear and tear. These can include toys used with food incentives (e.g., Kong toys), puzzle toys to create complexity for daily rations (including toys with hidden compartments for treats, or buttons and switches to obtain treats), or toys to promote chewing (Nylabone as an example, though care should be taken that no piece of the toy can be chewed off and swallowed). Toys with fleece, plush or rope could be ingested if the dog is left alone and should only be provided with supervision. If socially housed, care must be taken with high-value toys so that a fight does not occur between dogs. Multiple toys should be available in social housing situations. Separation may be necessary for food-based enrichment (Overall & Dyer, 2005).

Physical structure of housing space

There are many variations for the physical structure of the housing space for laboratory canines. The *Guide* states the space should be, at minimum, sufficient for full range of movement while standing, turning and lying down, using body weight as a guideline (0.74 to over 2.4 m^2 floor area, depending on the size of the dog). While the space guidelines can be met with caging for smaller dogs, it is always preferable to use much larger runs for all dog sizes (National Research Council, 2011).

Flooring can be a variety of substrates: slate flooring with drains beneath, solid flooring (usually concrete), sawdust, or access to the outdoors. Solid flooring with elevated rest areas and a separate space for elimination is the ideal option. Wire or slatted flooring can create pressure sores on the pads of dogs' feet (Field & Jackson, 2007). This type of flooring can also entrap toes and should be avoided. This issue can be partially addressed

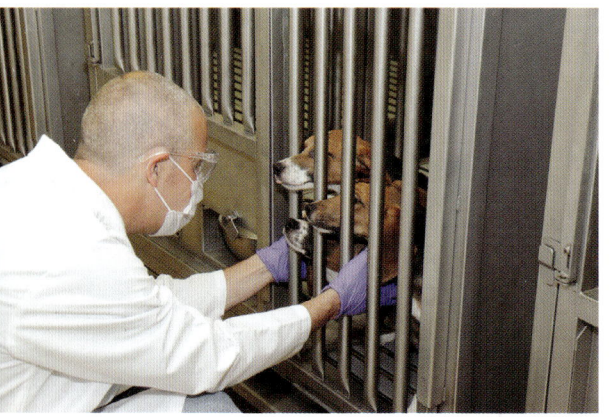

with resting pads, frequent exercise out of the enclosure, and sawdust used sparingly for moisture control. Regardless of flooring type, it is important to maintain a clean and dry environment. Housing space should also be large enough to allow the dogs to avoid areas that have been urinated or defecated in. This is a particularly important point when dogs are socially housed.

When possible, elevated resting surfaces should be used. Elevation has multiple benefits, as it adds complexity to the housing space and helps keep the resting surfaces clean and dry. When constructing raised platforms, care should be taken to ensure that the platforms and areas underneath them can be sanitized. The platforms should be large enough to accomodate the size and number of dogs who might use them (Anonymous, 2013b; Hubrecht, 1993).

Diet and husbandry care

Most laboratory canines are fed a dry kibble diet that is meat based and measured according to size, age and physical activity (Hubrecht, 1995). This can be fed from a bowl on the ground, or a feeder attached to the cage or run door to maintain cleanliness. It is preferable to keep food and water bowls off the floor, to prevent soiling and accidental spillage. When dogs are housed in groups, care must be taken to ensure that all are eating an appropriate amount. If aggression over food is noticed, then group-housed dogs should be separated for feeding. Treats must be approved by veterinary and laboratory staff to make sure they do not affect study goals or impact the health of the dog. Small treats for training or insertion in puzzle toys are often used.

Water can be provided either in a bowl or Lixit attached to a constant water source. While the Lixit is a convenient way to prevent contamination of the water supply, the bowl allows for a more natural drinking motion. If Lixits are used, care should be taken to ensure there are enough to support all dogs in the enclosure and that all dogs are drinking enough to be properly hydrated. The Lixits must be checked daily to make sure they are in working order.

Husbandry care includes cleaning of enclosures, human interaction and socialization, positive reinforcement training, nail trimmings (monthly, or as needed), and daily health checks. Positive interactions with the dogs are paramount to thorough health checks, as visual and tactile observations lead to a more thorough inspection. Technicians should be well versed in the common signs of distress and pain in canines (National Research Council, 1994). It is a common practice to spray down housing enclosures for daily cleaning. If this is the practice, the occupants must be taken out of the enclosure and the enclosure should be allowed to dry before they are returned. Further, it is not uncommon for dogs to urinate or defecate in a newly cleaned enclosure. Care staff should be aware of this habit so that they can remove the fecal matter.

Noise levels

The noise levels in kennels can reach 100 decibels. This level can be damaging to human and canine hearing (Hubrecht, 1995). Some facilities try to alleviate this by playing soothing music or using white noise in the kennel, which may decrease barking and agitation (Kilcullen-Steiner & Mitchell, 2001; Wells et al., 2002). Debarking should not be considered an option to control noise levels, as it can have significant adverse consequences (i.e., swelling, bleeding, and infection) with little gain (bark is only muffled, not removed) ("Alternatives to Debarking," 2014).

Retirement and rehoming of laboratory canines

It has previously been considered an extremely difficult feat to rehome laboratory canines, especially purpose-bred hounds and beagles. While laboratory canines do necessitate a unique degree of care and ability, adoption is increasingly viewed as an acceptable alternative if euthanasia is not necessary for the studies conducted (Simons, 2014). Many facilities have in-house adoption programs that release canines primarily to staff members, many of whom form attachments during their work with a particular dog (Anonymous, 2013c). There are often terms associated with such adoptions, including the dogs passing behavioral and medical screenings and having minimal surgical interventions during their studies. For facilities with a large number of canines being rehomed (or with a particular study group that will be available) it is advised to have a special socialization and introduction program before the dogs' release to increase the likelihood of adaptation to living in a house and outdoor setting (Evans et al., 1999; Burgess et al., 2010.) Some rehoming groups will work with the dogs on socialization and preparation for life in a home.

There are also some rescue groups who specialize in rehoming the most common laboratory breed, the beagle. For example, one of the largest and longstanding groups, the Beagle Rescue League, operates out of New Jersey and will work with laboratories to ensure proper socialization and successful home placement. If an outside group is used, research should be done into the group's success in providing smooth transitions and successful placements, as well as the group's history and possible political and public agendas. If facilities wish to remain anonymous or specify any other terms, these should be clearly spelled out in a written agreement.

Conclusion

A compassionate approach to laboratory canine housing and care can lead to better research and better animal welfare. Dry, comfortable and spacious housing with adequate food and cleanliness are the basis from which to build a comprehensive care program for canines. This should include attention to their social, behavioral and training needs. This also includes encouraging rehoming after research when possible.

REFERENCES

Alternatives to debarking surgery (2014, April 10) Retrieved from https://vetmed.tamu.edu/news/pet-talk/alternatives-to-debarking-surgery

Andrews-Kelly G 2010 Canine socialization through the use of "playrooms" or exercise rooms. *Enrichment Record 5*: 7–9. http://enrichmentrecord.com/canine-socialization-through-the-use-of-"playrooms"-or-exercise-rooms/

Anonymous 2013a Dealing with emotional fatigue. In: Reinhardt V (ed) *Compassion Makes a Difference—Discussions by the Laboratory Animal Refinement & Enrichment Forum, Volume III* pp 180–182. Animal Welfare Institute: Washington, DC. http://awionline.org/sites/default/files/CompassionMakesADifference.pdf

Anonymous 2013b Elevated resting surfaces for dogs. In: Reinhardt V (ed) *Compassion Makes a Difference—Discussions by the Laboratory Animal Refinement & Enrichment Forum, Volume III* pp 10–12. Animal Welfare Institute: Washington, DC. http://awionline.org/sites/default/files/CompassionMakesADifference.pdf

Anonymous 2013c Retiring and adopting animals who are no longer needed for research. In: Reinhardt V (ed) *Compassion Makes a Difference—Discussions by the Laboratory Animal Refinement & Enrichment Forum, Volume III* pp 169–176. Animal Welfare Institute: Washington, DC. http://awionline.org/sites/default/files/CompassionMakesADifference.pdf

Associated Press (2014, March 13) Americans Spent a Record $56 Billion on Pets Last Year. Retrieved from http://www.cbsnews.com/news/americans-spent-a-record-56-billion-on-pets-last-year/

Burgess W, French ED and Kendall LV 2010 Socialization program to improve research dog adoption rates. *American Association for Laboratory Animal Science Meeting Official Program*: 163

Evans EI, Gates GR and Green VD 1999 A "puppy playroom" as opportunity for exercise and learning prior to adoption. *American Association for Laboratory Animal Science Meeting Official Program*: 32

Field G and Jackson TA 2007 *The Laboratory Canine*. Taylor and Francis Group: Boca Raton, FL

Freedman DG, King JA and Elliot O 1991 Critical period in the social development of dogs. *Science 133*: 1016–1017

Fox MW 1975 Evolution of social behavior in canids. In: Fox MW (ed) *The Wild Canids* Van Nostrand Reinhold: New York, NY

Haug, LI 2006 *Environmental Enrichment for Dogs*. Texas Veterinary Behavior Services: Sugar Land, TX. http://texasvetbehavior.com/Canine_Enrichment.pdf

Hubrecht RC 1993 A comparison of social and environmental enrichment methods for laboratory housed dogs. *Applied Animal Behaviour Science 37*: 345–361

Hubrecht RC 1995 Dogs and dog housing. In: Smith CP and Taylor V (eds) *Environmental Enrichment Information Resources for Laboratory Animals: 1965—1995: Birds, Cats, Dogs, Farm Animals, Ferrets, Rabbits, and Rodents; AWIC Resource Series No. 2* pp 49–62. United States Department of Agriculture, Beltsville, MD and Universities Federation for Animal Welfare, Potters Bar, UK. http://pubs.nal.usda.gov/sites/pubs.nal.usda.gov/files/Enrichment1995.pdf

Hubrecht RC 2002 Comfortable quarters for dogs in research institutions. In: Reinhardt V and Reinhardt A (eds) *Comfortable Quarters for Laboratory Animals, Ninth Edition* pp 57–65. Animal Welfare Institute: Washington, DC. http://www.awionline.org/pubs/cq02/Cq-dogs.html

Hussain M, Leach A and Hardy C 2006 A refined method of restraint for dogs used in inhalation studies. *Animal Technology and Welfare 5*(3): 179–181

Kilcullen-Steiner C and Mitchell A 2001 Quiet those barking dogs. *American Association for Laboratory Animal Science Meeting Official Program*: 103

Laule G 1999 Training laboratory animals. In: Poole T (ed) *The UFAW Handbook on the Care and Management of Laboratory Animals, Seventh Edition* pp 21–27. Wiley-Blackwell: Oxford, UK

Loveridge GG 1994 Provision of environmentally enriched housing for dogs. *Animal Technology 45*: 1–19

National Research Council 1994 *Laboratory Animal Management: Dogs.* National Academy Press: Washington, DC

National Research Council 2011 *Guide for the Care and Use of Laboratory Animals, Eighth Edition.* National Academies Press: Washington, DC

Overall KL and Dyer D 2005 Enrichment strategies for laboratory animals from the viewpoint of clinical veterinary behavioral medicine: Emphasis on cats and dogs. *Institute for Laboratory Animal Research Journal 46*(2): 202–216

Roddis D 2005 How to teach an old dog new tricks. *Animal Technology and Welfare 4*: 181–184

Savastano GM 2013 Operant conditioning with laboratory beagles. *American Association for Laboratory Animal Science Meeting Official Program*: 144

Shulder L and Ogbin J 2010 Zen pen. *Enrichment Record 5*: 8. http://www.gr8tt.com/flipbooks/uniflip_ER_1010Folder/uniflip.swf

Simons A (2014, May 21) Beagle Freedom Law makes history in Minnesota. *Star Tribune.* Retrieved from http://www.startribune.com/politics/statelocal/260128011.html

Tabers BC, Corten DJ and Mehrtens AC 2009 Improvement of canine gavage through behavioral refinement. *American Association for Laboratory Animal Science Meeting Official Program*: 108–109

Trussell BA, King J and Smith D 1999 Application of environmental enrichment routines to regulatory toxicology studies in the beagle dog. *Animal Technology 50*: 131–133

Wells DL, Graham L and Hepper PG 2002 The influence of auditory stimulation on the behaviour of dogs housed in a rescue shelter. *Animal Welfare 11*: 385–393

Wolfle TL 1992 Socialization of dogs. In: Krulisch L (ed) *Implementation Strategies for Research Animal Well-being: Institutional Compliance with Regulations* pp 15–21. Scientists Center for Animal Welfare: Bethesda, MD

Cats

Cats

Irene Rochlitz, BVSc, MSc, PhD, FHEA, MRCVS

CENTRE FOR ANIMAL WELFARE AND ANTHROZOOLOGY, DEPARTMENT OF VETERINARY MEDICINE,
UNIVERSITY OF CAMBRIDGE

Implementation of the "Three Rs" (replacement, reduction and refinement) should be considered whenever animals are used in biomedical research (Russell & Burch, 1959). Refinement applies both to experimental procedures and to the way animals are housed and looked after. Housing conditions have a major impact on animal welfare so they should be well regulated to the highest standards. This is especially relevant for domestic cats (*Felis silvestris catus*), who cannot usually be taken out of their enclosures for walks in the way that dogs can.

Keeping cats in an enriched, stimulating environment that encourages a wide range of normal behaviors—and providing them with ongoing positive interactions with people—will result in enhanced health and welfare. In addition, when these cats are no longer required for research and are rehomed, they will be more likely to adapt successfully to their new environment (DiGangi & Levy, 2006).

Species-typical characteristics of cats

In order to understand and appreciate the species-typical characteristics of cats, knowledge of their evolutionary history is helpful. The domestic cat is descended from the North African wildcat (*Felis silvestris libyca*), a largely solitary-living, territorial species. This carnivorous predator lives and hunts alone; it is also the prey of larger carnivores. The species can be described as semi-arboreal, spending much of the day hiding in bushes and trees. Members of the species are active primarily at night, and come into direct contact with conspecifics mainly for mating.

Studies of feral cats, defined as free-living domestic cats with limited or no contact with humans, are also informative. Feral cats also hunt alone. But while they may

live singly, when there are sufficient food sources they often form groups (colonies) of mainly female relatives (Macdonald et al., 1987; Macdonald et al., 2000). Female feral cats may cooperate in the rearing of kittens (Macdonald et al., 2000), especially if they are related (e.g., mother-daughter, sister-sister). Cats can form strong social bonds with other individuals ("preferred affiliates," Curtis et al., 2003; Crowell-Davis et al., 2004), particularly if they are related: e.g., between kittens in the same litter and between kittens and their mother. They may also form close bonds with unrelated but familiar cats; this is likely to develop if they have been together from a young age. Cats who are members of the same group are often in tactile contact, exhibiting behaviors such as allorubbing (rubbing their bodies against each other), allogrooming (grooming each other), touching noses (a greeting response), and resting together (Curtis et al., 2003; Crowell-Davis et al., 2004; Rochlitz, 2009).

The wildcat relies primarily on olfactory communication, which is particularly well adapted to a solitary lifestyle, as it is long-lasting, individual, specific, and effective over long distances. Like the wildcat, both feral and nonferal domestic cats have an excellent sense of smell and rely primarily on olfactory communication. This consists of the deposition of scent (scent marking) mainly by urine spraying, claw scratching, and tactile rubbing (against objects or other cats), and less commonly by elimination (urination and defecation). A cat's nasal epithelium is dense with olfactory receptors, and there is a vomeronasal organ behind the upper incisors that is used in the "flehmen" response (Bradshaw et al., 2012). This response, where the upper lip is raised and air inhaled, is seen primarily in reaction to odors from other cats and is presumed to gather social information.

In addition to olfactory communication, cats use visual (i.e., posture, tail position, and facial expression) and vocal communication. Interestingly, they are more vocal towards humans than towards other cats (Brown & Bradshaw, 2014). It is thought that cats lack the facial musculature typical of social species (van den Bos, 1998) and do not have as large a behavioral repertoire for visual communication as, for example, the highly social, group-living dog. Nevertheless, while acknowledging that some signs may be subtle, much can be learned about a cat from careful observation.

Addressing the species-typical characteristics of cats in the research laboratory

This chapter will describe ways of addressing the species-typical characteristics of cats in research institutions, with the aim of meeting their needs and ensuring their good welfare. Much of this information is derived from reviews on environmental enrichment for domestic cats in the home, in catteries or in shelters (e.g., Rochlitz, 1999, 2000a, 2005; Ellis, 2009; Herron & Buffington, 2010; Ellis et al., 2013: Stella & Buffington, 2014), as well as in laboratories (e.g., Rochlitz, 1999, 2000a, 2000b; Overall & Dyer, 2005; McCune et al., 2014). The five main elements to the design of cat housing in research laboratories (adapted from Bloomsmith et al., 1991) are as follows:

1. Physical: size of enclosure, internal structure and complexity
2. Social: socialization—contact with conspecifics and humans
3. Sensory: olfaction and marking behaviors, visual and auditory stimuli, surrounding area within the cat's sensory range
4. Occupational: opportunities to explore the environment, exercise and play
5. Nutritional: provision of food and water, toileting (elimination)

Physical

Having evolved from a solitary-living species, cats have not been subject to selection pressures to develop a wide range of social communication behaviors or formal group structures such as strict dominance hierarchies. In addition, unlike the dog, they lack mechanisms to convey appeasement or reconciliation after conflicts (van den Bos, 1998; Casey & Bradshaw, 2007). The main way they avoid conflict is by avoiding each other. Studies have shown that a distance of 1 to 3 meters between cats is required to reduce stress and maintain harmony (Kessler & Turner, 1999; Barry & Crowell-Davis, 1999). If cats are unable to establish sufficient distance, enforced proximity may lead to a notable reduction in activity and increased stress, possibly resulting in overt aggression.

The minimum space requirements for the housing of cats in laboratories in Europe can be found in Appendix A of the European Convention ETS 123 (Council of Europe, 2006) and are presented in the table on the following page. Enclosures should be "walk-in" (2 m high or more) to allow caretakers to enter, as this makes it easier to interact with the cats in a normal manner and conduct maintenance activities effectively. These European dimensions are considerably larger than the minimum dimensions stated for the housing of cats in research facilities in the United States, where a 4 kg cat can be housed in an enclosure with 4 square feet (0.37 m^2) of floor space and a height of 2 feet (0.6 m) (National Research Council, 2011). In the author's opinion, the National Research Council minimum dimensions are too small; the minimum floor space requirement for cats should be determined by their socio-spatial needs rather than by their body weight. Research facilities should aim to exceed these dimensions in order to be able to create enclosures that are well designed to meet the cats' needs.

Minimum space requirements for the housing of cats, Appendix A to the European Convention ETS 123 (Council of Europe, 2006)

	Floor area (excluding shelves) m^2	Shelves m^2	Height m
Minimum for 1 adult cat	1.5	0.5	2
For each additional cat add	0.75	0.25	

The minimum space in which a queen and litter may be held is the space for a single cat, which should be gradually increased so that by 4 months of age litters have been re-housed to conform with the above space requirements for adults.

Once sufficient floor space is provided, it is the overall complexity and vertical space availability that are of greatest importance. Cats are agile and athletic animals who enjoy exploring, climbing, running, and jumping—spending more time off than on the floor. Structures that enable cats to make maximal use of the vertical dimension include climbing towers, climbing frames, raised walkways, "cat aerobic centers," and platforms or shelves placed at different heights. Slanting boards and steps will help kittens and small cats to reach the raised areas.

Cats spend long periods of time sleeping, dozing or resting. They prefer soft resting substrates, such as pillows or fleece beds (Crouse et al., 1995; Hawthorne et al., 1995) and materials that maintain a constant temperature (Roy, 1992).

Hiding is a coping behavior that cats frequently display in response to changes in their environment or to avoid interactions with other cats or people. Being potential prey as well as predator, and in order to avoid conflict and reduce stress, they prefer to be partially or completely hidden and, wherever possible, off the floor and preferably in a corner. This allows them to monitor their surroundings without being exposed or attacked from behind. These hiding and resting places, or "safe havens," must therefore provide concealment and height as well as comfort. Hammocks, high-sided trays, "igloos," cardboard or plastic boxes and similar structures, placed securely on raised shelves and ledges as well as on the floor, are all suitable. Various types of plastic children's furniture are often effective, inexpensive and can easily be cleaned. While cats don't particularly favor this substrate, it can be improved by the addition of soft bedding.

There should be more resting and hiding places in the enclosure than there are cats, in order to minimize competition and also because some cats like to change places. Cats will show individual preferences, and these should be identified and met wherever possible. Visual barriers, such as vertical panels, are very useful to divide the enclosure into separate spaces, enabling cats to make choices and to avoid others.

Whenever possible, cats should be housed as pairs or in groups. In some instances, it may be necessary to house a cat singly for recovery after a clinical procedure or for metabolic studies. This should be for as short a time as possible. The cage should still have at least 1.5 square meters of floor space, but it does not have to be walk-in. Ideally, it should be no less than 1 meter high so that the cat can stretch fully in the vertical direction, and a shelf must be installed in such a way that the cat can comfortably rest on it and freely move on and under it. A hiding box should also be provided, placed under or securely on the shelf. If there isn't room for a hiding box, a towel attached to the shelf so that it acts as a curtain for the space below, or a towel covering half of the front of the cage, are ways to create a hiding space for the cat.

Cages should not be stacked one on top of the other, and placing them on a shelf at waist height will make access easier for the caretaker. Methods to join two small cages together via a porthole to make a larger single cage have been described (UC-Davis Koret Shelter Medicine Program, 2010). Cats who have to be housed singly for more than 2–3 days should be allowed daily access to a larger enclosure where they can explore, play, and interact more naturally with the caretaker. Queens in the last 2 weeks of pregnancy, and queens with unweaned kittens should not be housed with other cats.

The floor of all cat enclosures should be smooth, nonslip and easy to clean. Wire-mesh or grid floors are not suitable, as they are uncomfortable for cats and may trap and injure their extremities (toes, paws and tails).

Social

Positive interactions with humans and other cats should continue beyond the sensitive period of social development (typically between 2 and 8 weeks of age) (Karsh & Turner, 1988) throughout the cats' lives. Periods of time each day should be set aside for interaction among cats and their caregivers (Loveridge, 1994; Rochlitz, 2005).

Most cats should be housed in groups providing that there is sufficient space, easy access to feeding and elimination areas, and an adequate number of hiding retreats and resting places. Many factors will determine the ideal group size, but it seems that 20 to 25 individuals is the maximal number for cats in laboratories (Hubrecht & Turner, 1998), although European guidelines recommend groups of up to 12 cats (Council of Europe, 2006). Cats who fail to adapt satisfactorily to living in groups should be identified and housed in compatible pairs or, if incompatible, singly. It is particularly important that singly housed cats receive additional daily human contact and, if judged to be beneficial, visual contact with other cats.

Neutered cats can be kept together in groups, as can intact females. While some authors suggest that intact males should be housed singly, others have shown that they can be housed successfully with other intact and neutered males (Podberscek et al., 1991); they can also be kept with neutered females.

The introduction of a new cat to a group should be done slowly and under supervision (Rochlitz, 2009). Initially, the cat should

be kept in a small cage placed within the group's enclosure. It is vital for the cat to have a box to retreat to, in order to escape the attention of the resident cats. Usually within 2 weeks, the newcomer can safely be released into the enclosure. The use of a synthetic analogue of naturally occurring feline facial pheromones, such as Felifriend (CEVA Animal Health Ltd.), can facilitate the introduction of an unfamiliar cat into an established group (Pageat & Tessier, 1997; Mills, 2005).

Sensory

Marking behaviors: Scratching on surfaces helps to maintain the claws, stretch the body, and exercise muscles and tendons in the legs. It is also a means of visual and olfactory communication. In addition to leaving visual striations or lines, claw scratching deposits scent from sebaceous glands in the cat's paws and is a form of marking behavior. Most cats are strongly motivated to scratch, so scratching posts, disks, rush matting, hessian, wood, or similar surfaces should be provided (Rochlitz, 2009). They should be placed in a number of locations throughout the enclosure, especially near entry/exit points and resting places, and replaced when worn. Some cats prefer to scratch on vertical surfaces, while others prefer horizontal surfaces. Scratching surfaces should be long enough so the cat can stretch fully.

Cats also scent mark by rubbing the sides of their face against protruding vertical structures such as corners and edges of boxes (as well as allorub with conspecifics). The deposition and exchange of scent are thought to be important in maintaining group cohesion and conveying other information. Excessive cleaning of scent-marked areas can disrupt these familiar and reassuring smells, so spot cleaning may be preferable, providing that adequate hygiene is maintained (Rochlitz, 2009). Another

way to maintain olfactory continuity in the enclosure is to wash only a portion of the bedding at a time.

Catnip, provided as a dry herb or in toys, may elicit positive behavioral responses (play, sniffing, pawing), which are seen in 50–70% of cats, as sensitivity to catnip is inherited (Ellis, 2007). As mentioned previously, synthetic pheromone products have been developed for use in cats. Felifriend is used to reduce fear and promote positive interactions among cats, and between cats and humans, while Feliway (also from CEVA) is used to reduce anxiety- and stress-related behaviors such as urine spraying (Mills, 2005; Ellis, 2009).

Visual and auditory stimuli: Seeing what is going on outside the cat's immediate enclosure gives the cat a sense of predictability and control and is often a rich source of stimulation, leading to improved welfare. DeLuca and Kranda (1992) found that cats in a colony spent almost all their time sitting on an indoor window ledge and watching activities in the hallway. External

windows bring in natural light, and outdoor activities are also a source of interest and stimulation. Windows with deep ledges or raised, resting platforms will facilitate comfortable viewing, whether it is towards the indoor or outdoor environment. Visual stimulation using television may hold the cat's attention, especially when prey items (e.g., small rodents, birds) and conspecfics (in amicable situations) are presented (Ellis & Wells, 2008).

Cats are very sensitive to noise and easily startled. It is important for kittens to be exposed to a range of normal sounds during their socialization period and beyond. Noises as familiar background sounds, such as low-volume music, may be reassuring (Loveridge, 1994; Newberry, 1995), but loud or unpredictable noises may cause stress. For this reason, cats should not be housed where they can hear dogs barking or other loud noises.

Occupational

An outdoor enclosure or pen will provide an area for exploration, exercise and stimulation. It should be secure and protected from contact with outdoor cats or other animals. Addition of climbing frames, tunnels, and other furniture will encourage activity. There should be more than one passageway (e.g., cat flap) leading to the outdoor enclosure so that the entry and exit point cannot be guarded and monopolized by one cat.

Play: The expression of play behaviors in cats, whether it is by playing alone or with humans, is usually interpreted as evidence of good welfare (table page 156). Uninterrupted contact periods for caretakers and their cats should be scheduled into the institution's daily routine. Individual cats often have specific likes and dislikes. For example, some enjoy being stroked on the head or being groomed, some prefer noncontact interaction

such as playing with humans via a toy, and some dislike being picked up (Soennichsen & Chamove, 2002; Rochlitz, 2005; Ellis et al., 2013). Toys that are small and mobile and have complex surface textures (e.g., fur, feathers) often elicit play (Hall & Bradshaw, 1998), as do simple objects like large paper bags or cardboard boxes (e.g., for "hide and seek" games). Toys should be regularly replaced to maintain interest and ensure their novelty effect (De Monte & Le Pape, 1997; Hall et al., 2001). There is a huge range of items available for playing with cats and some are more effective than others; the caretakers should be able to identify what their cats prefer. Because most cats play alone rather than in groups, the enclosure must be of sufficient size to ensure that cats can play safely without disturbing others.

Nutritional

Feeding: Cats prefer frequent small meals throughout the day and night (this would occur in the wild, as they catch prey sporadically) but are usually offered two meals a day. Ways of increasing the time the cat spends in feeding behaviors—in particular, in performing parts of the predatory sequences prior to consuming food—have been devised. For example, dry cat food ("kibble") can be hidden in the environment for the cat to find. Food can be put in puzzle feeders (or food balls), which are food-dispensing containers for the cat to manipulate. Flat activity boards, where the cat has to manipulate dry food out of obstacles on the device before it can be consumed (mimicking some predatory behaviors), are also commercially available. Because cats do not usually eat together, food bowls should be dispersed throughout the enclosure and placed on raised surfaces as well as on the floor. There should be a sufficient number of bowls so as to avoid conflict or guarding (monopolization). Consideration should be given to the provision of containers of grass for cats to chew, as this is thought to help eliminate fur balls (trichobezoars).

Drinking: Water should be available 24 hours a day, and changed daily to ensure freshness. Factors to consider include movement (e.g., water fountains), shape of container (some cats do not like their whiskers to touch the sides of the container when drinking), and position. Because cats often prefer to drink away from the feeding area and at times unconnected with feeding, there should be bowls of water in several locations, both near and away from food bowls (and outdoors as well as indoors, if cats have outside access).

Toileting (elimination): Cats are fastidious when it comes to toileting behavior. They urinate and defecate in locations that are safe, clean, not near resting and hiding places or areas of activity, and that are not under the constant scrutiny of other cats. Failure to use the litter tray should alert the caretaker to possible health or behavior problems in the group, or to inadequate provision of suitable toilet areas.

Litter boxes should be dispersed throughout the enclosure but, preferably, at least 0.5 to 1 meter away from feeding, drinking and resting/hiding areas, and easily accessible to kittens as well as adults. There should be at least one box per two cats and it should be cleaned at least once, and preferably twice, a day. More frequent cleaning will reduce the number of litter trays required, thereby freeing up available floor space. Ideally the litter box should be longer than 1.5 times the average body length of a cat, so that the cat can reach forward to rake clean litter back over their waste. Often, litter boxes are too short, so large plant trays may be a better choice. The space occupied by the litter box should not be considered as part of the overall floor space.

With regard to type of litter box (covered or not) and type of litter, cats have individual preferences and these should be met wherever possible. Within the litter box it is important that there is sufficient substrate to bury feces, and unscented, fine-grained clumping litter is often preferred. Fine-grained clumping litter should be completely removed weekly because scents and small particles of feces can coat the surfaces of individual grains (Overall & Dyer, 2005).

Refining husbandry and research procedures for cats

Handling: Socialization from a young age, daily positive interactions with familiar and unfamiliar people, and exposure to a complex environment promotes the development of kittens into adult cats that are easier to handle for research procedures. (Hoskins, 1995). Handlers should be empathetic, calm, gentle, and focused; speak in a quiet voice; and approach from the side rather than the front. They should avoid direct or unblinking eye contact with the cat. If the cat is frightened and uncooperative, covering the animal's head or wrapping the body in a towel will help to calm the cat down and make handling safer.

Scruffing (grasping folds of the cat's skin in the cervical area) is a behavior usually seen in cats in only limited circumstances, such as mothers carrying kittens and during mating (and, less commonly, fighting). It is not naturally used as an effective method of discipline (Rodan et al., 2011). Pinching the cat's skin along the back of the neck ("clipnosis") may also inhibit reactions, but it is unclear if this is due to pain or other effects (Tarttelin, 1991; Pozza et al., 2008). Neck scruffing or cervical skin pinching as methods of restraint by humans, therefore, should be avoided whenever possible, until more is known about how the cat experiences these methods. It is important to realize that when cats react negatively to handling it is usually due to fear, so handling methods should aim to reduce fear and gain the cat's trust rather than to escalate to aggression. Guidelines on the humane handling of cats can be found in Rodan et al. (2011).

Training: Most cats can be trained to interact in particular ways in order to make procedures safer and less aversive. Operant conditioning techniques are effective (Lockhart et al., 2013), providing there is sufficient time and the trainer is knowledgeable and patient, and only uses techniques based on reward. Most cats can be trained to tolerate routine procedures such as venipuncture, saliva collection (shown in photo above), nail trimming, tooth brushing, and having their temperature taken (Overall & Dyer, 2005). Fearful cats who react poorly to research procedures or to training should be removed from the program and put up for adoption.

Declawing: Declawing (onychectomy) is a surgical procedure performed in some countries (e.g., the United States) but regarded as a mutilation, and therefore banned, in others (e.g., the United Kingdom). It is likely to cause short-term pain and possibly also longer-term pain, and it may frustrate the cat's motivation to scratch. Cats can be handled safely and take part in a research program without being declawed, and declawing has not been shown to "improve" behavior or the chances for a successful adoption (DiGangi & Levy, 2006), so this procedure should not be performed.

Recognizing stress: Cats need predictability, familiarity and routine. When these are absent, cats' coping abilities are reduced, making them less resilient and more susceptible to stress (Herron & Buffington, 2010). This is highlighted in recent studies on sickness behaviors (Stella et al., 2011, Stella et al., 2014). Sickness behaviors refer to a group of nonspecific clinical and behavioral signs that normally occur in response to infection. These behaviors were noted in healthy colony-housed cats in response to environmental disturbances such as transient discontinuation of contact or interactions with the cats' primary caretaker, changes in time of day of routine husbandry, unfamiliar caretakers, and a delay in feeding time. The most common sickness behaviors observed were vomiting of hair, food, or bile; decreased appetite; and eliminating out of the litter tray. While some cats experiencing stress will show overt signs such as sickness behaviors, others will react differently, responding

Some behavioral measures of good and poor welfare in domestic cats in research institutions (modified from Rochlitz 2009, 2014)

Behavior	Good welfare	Poor welfare
Maintenance behaviors[1]	Normal levels	Reduced levels or absent; sickness behaviors
Activity, exploration and investigation of surroundings	Normal levels	Reduced levels or absent (rarely, high levels)
Social interactions with other cats in the group	Present; positive (affiliative) behaviors such as allorubbing, allogrooming, staying in proximity	Absent or negative; hostility, aggression, avoidance of each other
Interactions with caretakers	Initiates positive interactions with caretakers; positive response to initiation of interactions by caretakers	Failure to initiate interactions with caretakers; absence or negative response to initiation of interactions by caretakers
Types of behaviors shown	Shows a wide range of normal behavioral repertoire; friendly behaviors (e.g., tail-up position, rubbing, vocalization)	Persistent signs of timidity, anxiety, fear or aggression; hiding or attempting to hide for long periods; over-grooming; self-mutilation; excessive vocalization; excessive vigilance; feigned sleep[2]
Play	Presence of play (on own, with objects, with other cats, or with caretakers)	Absence of play

[1]Maintenance behaviors: feeding, drinking, grooming, claw scratching, resting, sleeping, urination, defecation.
[2]Feigned sleep: the cat appears to be asleep or resting (body is in sleep posture and eyes are closed or partly closed) but is awake and vigilant.

Comfortable Quarters for Laboratory Animals

to poor environmental conditions and disturbances by becoming inactive and reducing normal behaviors such as self-maintenance (feeding, drinking, grooming, claw scratching, resting, sleeping, urination, defecation), exploration, or play (McCune, 1992; Rochlitz, 1999; Casey & Bradshaw, 2007). Some behavioral measures of good and bad welfare in cats are summarized in the table on the previous page.

Source of cats
Cats should be obtained only from designated breeding establishments (class A dealers). Because the sensitive period of socialization is so early in cats (2 to 8 weeks of age), an informed and well-planned socialization program should be in place at the breeder. During this period it is particularly important that the kitten has social contacts with other kittens (e.g., litter mates) and with humans, and is exposed and habituated to the environmental conditions the kitten will subsequently encounter. In addition, the breeding program should take into account that friendliness (also described as boldness) to humans is, in part, genetically inherited from the father. Kittens from friendly fathers tend to react with greater boldness when faced with unfamiliar people and novel objects (McCune, 1995).

Final thoughts
Cats respond strongly to humans in their environment and organize their daily activity patterns around the caretakers' activity, preferring human contact over toys (Randall et al., 1990; DeLuca & Kranda, 1992). Caretakers should like cats, and be knowledgeable of their behavior and reactions to stress so that signs of poor welfare, whether overt or more subtle, are recognized promptly. However, it should be recognized that even though the caretakers are key components to a cat's welfare, they are not a substitute for a proper physical environment. High standards of housing, enrichment, management, and human interaction will create an optimal scenario for the cat in research.

The author would like to thank The WALTHAM Centre for Pet Nutrition (© Mars, Incorporated. All Rights Reserved) for providing the photographs on pages 149, 151, 154 and 155 in this chapter.

REFERENCES

Barry KJ and Crowell-Davis SL 1999 Gender differences in the social behaviour of the neutered indoor-only domestic cat. *Applied Animal Behaviour Science 64*: 193–211

Bloomsmith MA, Brent LY and Schapiro SJ 1991 Guidelines for developing and managing an environmental enrichment program for non-human primates. *Laboratory Animal Science 41*: 372–377

Bradshaw JWS, Casey RA and Brown SL 2012 *The Behaviour of the Domestic Cat, Second Edition*. CAB International: Wallingford, UK

Brown S and Bradshaw JWS 2014 Communication in the domestic cat: Within-and between- species. In: Turner DC and Bateson P (eds) *The Domestic Cat: The Biology of its Behaviour, Third Edition* pp 37–59. Cambridge University Press: Cambridge, UK

Casey RA and Bradshaw JWS 2007 The assessment of welfare. In: Rochlitz I (ed) *The Welfare of Cats* pp 23–46. Springer: Dordrecht, Netherlands

Crouse SJ, Atwill ER, Lagana M and Houpt KA 1995 Soft surfaces: A factor in feline psychological well-being. *Contemporary Topics in Laboratory Animal Science 34*(6): 94–97

Crowell-Davis SL, Curtis TM and Knowles RJ 2004 Social organization in the cat: a modern understanding. *Journal of Feline Medicine and Surgery 6*: 19–28

Curtis TM, Knowles RJ and Crowell-Davis SL 2003 Influence of familiarity and relatedness on proximity and allogrooming in domestic cats (*Felis catus*). *American Journal of Veterinary Research 64*: 1151–1154

De Monte M and Le Pape G 1997 Behavioural effects of cage enrichment in single-housed adult cats. *Animal Welfare 6*: 53–66

DeLuca AM and Kranda KC 1992 Environmental enrichment in a large animal facility. *Laboratory Animal 21*: 38–44

DiGangi BA and Levy JK 2006 Outcome of cats adopted from a biomedical research program. *Journal of Applied Animal Welfare Science 9*:143–163

Council of Europe 2006 *Appendix A of the European Convention for the Protection of Vertebrate Animals Used for Experimental and Other Scientific Purposes (ETS No. 123): Guidelines for Accommodation and Care of Animals (Article 5 of the Convention) Approved by the Multilateral Consultation*. Council of Europe: Strasbourg, France. http://conventions.coe.int/Treaty/EN/Treaties/PDF/123-Arev.pdf

Ellis SLH 2007 *Sensory enrichment for cats (*Felis silvestris catus*) housed in an animal rescue shelter* [dissertation]. Queen's University: Belfast, UK

Ellis S 2009 Environmental enrichment: Practical strategies for improving feline welfare. *Journal of Feline Medicine and Surgery 11*: 901–912

Ellis SLH and Wells DL 2008 The influence of visual stimulation on the behaviour of cats housed in a rescue shelter. *Applied Animal Behavior Science 113:* 166–74

Ellis SLH, Rodan I, Carney HC, Heath S, Rochlitz I, Shearburn LD, ... Westropp JL 2013 AAFP and ISFM Feline Environmental Needs Guidelines. *Journal of Feline Medicine and Surgery 15*: 219–230

Hall SL and Bradshaw JWS 1998 The influence of hunger on object play by adult domestic cats. *Applied Animal Behaviour Science 58*: 143–150

Hall SL, Bradshaw JWS and Robinson IH 2001 Object play in adult domestic cats: The roles of habituation and disinhibition. *Applied Animal Behaviour Science 79*: 263–271

Hawthorne AJ, Loveridge GG and Horrocks LJ 1995 The behaviour of domestic cats in response to a variety of surface-textures. In: Holst B (ed) *Proceedings of the Second International Conference on Environmental Enrichment* pp 84–94. Copenhagen Zoo: Copenhagen, Denmark

Herron ME and Buffington CA 2010 Feline focus: environmental enrichment for indoor cats. *Compendium on Continuing Education for the Practising Veterinarian 32*: E1–E5

Hoskins CM 1995 *The effects of positive handling on the behaviour of domestic cats in rescue centres* [thesis]. University of Edinburgh: Edinburgh, UK

Hubrecht RC and Turner DC 1998 Companion animal welfare in private and institutional settings. In Wilson CC and Turner DC (eds) *Companion Animals in Human Health* pp 267–289. Sage Publications: Thousand Oaks, CA

Karsh EB and Turner DC 1988 The human-cat relationship. In Turner DC and Bateson P (eds) *The Domestic Cat: The Biology of its Behaviour, First Edition* pp 159–177. Cambridge University Press: Cambridge, UK

Kessler MR and Turner DC 1999 Effects of density and cage size on stress in domestic cats (*Felis silvestris catus*) housed in animal shelters and boarding catteries. *Animal Welfare 8*: 259–267

Lockhart J, Wilson K and Lanman C 2013 The effects of operant training on blood collection for domestic cats. *Applied Animal Behaviour Science 143*: 128–134

Loveridge GG 1994 Provision of environmentally enriched housing for cats. *Animal Technology 45*: 69–87

Macdonald DW, Apps PJ, Carr GM and Kirby G 1987 Social dynamics, nursing coalitions and infanticide among farm cats, *Felis catus*. *Advances in Ethology* (supplement to *Ethology*) *28*: 1–66

Macdonald DW, Yamguchi N and Kerby G 2000 Group living in the domestic cat: Its socio-biology and epidemiology. In: Turner DC and Bateson P (eds) *The Domestic Cat: The Biology of its Behaviour, Second Edition* pp 95–115. Cambridge University Press: Cambridge, UK

McCune S 1992 *Temperament and the welfare of caged cats* [dissertation]. University of Cambridge, Cambridge, UK

McCune S 1995 The impact of paternity and early socialisation on the development of cats' behaviour to people and novel objects. *Applied Animal Behaviour Science 45*(1–2): 111–126

McCune S, Moesta A and Kruger K 2014 Housing and husbandry of cats. National Centre for Replacement, Refinement and Reduction of Animals in Research website. http://www.nc3rs.org.uk/our-resources/housing-and-husbandry/housing-and-husbandry-cats

Mills D 2005 Pheromonatherapy: Theory and applications. *In Practice 27*(7), 368–373

National Research Council 1996 *Guide for the Care and Use of Laboratory Animals, Seventh Edition*. The National Academy Press: Washington, DC. http://www.nap.edu/openbook.php?record_id=5140

Newberry RC 1995 Environmental enrichment: increasing the biological relevance of captive environments. *Applied Animal Behaviour Science 44*: 229–243

Overall KL and Dyer D 2005 Enrichment strategies for laboratory animals from the viewpoint of clinical veterinary behavioral medicine: Emphasis on cats and dogs. *ILAR Journal 46*(2): 202–216

Pageat P and Tessier Y 1997 Usefulness of the F4 synthetic pheromone for prevention of intraspecific aggression in poorly socialised cats. In: Mills D, Heath SE and Harrington LJ (eds) *Proceedings of the First International Conference on Veterinary Behavioural Medicine, Birmingham, UK, April 1–2* pp 64–72. Universities Federation for Animal Welfare: Potters Bar, UK. http://eprints.lincoln.ac.uk/1880/13/1_12.pdf

Podberscek AL, Blackshaw JK and Beattie AW 1991 The behaviour of laboratory colony cats and their reactions to a familiar and unfamiliar person. *Applied Animal Behaviour Science 31*: 119–130

Pozza ME, Stella JL, Chappuis-Gagnon AC, Wagner SO and Buffington CA 2008 Pinch-induced behavioral inhibition ('clipnosis') in domestic cats. *Journal of Feline Medicine and Surgery 10*: 82–87

Randall WR, Cunningham JT and Randall S 1990 Sounds from an animal colony entrain a circadian rhythm in the cat, *Felis catus* L. *Journal of Interdisciplinary Cycle Research 21*: 55–64

Rochlitz I 2014 Feline welfare issues. In: Turner DC and Bateson P (eds) *The Domestic Cat: The Biology of its Behaviour, Third Edition* pp 131–153. Cambridge University Press: Cambridge, UK

Rochlitz I 2009 Basic requirements for good behavioural health and welfare in cats. In: D Horwitz and DF Mills (eds) *BSAVA Manual of Canine and Feline Behavioural Medicine, Second Edition* pp 35–48. British Small Animal Veterinary Association: Gloucester, UK

Rochlitz I 2005 Housing and Welfare. In: Rochlitz I (ed) *The Welfare of Cats* pp 141–176. Springer: Dordrecht, Netherlands

Rochlitz I 1999 Recommendations for the housing of cats in the home, in catteries and animal shelters, in laboratories and in veterinary surgeries. *Journal of Feline Medicine and Surgery 3*: 181–91

Rochlitz I 2000a Feline welfare issues. In: Turner DC and Bateson P (eds) *The Domestic Cat: The Biology of Its Behaviour*. Cambridge University Press: Cambridge, UK

Rochlitz I 2000b Recommendations for the housing and care of domestic cats in laboratories. *Laboratory Animals 34*: 1–9

Rodan I, Sundahl E, Carney H, Gagnon AC, Heath S, Landsberg G, ... Yin S 2011 AAFP and ISFM feline-friendly handling guidelines. *Journal of Feline Medicine and Surgery 13*: 364–75

Roy D 1992 *Environmental enrichment for cats in rescue centres* [thesis]. University of Southampton: Southampton, UK

Russell WMS and Burch RL 1959 *The Principles of Humane Experimental Technique*. Methuen: London, UK

Soennichsen S and Chamove AS 2002 Responses of cats to petting by humans. *Anthrozoös 15*: 258–265

Stella JL, Lord LK and Buffington CA 2011 Sickness behaviors in response to unusual external events in healthy cats and cats with feline interstitial cystitis. *Journal of the American Veterinary Medical Association 238*: 67–73

Stella J, Croney C and Buffington C 2014 Environmental factors that affect the behavior and welfare of domestic cats (*Felis silvestris catus*) housed in cages. *Applied Animal Behaviour Science 160*: 94–105

Tarttelin M 1991 Restraint in the cat induced by skin clips. *Journal of Neuroscience 57*: 288

UC-Davis Koret Shelter Medicine Program 2010 Cat cage modifications: Making double compartment cat cages using a PVC portal. Retrieved from http://www.sheltermedicine.com/printpdf/68

van den Bos R 1998 Post-conflict stress-response in confined group-living cats (*Felis silvestris catus*) *Applied Animal Behaviour Science 59*: 323–330

Nonhuman Primates

Viktor Reinhardt, DVM, PhD

THE DESIGN OF species-adequate housing conditions and humane handling practices for nonhuman primates in research laboratories must take the following facts into account:

1. *In their natural habitat*, nonhuman primates maintain vocal, visual and/or tactile contact with other conspecifics; spend most of the day foraging, i.e., searching for, retrieving and processing food; show a vertical flight response during alarming situations, and retreat to high places during the night (Napier & Napier, 1994).
2. *In the research laboratory*, nonhuman primates experience anxiety (distress) before and intense fear (stress) during enforced restraint procedures (Reinhardt et al., 1995; Reinhardt, 1998).

Animal Welfare Concerns of Traditional Housing and Handling Practices

Alone in a boring enclosure: The inadequacy of housing a nonhuman primate alone in a boring enclosure is addressed by the following professional guidelines and legislative rules:

The International Primatological Society (1993 & 2007)
1. Pair or group housing in an enclosure must be considered the norm. For experimental animals, where housing in groups is not possible, keeping them in compatible pairs is a viable alternative social arrangement. A compatible conspecific probably provides more appropriate stimulation to a captive primate than any other potential environmental enrichment factor. Single caging should only be allowed where there is an approved protocol justification on *veterinary or welfare grounds* [emphasis added].
2. As monkeys and apes like to work for their food (Köhler, 1921; Yerkes & Yerkes, 1929; Murphy, 1976; Anderson & Chamove, 1984; Evans et al., 1989; Line et al., 1989; Menzel, 1991; Washburn & Rumbaugh, 1992; O'Connor and Reinhardt, 1994; Reinhardt, 1994a; Inglis et al., 1997; Taylor, 2002; de Rosa et al., 2003), increasing processing time, increasing foraging, or providing puzzle feeders or other feeding devices is encouraged.
3. The vertical dimension of the cage is of importance [because of the vertical flight response] and cages where the monkey is able to perch above human eye level are recommended.

The Primate Research Institute (2003) of Japan
1. Primates are very social animals. Physical contact, such as grooming, and noncontact communication through visual, auditory, and olfactory signals are vital elements of their lives. Providing animals with a satisfactory social interaction helps to buffer against the effects of stress, reduce behavioral abnormalities, increase opportunities for exercise and helps to develop physical and social competence.
2. Food presentation should satisfy the animal's interest in manipulating objects. In order to satisfy their requirement to interact with their environment, it is desirable to provide feeders that require complex handling or devices which in some way lead the animals to object manipulation.
3. Devices suitable for gross motor and behavioral patterns, such as perches and three-dimensional structures should be arranged to make as much use of the available space as is possible. Diversity is essential to the housing environment of laboratory animals. Windows through which the animals can see the outside world may help to alleviate some boredom.

The Medical Research Council (2004) of the United Kingdom

1. Primates should be socially housed as compatible pairs or groups. They should not be singly housed unless there is *exceptional* [emphasis added] scientific or veterinary justification.
2. The MRC will require *justification for the use of scientific procedures that restrict the opportunity to forage* [emphasis added].
3. The volume and height of the cage (or enclosure) are particularly important for macaques and marmosets, which flee upwards when alarmed. Their cages and enclosures should be floor-to-ceiling high whenever possible, allowing the animals to move up to heights where they feel secure. Cages and enclosures should be furnished to encourage primates to express their full range of behaviours. Depending on the species, this should normally include provision for resting, running, climbing and leaping.
4. Primates *must* [emphasis added] be provided with a complex and stimulating environment that promotes good health and psychological well-being and provides full opportunity for social interactions, exercise and to express a range of behaviours appropriate to the species.

The Canadian Council on Animal Care (1984 & 1993)

Any primate housed alone will probably suffer from social deprivation, the stress from which may distort processes, both physiological and behavioural. In the interest of well-being, a social environment is desired for each animal which will allow basic social contacts and positive social relationships. Social behaviour assists animals to cope with circumstances of confinement.

The United States Department of Agriculture (1991)

Dealers, exhibitors, and research facilities must develop, document, and follow an appropriate plan for environmental enhancement adequate to promote the psychological well-being of nonhuman primates. The plan *must* [emphasis added] include specific provisions to address the social needs of nonhuman primates of species known to exist in social groups in nature. The physical environment in the primary enclosures *must* [emphasis added] be enriched by providing means of expressing noninjurious species-typical activities. Examples of environmental enrichment include providing perches, swings, mirrors, and other increased cage complexities; providing objects to manipulate; varied food items; using foraging or task-oriented feeding methods; and providing interaction with the care giver or other familiar and knowledgeable person consistent with personnel safety precautions.

The National Health and Medical Research Council (1997) of Australia

For nonhuman primates social interaction is paramount for well-being. Social deprivation in all its forms *must* [emphasis added] be avoided. Animals that need to be individually caged, either for experimental or holding purpose (for example, aggressive adult males), must be given contact with conspecific animals. Accommodation should provide an environment which is as varied as possible. It should meet the behavioural requirements of the species being used. Emphasis *must* [emphasis added] be placed on environmental enrichment.

The Council of Europe (2006)

Because the common laboratory nonhuman primates are social animals, they should be housed with one or more compatible conspecifics. Single housing should only occur if there is justification on *veterinary or welfare grounds* [emphasis added]. The structural division of space in primate enclosures is of paramount importance. It is essential that the animals should be able to utilise as much of the volume as possible because, being arboreal, they occupy a three-dimensional space. To make this possible, perches and climbing structures should be provided.

Enforced restraint: Nonhuman primates experience intense fear when they are forcibly subjected to handling procedures. It may be true that procedures such as injection and blood sampling are simple, but they can be expected to "produce little or no discomfort" (Scientists Center for Animal Welfare, 1987) *only* if the subject is *not* forced to leave her or his cage and subsequently is *not* forced to hold still during such a procedure.

The problems associated with involuntary restraint are addressed by the following professional guidelines and legislative rules:

The International Primatological Society (2007)
Primates of many species can be quickly trained, using positive reinforcement techniques, to cooperate with a wide range of scientific, veterinary and husbandry procedures. Such training is advocated whenever possible as a less stressful alternative to traditional methods using physical restraint. Techniques that reduce or eliminate adverse effects not only benefit animal welfare but can also enhance the quality of scientific research, since suffering in animals can result in physiological changes which are, at least, likely to increase variability in experimental data and, at worst, may even invalidate the research. Restraint procedures should be used only when less stressful alternatives are not feasible.

The Primate Research Institute (2003) of Japan
Pain and other physical stress, such as physical or chair restraint, most definitely affect the behavior and psychology of laboratory animals. All possible measures to reduce their incidence should be taken. Animals should be trained to be as cooperative as possible to the procedures to facilitate the rapid completion of work and to alleviate stress in both the animals and people in charge.

The Home Office (1989) of the United Kingdom
The least distressing method of handling is to train the animal to co-operate in routine procedures. Advantage should be taken of the animal's ability to learn.

Enforced restraint is sometimes advocated with the assertion that nonhuman primates are unpredictable and readily scratch and bite handling personnel (Gisler et al., 1960; Ackerley & Stones, 1969; Valerio et al., 1969; Altman, 1970; Whitney et al., 1973; Henrickson, 1976; Wickings & Nieschlag, 1980; Robbins et al., 1986; Wolfensohn & Lloyd, 1994; Johns Hopkins University and Health System, 2001; Panneton et al., 2001; University of Arizona - IACUC Certification Coordinator, 2008; University of Minnesota - Investigators, and Animal Husbandry and Veterinary Staff, 2008). This contention overlooks the fact that the animals are not intrinsically "aggressive," but that enforced restraint *makes* them aggressive. Trying to bite or scratch the handling personnel is the biologically normal self-defense of any animal who is forcibly restrained. The very act of forceful restraint triggers, rather than prevents, aggressive self-defense. Gaining the animal's trust, and then training him or her to cooperate—instead of resist—during procedures eliminates the risks that are associated with self-defensive aggression. A cooperative animal is no longer given any reason to bite or scratch the investigator, animal technician, animal caretaker or veterinarian who is working *with*, rather than against, the animal during a procedure.

Refinement » Social enrichment

Animate enrichment addresses the social needs of nonhuman primates by promoting noninjurious contact and interaction with one (pair-housing) or several (group-housing) compatible conspecifics.

Establishing groups: Given that nonhuman primates are social animals, it is vital to address their need for compatible companionship when they are kept in research laboratories (United States Department of Agriculture, 1991). Housing the animals in compatible groups is the most species-appropriate strategy to address their social needs.

There are numerous reports on integrating animals into already established groups, but only a few on forming a new group of previously single-caged individuals. Fritz & Fritz (1979) and Fritz (1994) developed a protocol to introduce previously single-caged **chimpanzees** to unfamiliar peers. The newcomer is first moved into a specially designed social unit and kept next to the cage of a selected member of an already established group. The two chimpanzees have full olfactory, visual and auditory contact, as well as limited tactile contact. The selected group member is moved in as a cage mate for the newcomer as soon as friendly interactions through the separating cage mesh are consistently observed. After several days, another group member is introduced to the pair in this same way, then another is introduced to the trio, and so on until the newcomer has met all members of the group and is then fully integrated. A total of 59 of 60 chimpanzees of both sexes and all age classes were successfully socialized to compatible group-living in this manner, without a single incidence of serious fighting.

Kessel & Brent (2001) tranquilized adult single-caged **baboons** with ketamine and placed one trio of males in one enclosure and two trios of two females and one male in two other enclosures, where the animals regained consciousness. The formation of the three groups was accompanied by two incidences of wounding, which were superficial and required no medical treatment. Bourgeois & Brent (2005) confirmed these findings in a subsequent study with three adolescent male baboons. Group formation was accompanied by no overt aggression. Rough-and-tumble wrestling was observed and dominance was quickly established.

Bernstein & Mason (1963) released 11 **rhesus macaques** (three adult females, two adult males, one subadult female, one subadult male and four juveniles) simultaneously into a large enclosure. During the first hour, a total of 83 threats and 23 attacks were observed; injurious encounters were not recorded, but one of the two males soon showed signs of deteriorating health and died after 20 days. Gust et al. (1991) simultaneously introduced eight unfamiliar adult female rhesus macaques and one unfamiliar adult male in a large enclosure. There was no serious fighting, and in fact no contact aggression was recorded, even though firm dominance-subordinance relationships were established during the first 48 hours. Several females stayed in close proximity of the male, who copulated with two of them during the first day. The male's presence probably accounted for the females' tolerance of each other (Bernstein, 2007). Reinhardt (1991a) tried to form an isosexual rhesus macaque group consisting of six previously single-caged adult females and another group of six previously single-caged adult males. Future group members were first given ample opportunity to physically interact with each other on a one-to-one basis during a 1-week period. Dominance-subordinance establishment was ascertained in each dyad. The two groups were then formed by releasing the six animals simultaneously into a big cage. In both situations, aggressive incompatibility was heralded by certain subjects challenging other partners to whom they had been subordinate during the familiarization week. Aggressive harassment was intense and persistent. Alliances were quickly formed and several animals in union attacked selected targets. Victims were cornered, and they showed no resistance, except for fear-grinning and submissive crouching; they did so to no avail and the vicious attacks continued. Both groups were disbanded within the first hour to avoid fatal consequences.

Gust et al. (1996) introduced eight adult female and one adult male **pig-tailed macaques** simultaneously in a large enclosure: Group formation and the establishment of a social hierarchy were not associated with serious aggression; there was no contact aggression during the first 5 hours following the simultaneous release of the eight animals into the same enclosure. The presence of the male probably functioned as an aggression buffer between the females (Dazey et al., 1977).

Clarke et al. (1995) familiarized three single-caged adult male **long-tailed (cynomolgus) macaques** pairwise with each other in a noncontact housing arrangement for 2 weeks and subsequently released them as a trio in a large cage. No injurious fighting was recorded; the new group was compatible. Asvestas (1998) and Asvestas & Reininger (1999) established a group of 22 adult male long-tailed macaques by first forming 11 compatible pairs. After 9 months, all animals were sedated with ketamine and placed simultaneously in a big enclosure where they regained consciousness under careful supervision of the attending staff. The new group turned out to be compatible, even though four males were slightly injured during fighting.

Clarke et al. (1995) kept three male **lion-tailed macaques** in a housing arrangement that allowed all animals to see each other for a period of 2 weeks. The males were subsequently released simultaneously into a large cage. This event was not accompanied by serious fighting, but the group was disbanded because the three males avoided each other and were apparently sufficiently distressed that their well-being was compromised, especially that of the lowest-ranking animal, who did not obtain sufficient food. Stahl et al. (2001) released six unfamiliar adult lion-tailed macaques into well-structured, large living quarters and encountered no aggression-relation problems. The six males showed 91 noncontact agonistic interactions but no physical aggression during the first 6 hours.

King & Norwood (1989) released 11 single-caged female and 13 single-caged male **squirrel monkeys**, ranging in age from 1 to 18 years, without any preliminaries, into a well-structured room. The establishment of the new group was accompanied by two deaths—one male and one female—resulting from attacks by other monkeys.

No foolproof recipe is yet available for group formation of **capuchin monkeys**. Our knowledge of how to form capuchin groups does not come from systematic experimental study, but derives from husbandry problems faced occasionally by laboratories. Overall, group formation is a stressful procedure both for the animals and the caregivers, and although cumulative experience may help to reduce the risks of failure, the outcome can never be predicted with certainty (Visalberghi & Anderson, 1999).

Group-housing: Housing three or more nonhuman primates together in the same enclosure can bear substantial risks for individual members of the group, especially when mature animals of both sexes are present. The inherent constraints of confinement often make it impossible for individuals to keep appropriate social distance from each other, so as to avoid conflicts. Research-related and management-related interferences in the group's membership are bound to destabilize its social structure, thereby triggering rearrangements in the social hierarchy that are usually associated with overt aggression and social distress (Southwick, 1967; Kaplan et al., 1980; Kessler et al., 1985; Cohen et al., 1992; Visalberghi & Anderson, 1993; Alford et al., 1995; de Filippis et al., 2009).

Serious, sometimes fatal injuries resulting from aggression are not uncommon in captive groups of baboons (Rowell, 1967; Nagel & Kummer, 1974), chimpanzees (Alford et al., 1995), squirrel monkeys (Abee, 1985), marmosets (Poole, 1990), vervet monkeys (Knezevich & Fairbanks, 2004), pig-tailed macaques (Sackett et al., 1975; Erwin, 1977), and rhesus macaques (Kaplan et al., 1980; Kessler et al., 1985; Schapiro et al., 1994), Rolland (1991) notes that "By far the most common physical problem that I treat as clinical veterinarian is trauma sustained by macaques in group-housing situations."

No published report could be found of serious aggression problems in core groups of long-tailed macaques (cf., Aureli et al., 1993; Clarke et al., 1995; Ljungberg et al., 1997), stump-tailed macaques, mangabeys, capuchin monkeys (cf., Fragaszy et al., 1994), and tamarins (cf., Poole et al., 1999).

Successfully transferring single-caged primates to compatible group-housing can be an effective remedy for self-injurious biting. Fritz (1989) observed that an unspecified number of chimpanzees gradually stopped biting themselves after they were transferred from single- to compatible group-housing arrangements. Alexander & Fontenot (2003) noted that 10 adult male rhesus macaques, who engaged in self-injurious biting while they were single-caged, showed no signs of this behavioral pathology during the first four months after they were transferred to compatible isosexual group-housing.

Establishing pairs: "To enhance the life-style of a primate, one of the most effective, but often overlooked improvements is pair-housing" (Rosenberg & Kesel, 1994, p 469). Keeping nonhuman primates in compatible pairs is a good compromise to group-housing; it addresses the animals' basic social needs while providing more assurance of their safety, better access to individuals, and control over their reproduction. Initial, strong reservations against the transfer of single-caged animals to pair-housing arrangements have proven to be based on the erroneous idea of the aggressive and near-intractable monkey (Gisler et al., 1960; Bernstein et al., 1974; Wickings & Nieschlag, 1980; Line, 1987; Coe, 1991; Rosenberg & Kesel, 1994) and the disregard of basic ethological principles when establishing new pairs.

Adults with juveniles: Adults—both females and males—are normally inhibited from showing overt aggression toward juveniles. This circumstance makes it unproblematic to transition single-caged adults to compatible pair-housing arrangements: the naturally weaned juvenile is simply introduced with the adult in the adult's home cage. Typically, the adult will show parental responses, huddling with the young, spending much time grooming the young, and allowing the young to engage in often exuberant play behaviors. Even rhesus males, who have the reputation of being particularly aggressive (Wickings & Nieschlag, 1980), have the tendency to treat their little companions with gentleness and great tolerance. Reinhardt (1994b) transferred naturally weaned, surplus juveniles between the ages of 12 and 18 months from a rhesus macaque breeding colony without any preliminary precautions, pairwise to unfamiliar single-caged adults of both sexes. A total of 78 pairs were tested and pair compatibility ascertained during the first week in 96% (75/78) of cases: the adult did not injure the juvenile, the juvenile showed no signs of depression, and the adult shared food with the juvenile. Three pairs (4%) were incompatible. One female grabbed her female juvenile immediately upon her arrival; she continued to do this repeatedly during the next 30 minutes, after which the youth was removed. One male bit his male juvenile on the fourth day of introduction. The youngster was slightly injured, although not bleeding. When the juvenile started to consistently avoid the adult, the pair was split. Another male often grabbed his little male companion, even though he gently groomed him, and the two huddled with each other regularly. Gradually, however, the juvenile showed more and more avoidance behavior, and the two were finally separated after 9 days.

Juveniles with juveniles: Juveniles who have not yet reached the age when they become ambitious to dominate over others are usually compatible when they are introduced as pairs, even when they are strangers to each other. Reinhardt (1994b) transferred a total of 84 female and 22 male juvenile rhesus macaques to same-sex pair arrangements. All pairs were compatible throughout a 1-year follow-up period. Males were occasionally observed playfully wrestling with each other, but this never resulted in injurious aggression or depression.

Adults with adults: Adult primates have the tendency to react with hostility when they meet another adult conspecific with whom they are not familiar. Strangers first determine their dominance-subordinance relationship; this often involves potentially injurious fighting. To avoid this in the laboratory setting, single-caged adults assigned to be paired with another adult partner are usually first given the opportunity to get to know each other during a noncontact familiarization period.

Reinhardt et al. (1988) and Eaton et al. (1994) familiarized previously single-caged adult, female **rhesus macaques** in double cages with transparent partitions for 1 week, and then introduced them as pairs in a different double cage. Within the first 2 hours after introduction, dominance-determining fighting was witnessed in 27% (5/18) and 10% (2/21) of cases, respectively. The fights resulted in no serious injuries, but they were persistent and led to depression in the victim, in three dyads of the 1988 study and in two dyads of the 1994 study. These five pairs were classified as incompatible and the partners were permanently separated; consequently, pair compatibility during the first week was 83% (15/18) and 90% (19/21), respectively.

Reinhardt (1994b) made sure that the partners of 77 adult female rhesus dyads and 20 adult male rhesus dyads had established their dominance-subordination relationships during a noncontact familiarization period, before they were introduced in a different double cage.

This precaution was implemented in order to minimize the animals' need to engage in dominance-determining aggression upon being introduced to each other. The following gestures, reactions and behaviors were taken as indicators that one animal was subordinate and accepted the dominant position of his or her neighboring partner:

a. strictly unidirectional fear-grinning when being looked at by the neighbor,
b. withdrawing and/or looking away when being approached or looked at by the neighbor, and
c. quickly glancing at the neighbor followed by threatening against other animals of the room or against the observer.

Partners who had established such a relationship were then introduced to each other in a different double cage to avoid potential territorial antagonism (Reinhardt et al., 1988; Niemeyer et al., 1998). Newly formed pairs were regularly observed during the first week. Shortly after introduction, fighting took place in only 2 of the 97 dyads tested. Partners turned out to be compatible in 95% (73/77) of the female pairs and also in 95% (19/20) of the male pairs. Compatible partners did not engage in serious aggression, they shared both standard and supplemental food, and none of them became depressed. Within the first week, 5 of the 97 pairs turned out to be incompatible because of injurious aggression (two female pairs), depression (one male pair), or food monopolization (one female pair and one male pair). Doyle et al. (2008) demonstrated in eight biotelemetry-instrumented adult male rhesus macaques who were carefully familiarized in pairs that compatible partners showed no increased heart rate when they were introduced, suggesting that the pair-formation process was not a stressful experience for them.

Lynch (1998) tested 17 adult male **long-tailed macaque** dyads. Potential pairs had all established clear-cut dominance-subordinance relationships during a noncontact familiarization period; subsequent partner introduction was accompanied by fighting in only one incompatible pair. The other 16 pairs (94%) were compatible. Crockett et al. (1994) also familiarized the potential partners of 15 adult male and 15 adult female long-tailed macaques in a noncontact housing arrangement, but introduced the animals as pairs in the familiarization cage without prior verification that they had established dominance-subordination relationships. Under these circumstances, fighting occurred shortly after introduction in 67% (10/15) of the male pairs and in 13% (2/15) of the female pairs.

Reinhardt (1994c) transferred 10 adult female and six adult male **stump-tailed macaques** from single-housing to isosexual pair-housing by first allowing potential partners to establish dominance-subordinance relationships without risk of injury, during a noncontact familiarization phase. Following subsequent introduction in a new home cage, all eight pairs showed signs of compatibility. Female partners reconfirmed their rank relationships within 30 minutes with subtle gestures, never by overt aggression. Male partners engaged in hold-bottom rituals, whereby one puts both hands on the other's hips (de Waal & Ren, 1988) upon being introduced to each other. Two male pairs reconfirmed rank relationships within 30 minutes with gestures, while the third pair resorted to a brief noninjurious dominance-reconfirming fight, which was followed by another reconciliatory hold-bottom ritual.

Coe & Rosenblum (1984) introduced 10 unfamiliar, adult male **bonnet macaques** pairwise without any preliminaries. As usually occurs when unfamiliar males first meet, agonistic behaviors related to the establishment of dominance relations occurred at pair formation. The aggressive incidents were limited, usually involving threats and pursuit behavior; manual attacks occurred only infrequently. More typically, one animal submitted and indicated his subordinate status through communicative gestures. In the first week following pair formation, the occurrence of aggressive behavior subsided almost entirely. The males' response to this pairing procedure may reflect their reputation for showing the highest degree of male-male tolerance in the genus *Macaca*.

Bourgeois & Brent (2005) established four pairs of previously single-caged subadult male **baboons** by sedating potential companions and having them wake up together in the same cage. No serious aggression was witnessed during ten 30-minute observations conducted during the first 2 weeks. Jerome & Szostak (1987) allowed an unspecified number of adult female baboons to live pairwise with each other 4 hours a day, three times a week. The same pairs visited each other in either animal's cage. No overt aggression occurred during visits.

Majolo et al. (2003) checked the clinical records of 56 unfamiliar female **common marmosets** of different age classes who were paired with each other without familiarization. Overall, 22 (79%) out of 28 pairs were compatible; the other six pairs were split up within the first week after pair formation because one of the monkeys was subjected to intense aggression and/or was injured as a consequence of fighting.

Pair-housing: Reinhardt (1994b) formed 84 compatible pairs of juvenile female and 22 compatible pairs of juvenile male **rhesus macaques** and noted that the animals remained compatible for at least 12 months. There were 21 juvenile female pairs with cranial implants. Living together in the same cage did not constitute any specific risks for the animals (no local infections possibly caused by grooming the margins of the implantation site; cf., Anonymous, 2007) and no risk for the implants (no damage related to social interactions; cf., Anonymous, 2007).

Reinhardt (1994b) created 75 compatible adult-infant pairs who were allowed to stay together uninterruptedly. Compatibility was ascertained throughout a 12-month follow-up period. Incompatibility was noted after more than 1 year in two cases, when the now prepubertal young subjects started teasing their over-30-year-old companions, thereby creating excessive disturbance for these aged animals. Two of the infants lived with adult females who were tethered, and 32 paired infants had cranial implants. Both circumstances did not interfere with research protocols requiring remote sample collection and neuroendocrinological testing (cf., Reinhardt & Dodsworth, 1989).

Doyle et al. (2008) allowed eight, previously single-caged adult male rhesus macaques to live uninterruptedly as four compatible pairs. Over the course of 18 months, one bite laceration was incurred (after 3.5 months), but the pair remained compatible after the injury was treated and healed. Average fecal cortisol levels were significantly higher when the males lived alone than after they had lived with a companion for 20-39 weeks (83 ng/g versus 9 ng/g), indicating that long-term pair-housing was a less distressing situation than single-housing. Scan sampling revealed that the males groomed each other about 13% of the time.

Eaton et al. (1994) studied 12 newly formed, compatible adult female rhesus pairs over a 36-month period. During this time, only one pair became incompatible, when the two partners had a serious fight. Compatible companions groomed each other during about 30% of multiple 10-minute observation sessions. Reinhardt (1999) worked with three adult female and four adult male rhesus macaques who habitually bit themselves when they were caged alone. The provision of perches, gnawing sticks, and food puzzles did not alleviate this behavioral pathology, but when the seven animals were successfully paired with compatible partners, the self-biting stopped immediately in three cases and gradually in the remaining four cases. Baker et al. (2012) noted in 46 adult female and in 18 adult male rhesus macaques a statistically significant decrease in abnormal behavior (females: 54%; males: 18%) and anxiety-related behavior (females: 35%; males: 41%) 4 weeks after the animals had been transferred from single-housing to compatible pair-housing condition.

Reinhardt (1990b) assessed the clinical records of a rhesus macaque colony consisting of 237 single-housed and 382 pair-housed animals of both sexes and all age classes. The incidence of non-research-related veterinary treatment was 23% for single-caged animals, versus 10% for pair-housed animals, indicating that the animals' physical health was not jeopardized by sharing a cage with a companion. Schapiro & Bushong (1994) examined the clinical records of 98 juvenile rhesus macaques during 1 year when they were caged alone and the subsequent year when they lived in opposite-sex pairs. Individuals required veterinary treatment more than twice as often when they were single-housed (0.40 times/year) than when they were pair-housed (0.17 times/year). These findings were confirmed in a subsequent study with adult rhesus macaques in which pair-housed animals required significantly fewer medical interventions for diarrhea than did single- or group-housed animals (Schapiro et al., 1997).

Reinhardt (1990a & 1994b) formed 73 compatible adult female and 19 compatible adult male rhesus macaque pairs. The animals were allowed to live together uninterruptedly.
» Over a 12-month follow-up period, compatibility was 93% (68/73) for the female pairs and 84% (16/19) for the male pairs.
» During 60-minute video recordings of eight female pairs and four male pairs, females groomed each other 25% and hugged each other 4% of the time while males groomed each other 12% and hugged each other 2% of the time; the sex differences were statistically significant.
» Among the compatible pairs were four 30- to 35-year-old animals who were so old that they experienced a progressive loss in body weight. Living with a companion did not accelerate this biological process (Vertein & Reinhardt, 1993), suggesting that the permanent presence of a companion did not jeopardize their general health. These elderly rhesus macaques groomed each other, on average, during 22% of multiple 1-hour observations (Reinhardt & Hurwitz, 1993).
» Some animals were assigned to controlled food intake studies over the course of the first 2 years after pair formation. When this happened, they were allowed to stay in their home cage, where they were separated from their companions with a grated cage dividing panel during the day, and reunited for the night after food intake was recorded.
» The majority of the animals were assigned to a timed breeding program. All 18 females who gave birth during the first 2 years after pair formation were allowed to stay with their partners. The presence of offspring did not affect the compatibility between the two cage companions.
» There were 23 female pairs with one or both partners having cranial implants. This circumstance did not jeopardize the integrity of ongoing neurophysiological research of

one or both animals (cf., Truelove, 2009). Evidence shows that pair-housing provides a safe and practical social alternative to single-housing not only for juvenile and adult female macaques but also for adult male macaques with biomedical implants (Roberts & Platt 2004).

» When one partner had to be chair-restrained during an experiment, the companion was brought along in a mobile cage to provide emotional support (cf., Reinhardt & Dodsworth, 1989).

Crockett et al. (1994) established 15 compatible adult female and 12 compatible adult male **long-tailed macaque** pairs and housed them in such a way that partners were separated each day for 17 hours and subsequently reunited for 7 hours. While 100% of the female pairs successfully coped with this situation and remained compatible, only 50% of the male pairs adjusted; the other 50% became incompatible and had to be separated within 2 weeks of living together under these conditions. Lynch (1998) also formed 16 compatible adult male long-tailed macaque pairs, but partners could stay together without interruption. All pairs remained compatible throughout a 12-month follow-up period and longer.

Line et al. (1990) transferred five adult female long-tailed macaques to compatible pair-housing arrangements and observed each pair during daily 10-minute sessions throughout the first two weeks after pair formation. During these sessions companions groomed each other, on average, 31% of the time. The incidence of abnormal behaviors decreased significantly after the animals were transferred from single- to pair-housing; the five females had engaged in self-biting behavior when they were single-caged; all of them stopped this behavioral pathology once they were living with a companion.

Roberts & Platt (2005) studied one adult male long-tailed and eight adult male rhesus macaques who all had cranial implants and lived with compatible partners in a pair-housing arrangement. The presence of a social partner did not cause any problems with the implants, which lasted for an average of 21 months. Partners were separated daily for a few hours to participate in physiological experiments; this had no adverse effect on their compatibility which, depending on the length of the study, was confirmed for up to 40 months.

Murray et al. (2002) demonstrated the practicability of post-operative pair-housing in 15 female long-tailed macaques who were returned to their partners on the day of the operation. Change in dominance status, self-traumatic events, weight loss or diarrhea did not occur in any of these animals, and the incision sites healed unremarkably. The animals ate and drank normally, and they accepted their post-operative oral medication.

Coe & Rosenblum (1984) observed five adult male **bonnet macaque** pairs on four different days during 15-minute sessions in the course of the first week after the pairs were established. Subjects groomed each other on average 29% of the time.

Reinhardt (1994c) monitored five adult female and three adult male **stump-tailed macaque** pairs, who had lived together for 6 months, each pair for 60 minutes. On average, female partners groomed each other 19% and hugged each other 6% of the time; male partners groomed each other 13% and engaged in hold-bottom rituals 4% of the time.

Grooming-contact housing: Crockett et al. (1997) housed same-sex pairs of adult long-tailed macaques in double-cage units in which partners were separated 19 hours daily by a blind panel, and separated 5 hours daily by grooming-contact bars, allowing them to reach through with their arms. Of 16 female pairs tested, 100% were compatible and partners spent about 43% of scan sampling time grooming each other. Of 45 male pairs tested, 89% were compatible and partners spent about 7% of the time grooming each other.

The usefulness of grooming-contact bars, or woven wire panels with mesh openings large enough so that adjacent neighbors can groom each other (Coelho & Carey, 1990), has also been confirmed in adult heterosexual pairs of pig-tailed macaques (Crockett et al., 2001; Lee et al., 2005) and adult isosexual and heterosexual pairs of baboons (Coelho et al., 1991; Crockett & Heffernan, 1998). De Villiers & Seier (2010) transferred a single-caged subadult male baboon, who suffered from serious self-injurious biting, in a contact-grooming housing arrangement to a group of compatible females. The protected social contact resulted in healing of the self-inflicted laceration within 4 months. After 18 months, neither the self-injurious biting nor the wounds reoccurred.

Compared to other nonhuman primate species, rhesus macaques do not adjust well to the grooming-contact housing system (Crockett et al., 2006).

Social buffer: The compatible companion can serve as a social buffer during potentially stressful research-related situations, such as being transferred to a test room.

Coelho et al. (1991) measured blood pressure via arterial catheter implants of four tethered adult male **baboons** who were kept in an unfamiliar test room alone or in company of a familiar male baboon with whom they had visual, tactile and auditory contact through a wire mesh panel. Mean blood pressures were significantly lower when another baboon was present, suggesting that companionship mitigated the distress response (anxiety) to the test room environment.

Gust et al. (1994) transferred seven adult female **rhesus monkeys** from their group to an unfamiliar environment, either alone or together with a group member. During both conditions, the animals were initially equally distressed, as measured in alterations of cell-mediated immune parameters, but they recovered significantly quicker when they had the social support of a companion. Mason (1960) placed five infant rhesus macaques into a test room, either alone or as a pair with another infant. They showed significantly fewer signs of distress (crouching, self-clasping, vocalization, agitation) when they were tested in the company of another monkey, indicating that the companion had a calming, reassuring influence. Gunnar et al. (1980) confirmed these findings in 12 rhesus infants.

Similar observations were made by Hennessy (1984) in eight infant **squirrel monkeys** who vocalized significantly less when they were tested in an unfamiliar environment as a pair than when they were tested alone. A significant elevation of plasma cortisol was observed when the animals were exposed to the novel environment alone, not when a companion was present. Coe et al. (1982) noticed the same stress reducing effect in 14 adult male squirrel monkeys. Subjects showed significantly fewer distress reactions (vocalization, fear reactions, agitation) to a snake behind a mesh when another male was with them than when they were alone.

Recommendations: Compatible social companionship is probably the most important factor that influences the well-being of a nonhuman primate in the research laboratory. Unless there are veterinary and ethological reasons for exemptions, no nonhuman primate should be housed alone. Ideally, all animals should be housed with a companion, and at minimum, they should spend the whole night and most of the day with a companion. Animal care committees must be well informed about possibilities of allowing nonhuman primates to keep full or partial contact with another compatible companion during scientific investigations. The fact that historical data have been collected from single-caged primates is not an acceptable excuse for keeping animals alone during current research projects.

Refinement » Friendly contact with humans

Friendly contact with humans provides high-quality environmental enrichment by promoting and fostering mutual trust relationships between human and nonhuman primates.

Nonhuman primates respond to friendly attention from individual caretakers and individual investigators by gradually developing affectionate relationships with them, and overcoming their conditioned fear and distrust of humans. Positive interaction with monkeys and apes is essential for the well-being of the animals, data validity, and ease of handling (Wolfle, 1987). The behavior of a nonhuman primate during procedures depends on the confidence he or she has in the handler. This confidence is developed through regular human contact and, once established, should be preserved (Home Office, 1989). Good relations between the animals and personnel are important for animals to reduce stress and for personnel to obtain safer working conditions. Personnel who have gained the trust of animals can more easily perceive abnormal behaviors and the animals are more likely to cooperate with them during research procedures, such as restraint and blood sampling (The Primate Research Institute, 2003).

Animals who have developed a trust relationship with attending personnel give the impression that they like human contact. This suggests that human contact can have a relaxing, tension-releasing effect on them. Gantt et al. (1966) reports of a female rhesus macaque who was petted by a person several times on two different days. On both days a significant decrease of the animal's heart rate was noticed during the petting sessions.

Koban et al. (2005) exposed four male long-tailed macaques of unspecified age to daily 10-minute positive reinforcement training sessions for 2 months; four control subjects received no training sessions. The results indicate that the positive interaction with the human trainer made the animals feel at ease: serum cortisol concentration and heart rate were significantly lower in trained than in control subjects.

Baker (2004) increased the time from 2 to 4 hours that caretakers spent visiting—playing, grooming, talking, offering treats—seven adult female and five adult male chimpanzees housed in pairs and trios. Behavioral data were collected between the visits, allowing the carry-over effect of human interaction to be assessed. When the daily time of unstructured affiliation with personnel was doubled, the chimpanzees were more relaxed, spending more time grooming each other ($p<0.05$) and less time engaging in agonistic displays ($p<0.06$).

Recommendations: The quality of care provided by research personnel has a profound effect on the well-being of the animals and the quality of the science. Whenever possible, unstructured time should be set aside for personnel to spend time with the animals in their charge, so that human-animal interactions will be based upon affection and trust rather than apprehension, anxiety, and fear.

Refinement » Training to cooperate with humans during procedures

A friendly human-animal relationship based on mutual trust is the basic condition to obtain the cooperation of nonhuman primates during procedures that would otherwise require involuntary restraint and incur distress for the animal and risk for the human handler.

Training nonhuman primates to cooperate—rather than resist—during procedures achieves two goals at the same time:

1. Intellectual stimulation for the animal subject *and* for the human caregiver (enrichment);
2. Reduction of distress reactions of the animal and increase in safety of the personnel during husbandry- and research-related procedures (Refinement).

It has been documented for several species of nonhuman primates how individuals of both sexes, different age classes and different housing arrangements can be successfully trained to voluntarily cooperate with the attending personnel during the following research- and husbandry-related procedures:

Injection:
- » single-caged adult male mandrill (Priest, 1991);
- » single-caged adult male baboon (Levison, 1994),
- » single-caged adult male mustached guenon (Stringfield & McNary, 1998);
- » group-housed adult male lion-tailed macaques (Bayrakci, 2003);
- » single-caged male squirrel monkeys (Gillis et al. 2012;
- » single-caged, pair- and group-housed female and male chimpanzees of all age classes (Spragg, 1940; Videan et al.; 2005a; Russell et al., 2006) who required a mean total of about 87 minutes per animal to achieve the goal of the training.

Bentson et al. (2003) compared the stress response to injection in four single-caged rhesus macaques who were not trained with that of 17 single-caged rhesus macaques who had been trained to cooperate during this procedure. While serum cortisol concentrations did not increase in the trained subjects, cortisol increased significantly in the untrained subjects.

Blood collection:
- » single-caged adult male mandrill (Priest, 1990);
- » pair-housed adult female stump-tailed macaques who needed less than a cumulative mean total of 60 minutes per animal to achieve the goal of the training (Reinhardt & Cowley, 1992);
- » single-, pair- and group-housed chimpanzees of both sexes and all age classes (Laule et al., 1996; Schapiro, 2000 & 2005; Coleman et al., 2008);
- » single-caged and pair-housed rhesus macaques of both sexes and all age classes (Elvidge et al., 1976; Vertein & Reinhardt, 1989, Reinhardt, 1991b; Phillippi-Falkenstein & Clarke, 1992). Depending on the technique applied, a cumulative mean total of 40 to 160 minutes were invested to achieve the goal of the training (Reinhardt, 1991b; Pranger et al., 2006; Schapiro et al., 2007).

It has been shown in rhesus and stump-tailed macaques as well as in baboons that successfully trained animals show no behavioral and no physiological stress response—as measured in changes in serum cortisol concentration (Elvidge et al., 1976; Reinhardt, 1991b; Reinhardt & Cowley, 1992; Bentson et al., 2003)—when they cooperate during blood collection.

Blood pressure measurement:
» group-housed adult female and male woolly monkeys (Logsdon, 1995);
» single-caged adult male baboons (Mitchell et al., 1980; Turrkan et al., 1989).

Urine collection:
» group-housed adult male vervet monkeys (Kelly & Bramblett, 1981);
» group-housed adult female white-faced sakis (Shideler et al., 1994);
» single-caged and group-housed juvenile and adult chimpanzees (Laule et al., 1996; Lambeth et al., 2000);
» group-housed juvenile and adult marmosets of both sexes (Anzenberger & Gossweiler, 1993; McKinley et al., 2003; Smith et al., 2004);
» group-housed adult female tamarins (Snowdon et al., 1985; Smith et al., 2004).

Vaginal swabbing:
» group-housed stump-tailed macaques (Bunyak et al., 1982). After five training sessions of unspecified duration it was no longer necessary to net and restrain the females. Indeed, some of them began to voluntarily approach the researcher and present for vaginal swabbing.

Semen collection:
» group-housed gorillas (Brown & Loskutoff, 1998);
» group-housed chimpanzees (Perlman et al., 2003).

Oral drug administration:
» group-housed adult cotton-top tamarins of both sexes (Savastano et al., 2003);
» single-caged adult male baboons (Turrkan et al., 1989);
» single-caged and group-housed adult marmosets of both sexes (Peterson et al., 1988; Donnelly et al., 2007);
» single-caged adult male and females rhesus macaques (Winterborn, 2007; Anonymous, 2013).

Saliva collection:
» single-caged adult male rhesus macaques (Lutz et al., 2000);
» single-caged adult male squirrel monkeys (Tiefenbacher et al., 2003);
» single-caged and group-housed adult marmosets of both sexes (Cross et al., 2004);
» group-housed young baboons (Pearson et al., 2008).

Topical treatment:
» group-housed adult female gorillas (Segerson & Laule, 1995);
» group-housed female and male chimpanzees of all age classes (Perlman et al., 2001);
» pair-housed adult stump-tailed macaques of both sexes (Reinhardt & Cowley, 1990).

Weighing:

» pair-housed adult marmosets (McKinley et al., 2003); a cumulative mean total of about 1 hour per pair was needed to achieve the goal of the training.

Chairing:

» single-caged male pig-tailed macaques of unspecified age (Nahon, 1968);
» single-caged adult long-tailed macaques of both sexes (Skoumbourdis, 2008);
» single-caged juvenile and adult rhesus macaques of both sexes (Skoumbourdis, 2008); Bliss-Moreau et al., 2013):
» pair-housed juvenile rhesus macaques of both sexes (McMillan et al., 2014).

Capture:

» groups of bonobos (Bell, 1995);
» groups of Japanese macaques (Goodwin, 1997);
» groups of chimpanzees (Kessel-Davenport & Gutierrez, 1994; Boomsmith et al. 1998;
» groups of rhesus macaques (Reinhardt (1990c). In order to train a heterosexual group of 45 rhesus macaques to voluntarily cooperate during the routine one-by-one capture procedure, an average of 20 minutes was invested per group member, 15 hours for the whole group. It took about 15 minutes to catch all 45 animals without distressing them (Luttrell et al., 1994).

Checking faucet waterers: Checking automatic waterers can be time-consuming and cumbersome for staff, as well as intimidating for animals. Habe Nelsen et al. (2010) alleviated this by training eight rhesus and seven long-tailed macaques, using a laser pointer as the target, to check their Lixits. Seven of the rhesus and six of the long-tailed macaques were successfully trained; after several months, most of them longer needed the laser and checked the Lixits upon a verbal command and a simple hand gesture (Ferraro et al., 2013).

Recommendations: "Primates dislike being handled and are stressed by it; training animals to cooperate should be encouraged, as this will reduce the stress otherwise caused by handling. Training the animals is a most important aspect of husbandry, particularly in long-term studies. ... Training can often be employed to encourage the animals to accept minor interventions, such as blood sampling" (Council of Europe, 2006, p 48). "Positive reinforcement techniques should be used to train primates to cooperate with capture, handling, restraint and research procedures. The routine use of squeeze-back cages and nets should be actively discouraged" (National Center for the Replacement Refinement Reduction of Animals in Research, 2006, p 11). Training nonhuman primates to cooperate rather than resist during husbandry and common research procedures is perhaps the most effective approach to minimize or avoid data-biasing stress reactions and, hence, reduce the number of subjects needed to achieve statistical significance of the research results (cf., Brockway et al., 1993; Schnell & Gerber, 1997; National Center for the Replacement Refinement Reduction of Animals in Research, 2006). Research laboratories need to make more earnest and more consistent use of the animals' amazing potential to work with rather than against principal investigators and animal care personnel during handling and husbandry procedures.

Refinement » Feeding enrichment

Feeding enrichment promotes noninjurious food searching, food retrieving and/or food processing activities.

Vegetables and fruits: The following unprocessed produce has been fed to captive primates without any adverse side effects: apples, oranges, bananas, grapes, watermelons, pumpkins, squash, potatoes, carrots, string beans, corn on the cob, lettuce, celery, artichokes, bell peppers, sugar cane, cranberries, raspberries, coconuts, and peanuts in the shell (Bloomsmith et al., 1988; Spector & Bennett, 1988; Hayes, 1990; Beirise & Reinhardt, 1992; Nadler et al., 1992; Williams et al., 1992; Logsdon, 1994; Waugh, 2002).

When presented behind a barrier—for example behind the bars or mesh of the enclosure—whole fruits and vegetables promote not only food processing, but also skillful food retrieval behavior.

Beirise & Reinhardt (1992) distributed every week 1 kg whole peanuts and on a different day 32 ears of corn to a 16-member breeding group of rhesus macaques. After a habituation period of 8 weeks, 120-minute observations were conducted immediately after peanuts or corn were distributed in weeks 9, 10 and 11. Individual animals spent on average:
  » 77% of the time husking corn ears, chewing husks, and eating corn kernels, and
  » 47% of the time cracking peanut shells and eating peanuts.

Recommendations: Attending care personnel typically work under time pressure. To have them chop supplemental vegetables and fruits for the animals can be quite time-consuming and is unnecessary. The animals have all the time needed to process the material themselves, and they like to do it. Offering whole, rather than already processed, vegetables and fruits of the season provides effective feeding enrichment without extra time investment. It introduces variety into the monotonous standard feeding regimen of commercial, pelleted dry food and allows the animals to engage in species-typical food processing behaviors. The provision of fruits and vegetables are not extra "treats" but a part of the standard food ration; every animal should have access to at least one medium-size whole fruit or vegetable on a daily basis.

Standard food ration behind a barrier: Offering the daily food ration not freely accessible on the floor or in standard food boxes, but behind the bars or mesh wall/ceiling of the enclosure, is probably the easiest way of increasing the time that the animals can spend retrieving and processing their food. Reinhardt (1993a) distributed the daily biscuit ration of eight adult male, pair-housed rhesus macaques (a) first in their ordinary, freely accessible food boxes, and then (b) for a 2-week period on the cages' mesh ceilings with 2.2 cm^2 openings. Time spent retrieving biscuits was recorded for each animal during 4 hours following food distribution.
  » When the ration consisted of 66 small, bar-shaped biscuits, average foraging time increased 80-fold, from 17 seconds to 1,363 seconds.
  » When the ration consisted of 32 large, star-shaped biscuits, average foraging time increased 296-fold, from 12 seconds to 3,551 seconds.
Working for their daily biscuit ration did not affect the males' body-weight balances.

Reinhardt (1993b,1993c) observed eight pair-housed, adult male rhesus macaques and five female and seven male single-caged adult stump-tailed macaques, each individual for 30 minutes after their daily biscuit rations were distributed either in the ordinary food boxes with 7.3 x 4.7 cm access holes or in the same boxes remounted onto the 2.2 cm^2-opening mesh front panels of the cages a few centimeters away from the original access holes. All animals were habituated over a 30-day period to receiving their food in the food puzzles; their body weights did not change in the course of that time period.

» Rhesus macaques spent, on average, less than 1 minute collecting biscuits from the food box versus 18 minutes retrieving them from the food puzzle.

» Stump-tailed macaques also spent, on average, less than 1 minute collecting biscuits from the food box versus 19 minutes retrieving them from the food puzzle.

Bertrand et al. (1999) report of four single-caged rhesus macaques, of unspecified age and sex, who received their daily pellet ration in a freely accessible standard feeder, and four other single-caged subjects who received their pellet ration on 4 days in a foraging device fitted on the front of the cage. Manipulative skills were required to retrieve the pellets from this device. Over 90% of the food was eaten within the first 15 minutes with the standard feeder, whereas it took 60 minutes to reach this percentage using the foraging feeder. The amount of food waste was up to 17 times lower when the animals had to work for their food instead of collect it freely.

Murchison (1995) videotaped the behavior of 20 single-caged adult female pig-tailed macaques, each animal for 1 hour, when the ration of 40 biscuits was presented in the standard feeder with one big access hole (5 cm diameter) versus the same feeder with four small access holes (3 cm diameter). The animals spent, on average, 7 minutes using hands, teeth and feet to remove biscuits from the feeder with small holes versus less than 1 minute to collect biscuits from the standard feeder with one big access hole; the difference was statistically significant. Unlike with the standard feeder, the animals consumed most of the biscuits they retrieved from the test feeder; this implied that they dropped fewer pieces of biscuits on the floor and less food was wasted.

Bloom & Cook (1989) mounted a commercial puzzle feeder on the front panel of the cages of two adult male rhesus macaques and habituated the animals to retrieving their daily single portion of biscuits from this device. It took the two males, on average, 25 minutes to retrieve their food.

Expanded feeding schedule: Taylor et al. (1997) expanded the feeding schedule of a group of four adult female and one adult male bonnet macaques by dispersing one-half of the daily ration of 150 biscuits and 1 cup of sunflower seeds on the woodchip litter at the usual time in the morning, and the other half in the afternoon. Over a period of 10 weeks, the animals were observed during several 10-minute sessions starting 1 hour after food distribution. When they received their daily food ration in two small portions (weeks 6-10), rather than in one big portion (weeks 1-5), they spent 52% versus 26% of the observation periods foraging.

Special food in and on gadgets: Numerous devices, baited with food treats rather than the standard food have been developed to encourage foraging-related activities in captive primates.

Brent & Eichberg (1991) attached one Plexiglas sheet with holes on the mesh ceilings of the enclosures of eight heterogeneous groups of three or four **chimpanzees**. After a 7-day habituation period, commercial food treats were placed on these puzzle boards on four different occasions and the animals' response was recorded. During 1-hour observation sessions the chimpanzees spent, on average, 17% of the time retrieving treats from the puzzle.

Maki et al. (1989) designed metal pipe-feeder puzzles containing sticky foods—such as applesauce, mashed bananas, spaghetti sauce, and dry fruit drink powder. Four adult chimpanzees, living with other companions in pairs or trios, were observed during eight 30-minute sessions distributed over a period of 1 month, when a daily-filled pipe feeder was permanently mounted from outside on the chain-link fencing of the home quarters. The four subjects spent, on average, 23% of the sessions manufacturing dipping sticks from branches, and an additional 29% fishing with these tools for the moist foodstuff in the box.

Celli et al. (2003) mounted an open transparent polyethylene bottle, which was filled daily with honey, in front of the cages of three pairs of adult female chimpanzees and offered them plastic brushes, wires, chopsticks and rubber tubes from which they could chose suitable tools for retrieving honey from the bottle. During daily 60-minute observations—probably right after the bottle was filled—individual animals spent about 9 % of the time checking out suitable fishing tools, and 31% of the time retrieving honey from the bottle.

Gilloux et al. (1992) monitored a heterogeneous group of seven chimpanzees for twelve 120-minute sessions when a 15-cm-diameter plastic pipe filled with fruits, vegetables and biscuits was attached outside onto the welded mesh of the enclosure. The apes could manipulate food items to the open end of the pipe by inserting bamboo canes or willow twigs through holes drilled along the side of the pipe facing them. Individuals used the filled feeder, on average, during 18% of the time.

Lambeth & Bloomsmith (1994) conducted six 30-minute observations of eight adult female and six adult male chimpanzees, living in pairs or groups of four, after a PVC pipe cut in half and planted with rye grass was attached to the front panel of the chain link fencing of the subjects' enclosures. Individual animals spent, on average, 4% of the time picking grass with their fingers through the fencing; when sunflower seeds were added to the grass, they spent 20% of the time searching for and picking up seeds.

Bayne et al. (1992) secured Plexiglas boards covered with artificial turf inside the cages of eight single-housed adult male **rhesus macaques**. Commercial, flavored food particles were sprinkled on the turf boards daily 2 hours after the morning feeding; this was followed by 30-minute observations of each subject on 20 days over the course of a 6-month period. The males foraged, on average, during 52% of the observation periods; there were no signs that they lost interest in foraging from the turf boards over time.

Lutz & Farrow (1996) mounted turf boards to the outside of the front panel of the cages of ten adult female **long-tailed macaques** and sprinkled sunflower seeds on the turf every morning, after the animals had received their daily biscuit ration. During three weekly 30-minute observations conducted at random times over a period of 8 weeks, the animals spent an average 11% of the time contacting the board. The boards were used by the animals with consistency; there was no indication that they lost interest in them over time.

Bryant et al. (1988) released six individually caged, adult male long-tailed macaques, one animal at a time, for 30 minutes into a playpen on 12 days, distributed over a 3-week period. The playpen was furnished with a nylon ball, a telephone directory, a nylon rope and a tray placed below the grid floor of the cage, containing woodchips scattered with sunflower seeds and peanuts. The animals showed little interest in the nonfood enrichment items but spent a considerable amount of the time reaching through the wire mesh of the cage floor to retrieve seeds and peanuts.

Fekete et al. (2000) mounted a turf board inside, on a shelf of the cages of 10 pair-housed adult female **squirrel monkeys** and sprinkled a mixture of nuts, seeds and dried fruits onto the board on 11 consecutive days, right after the normal food was distributed. During the first 20 minutes, individuals spent approximately 36% of the time foraging.

Chamove & Scott (2005) made 360-minute video recordings of four family groups (5–11 individuals) of cotton-top **tamarins** after they were presented with a forage box to which they were habituated. The box was filled with a mixture of sawdust and small food items. Over the 6 hours, any given monkey was engaged in searching for and retrieving food from the box approximately 7% of the time.

Roberts et al. (1999) injected acacia gum into 2.5-cm-deep holes of 30-cm-long branch segments and placed one gum feeder each in the cages of 28 adult **marmosets**. The feeders were left in the cages for 5 days and the animals tested right after gum was injected into the branch on day 1 and day 5. During the 30-minute test sessions, individual animals spent, on average, 43% of the time gum-foraging on day 1, and 10% of the time gum-foraging on day 5. The branches were already heavily gouged on day 5.

Special food mixed with a substrate: Anderson & Chamove (1984) observed eight group-housed young stump-tailed macaques who were kept on woodchip litter (a) during 2 days before, and (b) during 2 days after 350 grams of mixed grain was distributed on the litter. In the course of 220 one-minute scan sessions conducted in both conditions, individual animals spent, on average, 6% of the time foraging on plain litter versus 30% of the time foraging on litter mixed with grain. Blois-Heulin & Jubin (2004) studied red-capped mangabeys and reported similiar findings.

Bryant et al. (1988) observed six adult male long-tailed macaques alone in a test cage each day for 30 minutes. The cage had a tray placed below the grid floor containing woodchips mixed with sunflower seeds and peanuts. Individuals spent approximately 37% of the time reaching through the grid floor, searching for and retrieving food from the woodchip litter. The interest in this activity increased over the course of a 12-day study period.

Recommendations: Nonhuman primates are biologically programmed to spend a major portion of their time foraging: searching for, retrieving and processing food. Allowing them to engage in foraging activities, rather than pick up freely accessible food in the research laboratory setting is an easy option for environmental enrichment; it should be a default practice in every primate research institution. The least expensive, yet very effective feeding enrichment is the distribution of the daily ration in such a way that the animals have to engage in skillful manipulation techniques to retrieve and process the food.

Refinement » Inanimate enrichment
Inanimate enrichment increases the complexity of the living quarters and promotes noninjurious contact and interaction with objects.

Structural enrichment » perches: The spatial limitation of the legally minimum-size standard cage can make it quite a challenge to open up the vertical dimension for the confined animal in a species-appropriate manner. This applies particularly for animals caged in the lower row of the prevailing double-cage arrangements. These animals are forced to live in a shady, cave-like environment close to the ground. Not surprisingly, when given the choice to stay in a bottom-row or in a top-row cage, macaques show a strong preference for the upper row cage (Westlund, 2002; MacLean et al., 2009).

A high perch opens up the vertical dimension, thereby increasing the usable cage space and promoting species-adequate behaviors, such as climbing, leaping (if the cage is large enough), balancing, bouncing, perching, sleeping, looking out, retreating to a safe place during alarming situations, and retreating to a dry place during the cage-cleaning procedure. Access to a high resting site has survival value for nonhuman primates. This explains why they do not lose interest in high resting surfaces over time.

Reinhardt (1989) assessed the time budgets of 25 adult male rhesus macaques who were housed in single-cages, each equipped with a 120-cm-long polyvinyl chloride (PVC) pipe that had a diameter of 5 cm and was installed diagonally, with a slope of 15°, about 40 cm above the floor. The males had been exposed to these perches for 12 months. There were 14 males in upper-row cages and 11 males in lower-row cages. During 120 minutes of observations, the average amount of time spent on the perch was:
  » 45% for the males in lower-row cages, versus
  » 15% for the males in upper-row cages; the difference was statistically significant.
Lower-row cage individuals were probably more attracted by their perch because they lived closer to the ground and at a greater distance from the light source. Sitting on an elevated surface was more advantageous for them than for individuals in the high and relatively bright upper-row cages.

Woodbeck & Reinhardt (1991) confirmed these findings in 28 pairs of adult female rhesus macaques who lived in double cages, each furnished with two 120-cm-long PVC pipes, located either in the bottom row (n=14 animals) or in the top row (n=14 animals). The females had been exposed to these perches for more than 24 months. During seven 30-minute observations conducted in the late afternoon when personnel were no longer in the building, average amount of time spent on the perch was:
  » 33% for the females in lower-row cages, versus
  » 7% for the females in upper-row cages; this difference was also statistically significant.

Similar findings were reported by Shimoji et al. (1993), who attached four parallel-connected PVC pipes, 5 cm in diameter, to the back of the cage 27 cm off the floor, of 10 female and 10 male adult long-tailed macaques for a 3-day study period. Remote video recordings revealed that animals caged on the bottom row of the rack spent, on average, 26% of the day on the perch, while animals caged on the top row spent only 14% of the day on the perch; the difference was not statistically different, but it was consistent on each of the 3 days.

Elevated structures not only make the vertical dimension accessible to the animals, but they also provide them with easy ways of quickly getting away from each other in situations of potential antagonistic conflict. Kitchen & Martin (1995) observed five pairs of common marmosets, each for a total of 20 hours, when their cages were barren versus equipped with three perches, 2.5 cm in diameter. When they had access to perches, the marmosets stopped showing startle responses and the incidence of aggressive interaction was significantly reduced. Neveu & Deputte (1996) recorded the behavior of a breeding troop of gray-cheeked mangabeys, consisting of three adult and two juvenile females and one adult and one subadult male, during 30-minute sessions when they lived in a barren cage versus a cage of the same dimensions but fitted with four perches at different heights. Access to perches decreased agonistic behaviors from about 25% to 0 % of all interactions; at the same time it increased socially positive behaviors significantly from about 2% to 10% of all interactions. Nakamichi & Asanuma (1998) tested a group of four adult female Japanese macaques in two identically sized enclosures that were either unstructured or furnished with eight wooden perches at different heights. Several 15-minute observation sessions showed that the average number of agonistic interactions was significantly lower in the furnished cage than in the unfurnished cage.

Structural enrichment » swings: In their natural habitat, nonhuman primates usually do not swing on branches or lianas. It is, therefore, not surprising that they have little use for swings in research labs, especially since the small size of their living quarters does not provide sufficient space to actually swing back and forth.

Bryant et al. (1988) observed six single adult male long-tailed macaques daily for 30-minutes in a play cage that was equipped with a swing suspended 60 cm from the ceiling. During a period of 12 days, two males never used the swing; the four others spent, on average, less than 2% of the time on it.

Kopecky & Reinhardt (1991) installed a PVC perch in one section and a PVC swing at the same height in the other section of upper-row double cages of 14 adult, pair-housed rhesus macaques and observed each animal after 1 month for 60 minutes. Subjects spent, on average, 7 minutes on the perch, but only half a minute on the swing. It was concluded that the animals' statistically significant preference for the perch was probably related to the fact that the perch, unlike the swing, was a fixed structure permitting continuous relaxed postures rather than brief balancing. Moreover, the perch, unlike the swing, allowed the animals to sit right in front of the cage within sight of the events going on in the room.

Dexter & Bayne (1994) tested nine adult single-caged rhesus macaques of both sexes in the presence of either two types of PVC swings, a hemp rope swing or a swing made of artificial vine. Each animal was exposed to the swings for a three-week period and observed three times for 30-minute sessions. The animals manipulated the swings but showed little inclination to actually use them for swinging. Altogether, swinging was witnessed only six times in the course of 360 minutes of observation; the average time that a monkey was actually swinging was less than 1 minute.

Also in relatively large group-enclosures, adult primates show hardly any interest in movable structures such as swings, ropes, suspended barrels or Ferris wheels, but they will spend most of the day and all night on fixed structures such as platforms, shelves, ladders, benches, and perches well above the floor area (langurs: Schwenk, 1992; rhesus macaques: Lehman & Lessnau, 1992; baboons: Kessel & Brent, 1996; chimpanzees: Howell et al., 1997).

Structural enrichment » visual barriers:

The spatial constraint of the cage makes it difficult to add structures in which an animal can take visual refuge from a dominant cage mate, but vertical blinds can readily be installed without occupying part of the floor area.

Reinhardt & Reinhardt (1991) inserted a privacy panel, consisting of a sheet of stainless steel with a rectangular 23 x 32 cm large passage hole close to the back wall of the cage, between the two halves of each double cage of 15 adult female rhesus pairs. One-hour observations before, and 7 days after placement of the privacy panels revealed that companions:
  » spent significantly more time in the same half of the cage (46 versus 37 minutes),
  » spent significantly more time engaged in affiliative interactions (22 versus 16 minutes), and
  » had fewer agonistic disputes (0.3/h vs. 2.2/h; difference statistically not different) when they had the option of visual seclusion.

Basile et al. (2007) observed 18 male/male pairs, 2 female/female pairs, and 5 male/female pairs before and 1 week after a privacy divider was placed in their double cages. The blind was oriented in such a way as to physically divide the front half of the cage, while leaving open access through the rear half. With the privacy divider in place, the animals spent significantly more time in the same half of the cage than without the divider. It was concluded that the privacy divider may provide a safe haven and give monkeys the ability to diffuse hostile situations before they escalate. McCormack & Megna (2001) placed privacy panels into the enclosure of a 126-animal breeding troop of rhesus macaques and noted a significant decrease in threatening, chasing, fear grinning, and screaming. Estep & Baker (1991) observed a breeding troop of 26 stump-tailed macaques during 90-minute sessions both before and after two solid temporary walls were erected within the animals' enclosure. The incidence of contact aggression was significantly lower when the monkeys had the option of breaking visual contact with other group members by moving behind these walls. Maninger et al. (1998) installed visual barriers in the living quarters of two breeding groups of 23 pig-tailed macaques and noted that the option of visual seclusion significantly reduced instances of biting, grabbing and chasing.

Erwin et al. (1976) studied agonistic interactions between adult female pig-tailed macaques who lived in four breeding groups; there were approximately 12 females in each group. Daily 20-minute observations were conducted of each group (a) during a 5-day control period and (b) during a 5-day experimental period when a concrete cylinder, approximately 1 m in length and 50 cm in diameter, was firmly placed in each enclosure. The mean incidence of agonistic interactions per 20-minute observation session was 94 during the control condition versus only 45 during the experimental condition; the difference was statistically significant. The monkeys used the cylinders as escape routes to hide from potential aggressors.

Recommendations: Nonhuman primates spend the night and a great portion of the day on elevated sites at a safe distance from terrestrial threats. "Even macaques, which some describe as semiterrestrial, spend most of the day in elevated locations and seek the refuge of trees at night. These animals might perceive the presence of humans above them as particularly threatening" (National Research Council, 1998, p 92 & 118). It is very important that their enclosures in research facilities are furnished with elevated structures allowing the animals to retreat to and rest on relatively safe surfaces. The usefulness of such structures depends on their placement in the cage. Any resting surface should be installed in such a way that an animal can:

> » sit right in front of the cage and check out what is going on in the room,
> » retreat to the back of the cage when frightened, for example, when a fear-inducing investigator enters the room,
> » sit on it without touching the ceiling with the head and without touching the floor with the tail, and
> » use the space beneath it for normal postural adjustments.

High resting surfaces are not really enriching the environment of the animals; they are a *necessity* and, therefore, should be mandatory furniture in every primary enclosure of nonhuman primates.

Visual barriers and/or safe escape options—with entrance *and* exit—should be mandatory provisions of any primary living quarters of pair-housed and group-housed primates in order to minimize conflicts triggered by the unnatural spatial constraint of confinement.

Toys: Nonhuman primates are too intelligent not to quickly get bored by toys, unless these can gradually be destroyed. Not surprisingly, they are much more interested in destructible than in durable toys (chimpanzee: Shefferly et al., 1993; Brent & Stone, 1998; Videan et al., 2005b; orangutan: Heuer & Rothe, 1998; pig-tailed macaques: Cardinal & Kent, 1998).

A conspicuous habituation to most commercial toys has been documented in chimpanzees (rubber and plastic toys for small children: Paquette & Prescott, 1988; Kong toys: Pruetz & Bloomsmith, 1992; indestructible toy ball: Shefferly et al., 1993), rhesus macaques (nylon balls: Ross & Everitt, 1988; plastic toys for small children: Hamilton, 1991; nylon balls and rings, Kong toys: Weick et al., 1991; Kong toys: Bayne et al., 1993), baboons (nylon bones: Brent & Belik, 1997), long-tailed macaques (Kong toys: Crockett et al., 1989) and pig-tailed macaques (plastic toys for small children: Cardinal & Kent, 1998; rubber and rawhide balls: Kessel & Brent, 1998). To be of some value for the animals, most commercial toys need to be replaced on a regular basis to make use of their short-lived novelty effect.

Gnawing sticks: Unlike many commercial toys, dry deciduous tree branches cut into gnawing sticks do not lose their novelty effect over time, since they steadily change their configuration and texture due to wear and progressive dehydration. The animals use the sticks for gnawing, nibbling, chewing, manipulating and playing. Long-term use of gnawing sticks by several hundred rhesus macaques resulted in no recognizable health hazards (Reinhardt, 1997a).

Comfortable Quarters for Laboratory Animals

Reinhardt (1990d) had provisioned 20 adult pair-housed stump-tailed macaques each with gnawing sticks for 2 months. During a 60-minute observation session, 16 of the animals gnawed the wooden material, on average, 8% of the time.

Reinhardt (1990a) assessed the time budgets of 60 pair-housed rhesus macaques of both sexes. Each pair had continuous access to one regularly replaced gnawing stick for 18 months or longer. During two 30-minute remote video recordings, the gnawing stick was used by 94% (17/18) of the subadult animals versus 64% (27/42) of the adult animals. On average, subadults spent 10% and adults spent 3% of the recording time in direct contact with the stick. The sexes did not differ significantly in their use of the wooden sticks; these were not only gnawed but also carried around and manipulated.

Sticks of sun-dried red oak branches are particularly suitable because they gradually wear into flakes that are so small that even large quantities pass through the sewer drains without clogging them (Reinhardt, 1992).

Paper and cardboard boxes: Recycled paper and cardboard boxes are not expensive, but they can offer effective environmental enrichment for primates in small cages or larger enclosures.

Kessel et al. (1995) scattered shredded paper once a week throughout the room of a group of five young male chimpanzees. After a habituation period of 1 week, 54-minute daily observation sessions were carried out during 2 weeks. The animals spent, on average, 27% of the time playing with the paper.

Bryant et al. (1988) transferred six adult male long-tailed macaques from their standard home cages to a play pen, furnished with a telephone directory and a nylon ball, each day for 30 minutes over a 12-day test period. The animals had very little or no use for the nylon ball, but they spent, on average, 10% of the time examining and shredding the telephone directory. Their interest in the paper material remained fairly constant; there was no indication that they lost interest in it over the course of time.

Beirise & Reinhardt (1992) placed a cardboard box into the pen of a 16-member breeding group of rhesus macaques once a week. After a habituation period of 8 weeks, the animals were observed for 120 minutes after placement of the cardboard box during weeks 9, 10 and 11. Individuals spent, on average, 65% of the time with the box, tearing it apart, shredding it and chewing pieces of it.

Water: Basins filled with water for swimming, diving for food items, fishing for food items, and playing have been employed for caged and group-housed long-tailed macaques (Gilbert & Wrenshall, 1989), squirrel monkeys (King & Norwood, 1989), and rhesus macaques (Anderson et al., 1992; Rawlins, 2005) without adverse effects, other than much splashing.

Mirrors: Both apes and monkeys are fascinated by their own reflections, and they use a mirror to check out the immediate environment without directly looking at it (Gallup, 1970; Lethmate & Dücker, 1973; Eglash & Snowdon, 1983; Platt & Thompson, 1985; Anderson,

1986; Lambeth & Bloomsmith, 1992; O'Neill et al., 1997; Chiappa et al., 2004; de Waal et al., 2005; Schultz, 2006).

Mirrors that can be manipulated are particularly useful for animals who are housed alone, while socially housed animals tend to focus their attention more on the social partner than on the mirror. Harris & Edwards (2004) hung stainless steel, 15-cm-diameter mirrors on the cages' front panels of 25 single male vervet monkeys, and observed each subject during four 30-minute sessions, 10 months, and again 16 months after the initial introduction of the mirrors. The average amount of time spent contacting the mirror and looking into the mirror was consistent at 5%, indicating that the animals had a sustained interest in them.

Windows: Whenever possible, rooms housing nonhuman primates should be provided with windows, since they are a source of natural light and can provide health benefits as well as environmental enrichment (International Primatological Society, 2007).

Pairs of male long-tailed macaques, transferred regularly for 1.5 hours to a playroom with windows, spend about 67% of the time looking out the windows (Lynch & Baker, 2000).

Light: There seems to be an international regulatory and professional consensus that lighting must be uniformly diffused throughout animal facilities and *must* provide sufficient illumination to facilitate housekeeping, cleaning, and inspection of animals, and maintain the well-being of the animals (United States Department of Agriculture, 1991; cf., Institute of Laboratory Animal Resources, 1980; National Research Council, 1996; Fortman et al., 2002; International Primatological Society, 2007). These important stipulations are meaningless as long as the traditional double-tier caging system prevails in some countries (e.g., Rosenberg & Kesel, 1994). Sanitation trays beneath the upper tier of cages, reduce significantly the amount of light from ceiling-mounted fixtures that can penetrate to the lower-cage tier; animals in the lower tier are thus relegated to a permanent state of semi-gloom (Mahoney, 1992). Illumination is often so poor that flashlights are needed to identify animals, check their well-being and make sure that the floor and the corners of the cage are adequately cleaned (Reinhardt, 1997b; Reasinger & Rogers, 2001; Savane, 2008).

Rotating cage positions relative to the light source—as is sometimes recommended (Canadian Council on Animal Care, 1993; National Research Council, 1996) and practiced (Ott, 1974; Ross & Everitt, 1988; Shively, 2001; Buchanan-Smith et al., 2002)—rotates the inherent problem, but it does not solve it: There will always be half of a population of double-tier caged animals who live in the lower tier in the shade cast by the cages of the upper tier.

Recommendations: "A two-tiered system is not recommended as these cages are usually too small. The lower tiers do not allow primates to engage in their vertical flight response, are often darker, and animals in the lower cages tend to receive less attention from attending personnel" (International Primatological Society, 2007, p 12). Keeping nonhuman primates in single-tier, rather than multi-tier caging systems in tall cages equipped with high resting surfaces is, at the moment, the only satisfactory refinement option to deal with the problems associated with the lower-row cage situation. It:
  » provides all animals of the room uniform illumination,
  » creates uniform illumination to aid in maintaining good housekeeping practices, adequate cleaning, and adequate inspection of animals, and

» allows the animals to access the arboreal dimension of their enclosures and retreat to relatively safe vantage points above eye-level of attending personnel.

Wall-mounted lights illuminating lower-row cages from behind can possibly even out the illumination differences between upper and lower row (MacLean et al., 2009), but they will keep the occupants of the lower row cages restricted to the terrestrial dimension.

Videos, television and music: Schapiro & Bloomsmith (1995) presented 49 single-caged yearling rhesus macaques with videotapes of chimpanzees and rhesus macaques in natural settings most of the day for a period of 3 months. During 15-minute observation sessions, subjects were looking at the monitor about 7% of the time. The possibility was not ruled out that the animals would have shown the same interest in the blank monitor.

Markowitz & Line (1989) mounted a radio device on the cages of five single adult female rhesus macaques. The radio had been available for a 14-week period and was preset to a soft rock music station; the animals could turn the radio on and off by touching two different bars. When they were tested during weeks 8–14, individual animals turned on the radio for 0–24 hours per day; on average the radio was turned on for about 12 hours per day. The monkeys showed no signs of losing interest in listening to the music.

Brent & Weaver (1996) noted in four single-caged baboons that the animals' mean heart rate was significantly lower when they could listen to a radio station playing oldies than when the radio was turned off. This calming effect may have been indirect, with the music masking the noise coming from other animal rooms, the ventilation system, and the caretaking staff.

McDermott & Hauser (2007) gave four adult cotton-top tamarins and four adult common marmosets the choice of listening to various noises and various kind of music. The animals showed a significant preference for soft over loud noise and for slow-tempo over fast-tempo music. Both tamarins and marmosets strongly and consistently preferred silence over musical stimuli (flute lullaby: $p<0.0001$; sung lullaby: $p<0.003$; Mozart concerto: $p<0.0001$), suggesting that they did not find such stimuli pleasurable or relaxing.

Recommendations: Before an institution plans to implement video, television or music enrichment programs for its nonhuman primates, it is advisable to first check if the animals actually benefit from such an investment. Playing music, videos or television programs in animal rooms may entertain the attending personnel but not necessarily the caged animals; they may have different preferences—including silence—which need to be respected to safeguard the animals' well-being.

Conclusions

Species-adequate, effective and practicable options for providing social enrichment, feeding enrichment and inanimate enrichment, and practicable options of training nonhuman primates to cooperate during common procedures have been described, tested and documented in the scientific and professional literature. Making life easier for nonhuman primates in research laboratories is not only a very basic ethical responsibility of the biomedical research industry and an important animal welfare issue, but it is also a fundamental condition for the scientific validity of the research data collected from these animals (Animal Welfare Institute, 1979; National Research Council, 1985; Meyerson, 1986; Donnelley, 1990; Morton, 1990; Novak & Bayne, 1991; Schwindaman, 1991; Institute for Laboratory Animal Research, 1992; Chance & Russell, 1997; Fuchs, 1997; Öbrink & Rehbinder, 1999; Richmond, 2002; Reinhardt & Reinhardt, 2002; Russell, 2002).

"Animals should be housed with the goal of maximizing species-specific behaviors and minimizing stress-induced behaviors" (National Research Council, 1996, p 21-22). "The maintenance and use of nonhuman primates should only be permitted in facilities which can truly provide the high quality of housing, and care and attention which these animals require, if their normal physiology and behavior are to be maintained" (Balls, 1995, p 286).

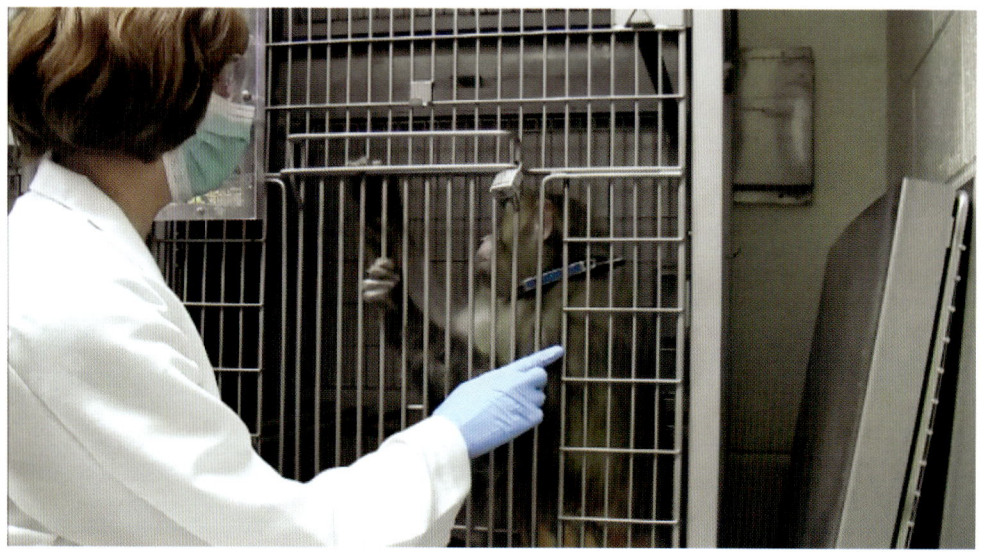

REFERENCES

Abee CR 1985 Medical care and management of the squirrel monkey. In: Rosenblum LA and Coe CL (ed) *Handbook of Squirrel Monkey Research* pp 447–488. Plenum Press: New York, NY

Ackerley ET and Stones PB 1969 Safety procedures for handling monkeys. *Laboratory Animal Handbooks 4*: 207–211

Alexander S and Fontenot MB 2003 Isosexual social group formation for environmental enrichment in adult male *Macaca mulatta. AALAS 54th National Meeting Official Program*: 141

Alford PL, Bloomsmith MA, Keeling ME and Beck TF 1995 Wounding aggression during the formation and maintenance of captive, multimale chimpanzee groups. *Zoo Biology 14*: 347–359

Altman NH 1970 Restraint of monkeys in clinical examination and treatment. *Journal of the American Veterinary Medical Association 159*: 1222

Animal Welfare Institute 1979 *Comfortable Quarters for Laboratory Animals, Seventh Edition*. Animal Welfare Institute: Washington, DC

American Association for Laboratory Animal Science 2001 *Cost of Caring: Recognizing Human Emotions in the Care of Laboratory Animals*. American Association for Laboratory Animal Science: Memphis, TN

Anderson JR 1986 Mirror-mediated finding of hidden food by monkeys *(Macaca tonkeana* and *Macaca fascicularis). Journal of Comparative Psychology 100*: 237–242

Anderson JR and Chamove AS 1984 Allowing captive primates to forage. *Standards in Laboratory Animal Management. Proceedings of a Symposium* pp 253–256. The Universities Federation for Animal Welfare: Potters Bar, UK

Anderson JR, Peignot P and Adelbrecht C 1992 Task-directed and recreational underwater swimming in captive rhesus monkeys *(Macaca mulatta). Laboratory Primate Newsletter 31*(4): 1–4

Anonymous 2003 Enrichment survey results. *Animal Keeper's Forum 30*: 513–515

Anonymous 2007 Pair-housed monkeys with head cap implants. In: Baumans V, Coke C, Green J, Moreau E, Morton D, Patterson-Kane E, Reinhardt A, Reinhardt V and Van Loo P (eds) *Making Lives Easier for Animals in Research Labs: Discussions by the Laboratory Animal Refinement & Enrichment Forum* pp 146–147. Animal Welfare Institute: Washington, DC

Anonymous 2013 Oral dosing off monkeys. In: Reinhardt V (ed) *Compassion Makes a Difference - Discussions by the Laboratory Animal Refinement & Enrichment Forum, Volume III* pp 108–115. Animal Welfare Institute: Washington, DC

Anzenberger G and Gossweiler H 1993 How to obtain individual urine samples from undisturbed marmoset families. *American Journal of Primatology 31*: 223–230

Asvestas C 1998 Pairing *Macaca fascicularis. Laboratory Primate Newsletter 37*(3): 5

Asvestas C and Reininger M 1999 Forming a bachelor group of long-tailed macaques *(Macaca fascicularis). Laboratory Primate Newsletter 38*(3): 14

Aureli F, Veenema H and van Eck C 1993 Coping with short-term crowding in long-tailed macaques. *American Journal of Primatology. 31*: 295

Baker KC 1997 Human interaction as enrichment for captive chimpanzees: A preliminary report. *American Journal of Primatology 42*: 92

Baker KC 2004 Benefits of positive human interaction for socially housed chimpanzees. *Animal Welfare 13*: 239–245

Baker KC, Bloomsmith MA, Oettinger B, Neu K, Griffis C, Schoof V and Maloney M 2012 Benefits of pair housing are consistent across a diverse population of rhesus macaques. *Applied Animal Behaviour Science 137*: 148–156

Basile BM, Hampton RR, Chaudhry AM and Murray EA 2007 Presence of a privacy divider increases proximity in pair-housed rhesus monkeys. *Animal Welfare 16*(1): 37–39

Bayne K 2002 Development of the human-research animal bond and its impact on animal well-being. *ILAR Journal 43*(1): 4–9

Bayne K, Dexter SL, Mainzer H, McCully C, Campbell G and Yamada F 1992 The use of artificial turf as a foraging substrate for individually housed rhesus monkeys *(Macaca mulatta). Animal Welfare 1*: 39–53

Bayne K, Hurst JK and Dexter SL 1992 Evaluation of the preference to and behavioral effects of an enriched environment on male rhesus monkeys. *Laboratory Animal Science 42*: 38–45

Bayne K, Dexter SL, Hurst JK, Strange GM and Hill EE 1993 Kong toys for laboratory primates: Are they really an enrichment or just fomites? *Laboratory Animal Science 43*: 78–85

Bayrakci R 2003 Starting an injection training program with lion-tailed macaques (Macaca silenus). *Animal Keeper's Forum 30*: 503–512

Beirise JH and Reinhardt V 1992 Three inexpensive environmental enrichment options for group-housed *Macaca mulatta. Laboratory Primate Newsletter 31*(1): 7–8

Bell BK 1995 Dealing with specific behavioral problems using operant conditioning with bonobos *(Pan paniscus). American Zoo and Aquarium Association (AZA) Regional Conference Proceedings*: 26–28

Bentson KL, Capitanio JP and Mendoza SP 2003 Cortisol responses to immobilization with Telazol or ketamine in baboons *(Papio cynocephalus/anubis)* and rhesus macaques *(Macaca mulatta). Journal of Medical Primatology 32*: 148–160

Bernstein IS 2007 Social mechanisms in the control of primate aggression. In: Campbell CJ, Fuentes A, MacKinnon KC, Panger M and Bearder SK (eds) *Primates in Perspective* pp 562–571. Oxford Univ Press: New York, NY

Bernstein IS and Mason WA 1963 Group formation by rhesus monkeys. *Animal Behaviour 11*: 28–31

Bernstein IS, Gordon TP and Rose RM 1974 Factors influencing the expression of aggression during introductions to rhesus monkey groups. In: Holloway RL (ed) *Primate Aggression, Territoriality, and Xenophobia* pp 211–240. Academic Press: New York, NY

Bertrand F, Seguin Y, Chauvier F and Blanquié JP 1999 Influence of two different kinds of foraging devices on feeding behaviour of rhesus macaques *(Macaca mulatta). Folia Primatologica 70*: 207

Bliss-Moreau E, Moadab G and Theil J 2013 Efficient cooperative chair training. *36th Meeting of the American Society of Primatologists Scientific Program*: Abstract # 118

Blois-Heulin C and Jubin R 2004 Influence of the presence of seeds and litter on the behaviour of captive red-capped mangabeys Cercocebus torquatus torquatus. Applied Animal Behaviour Science 85: 340–362

Bloom KR and Cook M 1989 Environmental enrichment: Behavioral responses of rhesus to puzzle feeders. Lab Animal 18(5): 25,27,29,31

Bloomsmith MA, Alford PL and Maple TL 1988 Successful feeding enrichment for captive chimpanzees. American Journal of Primatology 16: 155–164

Bloomsmith MA, Stone AM and Laule GE 1998 Positive reinforcement training to enhance the voluntary movement of group-housed chimpanzees within their enclosure. Zoo Biology 17: 333–341

Bourgeois SR and Brent L 2005 Modifying the behaviour of singly caged baboons: evaluating the effectiveness of four enrichment techniques. Animal Welfare 14: 71–81

Brent L and Eichberg JW 1991 Primate puzzleboard: A simple environmental enrichment device for captive chimpanzees. Zoo Biology 10: 353–360

Brent L and Weaver D 1996 The physiological and behavioral effects of radio music on singly housed baboons. Journal of Medical Primatology 25: 370–374

Brent L and Belik M 1997 The response of group-housed baboons to three enrichment toys. Laboratory Animals 31 : 81–85

Brent L and Stone AM 1998 Destructible toys as enrichment for captive chimpanzees. Journal of Applied Animal Welfare Science 1: 5–14

Brockway BP, Hassler CR and Hicks N 1993 Minimizing stress during physiological monitoring. In: Niemi SM and Willson JE (eds) Refinement and Reduction in Animal Testing pp 56–69. Scientists Center for Animal Welfare: Bethesda, MD

Brown CS and Loskutoff NM 1998 A training program for noninvasive semen collection in captive western lowland gorillas (Gorilla gorilla gorilla). Zoo Biology 17: 143–151

Bryant CE, Rupniak NMJ and Iversen SD 1988 Effects of different environmental enrichment devices on cage stereotypies and autoaggression in captive cynomolgus monkeys. Journal of Medical Primatology 17: 257–269

Buchanan-Smith HM, Shand C and Morris K 2002 Cage use and feeding height preferences of captive common marmosets (Callithrix j. jacchus) in two-tier cages. Journal of Applied Animal Welfare Science 5: 139–149

Bunyak SC, Harvey NC, Rhine RJ and Wilson MI 1982 Venipuncture and vaginal swabbing in an enclosure occupied by a mixed-sex group of stumptailed macacaques (Macaca arctoides). American Journal of Primatology 2: 201–204

Canadian Council on Animal Care 1993 Guide to the Care and Use of Experimental Animals, Volume 1, 2nd Edition. Canadian Council on Animal Care: Ottawa, Canada

Canadian Council on Animal Care 1984 Chapter XX: Nonhuman primates. In: Guide to the Care and Use of Experimental Animals, Volume 2 pp 163–173. Canadian Council on Animal Care: Ottawa, Canada

Cardinal BR and Kent SJ 1998 Behavioral effects of simple manipulable environmental enrichment on pair-housed juvenile macaques (Macaca nemestrina). Laboratory Primate Newsletter 37(1): 1–3

Celli ML, Tomonagaa M, Udonob T, Teramotob M and Naganob K 2003 Tool use task as environmental enrichment for captive chimpanzees. Applied Animal Behaviour Science 81: 171–182

Chamove AS and Scott L 2005 Forage box as enrichment in single- and group-housed callitrichid monkeys. Laboratory Primate Newsletter 44(2): 13–17

Chance MRA and Russell WMS 1997 The benefits of giving experimental animals the best possible environment. In: Reinhardt V (ed) Comfortable Quarters for Laboratory Animals, Eighth Edition pp 12–14. Animal Welfare Institute: Washington, DC

Chiappa P, Ortiz-Sánchez V and Antonio-Garcés J 2004 Social status and mirror behaviour in Macaca arctoides. Folia Primatologica 75(Supplement): 364

Clarke AS, Czekala NM and Lindburg DG 1995 Behavioral and adrenocortical responses of male cynomolgus and lion-tailed macaques to social stimulation and group formation. Primates 36: 41–46

Coe CL 1991 Is social housing of primates always the optimal choice? In: Novak MA and Petto AJ (eds) Through the Looking Glass. Issues of Psychological Well-being in Captive Nonhuman Primates. pp 78–92. American Psychological Association: Washington, DC

Coe CL and Rosenblum LA 1984 Male dominance in the bonnet macaque: A malleable relationship. In: Barchas PR and Mendoza SP (eds) Social Cohesion. Essays Toward a Sociophysiological Perspective pp 31–64. Greenwood Press: Westport, CT

Coe CL, Franklin D, Smith ER and Levine S 1982 Hormonal responses accompanying fear and agitation in the squirrel monkey. Physiology and Behavior 29: 1051–1057

Coelho AM and Carey KD 1990 A social tethering system for nonhuman primates used in laboratory research. Laboratory Animal Science 40: 388–394

Coelho AM, Carey KD and Shade RE 1991 Assessing the effects of social environment on blood pressure and heart rates of baboons. American Journal of Primatology 23: 257–267

Cohen S, Kaplan JR, Cunnick JE, Manuck SB and Rabin BS 1992 Chronic social stress, affiliation, and cellular immune response in nonhuman primates. Psychological Science 3: 301–304

Coleman K, Pranger L, Maier A, Lambeth SP, Perlman JE, Thiele E and Schapiro SJ 2008 Training rhesus macaques for venipuncture using positive reinforcement techniques: A comparison with chimpanzees. Journal of the American Association for Laboratory Animal Science 47: 37–41

Council of Europe 2006 Appendix A of the European Convention for the Protection of Vertebrate Animals Used for Experimental and Other Scientific Purposes (ETS No. 123). Council of Europe: Strasbourg, France

Crockett CM and Heffernan KS 1998 Grooming-contact cages promote affiliative social interaction in individually housed adult baboons. American Journal of Primatology 45: 176

Crockett CM, Bielitzki JT, Carey A and Velez A 1989 Kong toys as enrichment devices for singly-caged macaques. Laboratory Primate Newsletter 28(2): 21–22

Crockett CM, Bowers CL, Bowden DM and Sackett GP 1994 Sex differences in compatibility of pair-housed adult longtailed macaques. American Journal of Primatology 32: 73–94

Crockett CM, Bellanca RU, Bowers CL and Bowden DM 1997 Grooming-contact bars provide social contact for individually caged laboratory primates. *Contemporary Topics in Laboratory Animal Science 36*(6): 53–60

Crockett CM, Koberstein D and Heffernan KS 2001 Compatibility of laboratory monkeys housed in grooming-contact cages varies by species and sex. *American Journal of Primatology 54*(Supplement): 51–52

Crockett CM, Lee GH and Thom JP 2006 Sex and age predictors of compatibility in grooming-contact caging vary by species of laboratory monkey. *International Journal of Primatology 27*(Supplement): 417

Cross N, Pines MK and Rogers LJ 2004 Saliva sampling to assess cortisol levels in unrestrained common marmosets and the effect of behavioral stress. *American Journal of Primatology 62*: 107–114

Dazey J, Kuyk K, Oswald M, Marenson J and Erwin J 1977 Effects of group composition on agonistic behavior of captive pigtailed macaques (*Macaca nemestrina*). *American Journal of Physical Anthropology 46*: 73–76

de Filippis B, Chiarotti F and Vitale A 2009 Severe intragroup aggressions in captive common marmosets (*Callithrix jacchus*). *Journal of Applied Animal Welfare Science 12*: 214–222

de Rosa C, Vitale A and Puopolo M 2003 The puzzle-feeder as feeding enrichment for common marmosets (*Callithrix jacchus*): a pilot study. *Laboratory Animals 37*: 100–107

de Villiers C and Seier JV 2010 Stopping self injurious behaviour of a young male Chacma baboon (*Papio ursinus*). *Animal Technology and Welfare 9*(2): 77–80

de Waal FBAF 1997 Conflict resolution and distress alleviation in monkeys and apes. *Annals of the New York Academy of Science 807*: 317–328

de Waal FBM and Ren RM 1988 Comparison of the reconciliation behavior of stumptail and rhesus macaques. *Ethology 78*: 129–142

de Waal FBM, Dindo M, Freeman CA and Hall MJ 2005 The monkey in the mirror: Hardly a stranger. *Proceedings of the National Academy of Sciences 102*: 140–147 http://www.pnas.org/cgi/content/full/102/32/11140#REF3

Dexter SL and Bayne K 1994 Results of providing swings to individually housed rhesus monkeys (*Macaca mulatta*). *Laboratory Primate Newsletter 33*(2): 9–12 http://www.brown.edu/Research/Primate/lpn33-2.html#bayne

Donnelly MJ, Wickham A, Kulick A, Rogers I, Stribling S, Strack A, Doerning B and Feeney W 2007 A refinement of oral dosing in the common marmoset (*Callithrix jacchus*). *AALAS 58th National Meeting Official Program*: 45

Donnelley S 1990 Animals in science: The justification issue. In: Donnelley S and Nolan K (eds) *Animal, Science and Ethics* pp 8–13. The Hastings Center Report: Briarcliff Manor, CT

Doyle LA, Baker KC and Cox LD 2008 Physiological and behavioral effects of social introduction on adult male rhesus macaques. *American Journal of Primatology 70*: 1–9

Eaton GG, Kelley ST, Axthelm MK, Iliff-Sizemore SA and Shiigi SM 1994 Psychological well-being in paired adult female rhesus (*Macaca mulatta*). *American Journal of Primatology 33*: 89–99

Eglash AR and Snowdon CT 1983 Mirror-image responses in pygmy marmosets (*Cebuella pygmaea*). *American Journal of Primatology 5*: 211–219

Elvidge H, Challis JRG, Robinson JS, Roper C and Thorburn GD 1976 Influence of handling and sedation on plasma cortisol in rhesus monkeys (*Macaca mulatta*). *Journal of Endocrinology 70*: 325–326

Erwin J 1977 Factors influencing aggressive behavior and risk of trauma in the pigtail macaque (*Macaca nemestrina*). *Laboratory Animal Science 27*: 541–547

Erwin J, Anderson B, Erwin N, Lewis L and Flynn D 1976 Aggression in captive pigtail monkey groups: Effects of provision of cover. *Perceptual and Motor Skills 42*: 319–324

Estep DQ and Baker SC 1991 The effects of temporary cover on the behavior of socially housed stumptailed macaques (*Macaca arctoides*). *Zoo Biology 10*: 465–472

Evans HL, Taylor JD, Ernst J and Graefe JF 1989 Methods to evaluate the well-being of laboratory primates. Comparison of macaques and tamarins. *Laboratory Animal Science 39*: 318–323

Fekete JM, Norcross JL and Newman JD 2000 Artificial turf foraging boards as environmental enrichment for pair-housed female squirrel monkeys. *Contemporary Topics in Laboratory Animal Science 39*(2): 22–26

Ferraro A, Brunelli R, Nelsen SL, Andrews-Kelly G and Schultz P 2013 Making use of a laser pointer as training and enrichment tool: a discussion by the Laboratory Animal Refinement & Enrichment Forum. *Animal Technology and Welfare 12*: 195–196

Fortman JD, Hewett TA and Bennett BT 2002 *The Laboratory Nonhuman Primate*. CRC Press: Boca Raton, FL

Fragaszy DM, Baer J and Adams-Curtis LE 1994 Introduction and integration of strangers into captive groups of tufted capuchins (*Cebus apella*). *International Journal of Primatology 15*: 399–420

Fritz J 1989 Resocialization of captive chimpanzees: An amelioration procedure. *American Journal of Primatology 19*(Supplement): 79–86

Fritz J 1994 Introducing unfamiliar chimpanzees to a group or partner. *Laboratory Primate Newsletter 33*(1): 5–7 http://www.brown.edu/Research/Primate/lpn33-1.html#jo

Fritz P and Fritz J 1979 Resocialization of chimpanzees: Ten years of experience at the Primate Foundation of Arizona. *Journal of Medical Primatology 8*: 202–221

Fuchs E 1997 Requirements of biomedical research in terms of housing and husbandry: Neuroscience. *Primate Report 49*: 43–46

Gallup GG 1970 Chimpanzees: Self-recognition. *Science 167*: 86–87

Gantt WH, Newton JEO, Royer FL and Stephens JH 1966 Effect of person. *Conditional Reflex 1*: 18–35

Gilbert SG and Wrenshall E 1989 Environmental enrichment for monkeys used in behavioral toxicology studies. In: Segal EF (ed) *Housing, Care and Psychological Well-being of Captive and Laboratory Primates* pp 244–254. Noyes Publications: Park Ridge, NJ

Gillis TE, Janes AC and Kaufman MJ 2012 Positive rreinforcement training in squirrel monkeys using clicker training. *American Journal of Primatology 74*: 712–720

Gilloux I, Gurnell J and Shepherdson D 1992 An enrichment device for great apes. *Animal Welfare 1*: 279–289

Gisler DB, Benson RE and Young RJ 1960 Colony husbandry of research monkeys. *Annals of the New York Academy of Sciences 85*: 758–568

Goodwin J 1997 The application, use, and effects of training and enrichment variables with Japanese snow macaques (*Macaca fuscata*) at the Central Park Wildlife Center. *American Zoo and Aquarium Association (AZA) Regional Conference Proceedings*: 510–515

Günther MM 1998 Influence of habitat structure on jumping behaviour in *Galago moholi. Folia Primatologica 69*(Supplement): 410

Gunnar MR, Gonzalez CA and Levine S 1980 The role of peers in modifying behavioral distress and pituitary-adrenal response to a novel environment in year-old rhesus monkeys. *Physiology and Behavior 25*: 795–798

Gust DA, Gordon TP, Wilson ME, Ahmed-Ansari A, Brodie AR and McClure HM 1991 Formation of a new social group of unfamiliar female rhesus monkeys affects the immune and pituitary adrenocortical systems. *Brain, Behavior, and Immunity 5*: 296–307

Gust DA, Gordon TP, Brodie AR and McClure HM 1994 Effect of a preferred companion in modulating stress in adult female rhesus monkeys. *Physiology and Behavior 55*: 681–684

Gust DA, Gordon TP, Wilson ME, Brodie AR, Ahmed-Ansari A and McClure HM 1996 Group formation of female pigtail macaques (*Macaca nemestrina*). *American Journal of Primatology 39*: 263–273

Haba Nelsen SL, Bradford D and Houghton P 2010 Laser lixit™ training: an alternative form of target training that can be utilized in the daily husbandry care of rhesus macaques (*Macaca mulatta*) and cynomolgus macaques (*Macaca fascicularis*). *American Journal of Primatology 72*(Supplement): 27

Hamilton P 1991 Enrichment toys and tools in recent trials. *Humane Innovations and Alternatives in Animal Experimentation 5*: 272–277

Harris HG and Edwards AJ 2004 Mirrors as environmental enrichment for African green monkeys . *American Journal of Primatology 63*: 459–467

Hayes SL 1990 Increasing foraging opportunities for a group of captive capuchin monkeys (*Cebus capucinus*). *Laboratory Animal Science 40*: 515–519

Hennessy MB 1984 Presence of companion moderates arousal of monkeys with restricted social experience. *Physiology and Behavior 33*: 393–398

Henrickson RV 1976 The nonhuman primate. *Lab Animal 5*(4): 60–62

Heuer A and Rothe H 1998 Environmental enrichment for four subadult orangutans at the Hannover Zoo [article in German]. *Der Zoologische Garten 2*: 119–133

Home Office 1989 *Animals (Scientific Procedures) Act 1986. Code of Practice for the Housing and Care of Animals Used in Scientific Procedures.* Her Majesty's Stationery Office: London, UK

Howell SM, Miteva E, Fritz J and Baron J 1997 The provision of cage furnishings as environmental enrichment at the Primate Foundation of Arizona. *The Newsletter 9*(2): 1–5

Inglis IR, Forkmann B and Lazarus J 1997 Free food or earned food? A review and fuzzy model of contrafreeloading. *Animal Behaviour 53*: 1171–1191

Institute for Laboratory Animal Resources 1980 *Laboratory Animal Management: Nonhuman Primates.* National Academy Press: Washington, DC

Institute for Laboratory Animal Research 1992 *Recognition and alleviation of pain and distress in laboratory animals.* National Academy Press: Washington, DC

International Primatological Society 1993 IPS International guidelines for the acquisition, care and breeding of nonhuman primates, Codes of Practice 1–3. *Primate Report 35*: 3–29

International Primatological Society 2007 *IPS International Guidelines for the Acquisition, Care and Breeding of Nonhuman Primates.* International Primatological Society: Bronx, NY

Jerome CP and Szostak L 1987 Environmental enrichment for adult, female baboons (*Papio anubis*). *Laboratory Animal Science 37*: 508–509

Johns Hopkins University and Health System 2001 Restraint techniques for animals - Non human primates. *Animal Care and Use Training* [Web site]: Accessed 02/11/2008

Kaplan JR, Manning P and Zucker E 1980 Reduction of mortality due to fighting in a colony of rhesus monkeys (*Macaca mulatta*). *Laboratory Animal Science 30*: 565–570

Kelley TM and Bramblett CA 1981 Urine collection from vervet monkeys by instrumental conditioning. *American Journal of Primatology 1*: 95–97

Kessel-Davenport AL and Gutierrez T 1994 Training captive chimpanzees for movement in a transport box. *The Newsletter 6*(2): 1–2

Kessel AL and Brent L 1995 An activity cage for baboons, Part I. *Contemporary Topics in Laboratory Animal Science 34*(6): 74–79

Kessel AL and Brent L 1996 Space utilization by captive-born baboons (*Papio* sp.) before and after provision of structural enrichment. *Animal Welfare 5*: 37–44

Kessel AL and Brent L 1998 Cage toys reduce abnormal behavior in individually housed pigtail macaques. *Journal of Applied Animal Welfare Science 1*: 227–234

Kessel AL and Brent L 2001 The rehabilitation of captive baboons. *Journal of Medical Primatology 30*: 71–80

Kessel AL, Brent L and Walljasper T 1995 Shredded paper as enrichment for infant chimpanzees. *Laboratory Primate Newsletter 34*(4): 4–6

Kessler MJ, London WT, Rawlins RG, Gonzales J, Martines HS and Sanches J 1985 Management of a harem breeding colony of rhesus monkeys to reduce trauma-related morbidity and mortality. *Journal of Medical Primatology 13*: 91–98

King JE and Norwood VR 1989 Free-environment rooms as alternative housing for squirrel monkeys. In: Segal EF (ed) *Housing, Care and Psychological Wellbeing of Captive and Laboratory Primates* pp 102–114. Noyes Publications: Park Ridge, NJ

Kitchen AM and Martin AA 1996 The effects of cage size and complexity on the behaviour of captive common marmosets, *Callithrix jacchus jacchus. Laboratory Animals 30*: 317–326

Knezevich M and Fairbanks L 2004 Tooth blunting as a wound reduction strategy in group living vervet monkeys (*Chlorocebus aethiops*). *American Journal of Primatology 62*(Supplement): 45

Koban TL, Miyamoto M, Donmoyer G and Hammar A 2005 Effects of positive reinforcement training on cortisol, hematology and cardiovascular parameters in cynomolgus macaques (*Macaca fascicularis*). *American Journal of Primatology 66*(Supplement): 148

Köhler, W 1921 *Intelligenzprüfungen an Menschenaffen.* Springer Verlag: Berlin, Germany

Kopecky J and Reinhardt V 1991 Comparing the effectiveness of PVC swings versus PVC perches as environmental enrichment objects for caged female rhesus macaques. *Laboratory Primate Newsletter 30*(2): 5–6

Lambeth SP and Bloomsmith MA 1992 Mirrors as enrichment for captive chimpanzees (*Pan troglodytes*). *Laboratory Animal Science 42*: 261–266

Lambeth SP and Bloomsmith MA 1994 A grass foraging device for captive chimpanzees (*Pan troglodytes*). *Animal Welfare 3*: 13–24

Lambeth SP, Perlman JE and Schapiro SJ 2000 Positive reinforcement training paired with videotape exposure decreases training time investment for a complicated task in female chimpanzees. *American Journal of Primatology 51*(Supplement): 79–80

Laule GE, Thurston RH, Alford PL and Bloomsmith MA 1996 Training to reliably obtain blood and urine samples from a diabetic chimpanzee (*Pan troglodytes*). *Zoo Biology 15*: 587–591

Lee GH, Thom JP and Crockett CM 2005 Factors predicting compatible grooming-contact pairings in four species of laboratory monkeys. *American Journal of Primatology 66*(Supplement): 83–84

Lehman SM and Lessnau RG 1992 Pickle barrels as enrichment objects for rhesus macaques. *Laboratory Animal Science 42*: 392–397

Lethmate J and Dücker G 1973 Studies on self-recognition in a mirror in orang-utans, chimpanzees, gibbons and various other monkey species [article in German]. *Zeitschrift für Tierpsychologie* [Ethology] *33*: 248–269

Levison PK, Fester CB, Nieman WH and Findley JD 1964 A method for training unrestrained primates to receive drug injection. *Journal of the Experimental Analysis of Behavior 7*: 253–254

Lindburg DG 1971 The rhesus monkey in North India: an ecological and behavioral study. In: Rosenblum LA (ed) *Primate Behavior: Developments in Field and Laboratory Research, Volume 2* pp 1–106. Academic Press: New York, NY

Line SW 1987 Environmental enrichment for laboratory primates. *Journal of the American Veterinary Medical Association 190*: 854–859

Line SW and Morgan KN 1991 The effects of two novel objects on the behaviour of singly caged adult rhesus macaques. *Laboratory Animal Science 41*: 365–369

Line SW, Markowitz H, Morgan KN and Strong S 1989 Evaluation of attempts to enrich the environment of single-caged nonhuman primates. In: Driscoll JW (ed) *Animal Care and Use in Behavioral Research: Regulation, Issues, and Applications* pp 103–117. Animal Welfare Information Center National Agricultural Library: Beltsville, MD

Line SW, Morgan KN, Markowitz H, Roberts J and Riddell M 1990 Behavioral responses of female long-tailed macaques (*Macaca fascicularis*) to pair formation. *Laboratory Primate Newsletter 29*(4): 1–5

Ljungberg T, Westlund K and Rydén L 1997 Ethological studies of well-being in two species of macaques after transition from single-cages to housing in social groups *EUPREN/EMRG Workshop* [Web site]: Accessed 01/11/2008

Logsdon S 1994 Enrichment for woolly monkeys. *The Shape of Enrichment 3*(1): 8

Logsdon S 1995 Use of operant conditioning to assist in the medical management of hypertension in woolly monkeys. *American Zoo and Aquarium Association Regional Conference Proceedings*: 96–102

Luttrell L, Acker L, Urben M and Reinhardt V 1994 Training a large troop of rhesus macaques to cooperate during catching: Analysis of the time investment. *Animal Welfare 3*: 135–140

Lutz CK and Farrow RA 1996 Foraging device for singly housed longtailed macaques does not reduce stereotypies. *Contemporary Topics in Laboratory Animal Science 35*(3): 75–78

Lutz C, Tiefenbacher S, Jorgenson MJ, Meyer JS and Novak MA 2000 Techniques for collecting saliva from awake, unrestrained, adult monkeys for cortisol assay. *American Journal of Primatology 52*: 93–99

Lynch R 1998 Successful pair-housing of male macaques (*Macaca fascicularis*). *Laboratory Primate Newsletter 37*(1): 4–5

Lynch R and Baker D 2000 Primate Enrichment: A room with a view. *Laboratory Primate Newsletter 39*(1): 12

MacLean EL, Roberts Prior S, Platt ML and Brannon EM 2009 Primate location preference in a double-tier cage: The effects of illumination and cage height. *Journal of Applied Animal Welfare Science 12*: 73–81

Mahoney CJ 1992 Some thoughts on psychological enrichment. *Lab Animal 21*(5): 27, 29, 32–37

Majolo B, Buchanan-Smith HM and Morris K 2003 Factors affecting the successful pairing of unfamiliar common marmoset (*Callithrix jacchus*) females: Preliminary results. *Animal Welfare 12*: 327–337

Maki S, Alford PL, Bloomsmith MA and Franklin J 1989 Food puzzle device simulating termite fishing for captive chimpanzees (*Pan troglodytes*). *American Journal of Primatology 19*(Supplement): 71–78

Maninger N, Kim JH and Ruppenthal GC 1998 The presence of visual barriers decreases antagonism in group housed pigtail macaques (*Macaca nemestrina*). *American Journal of Primatology 45*: 193–194

Markowitz H and Line SW 1989 Primate research models and environmental enrichment. In: Segal EF (ed) *Housing, Care and Psychological Wellbeing of Captive and Laboratory Primates* pp 202–212. Noyes Publications: Park Ridge, NJ

Mason WA 1960 Socially mediated reduction in emotional responses of young rhesus monkeys. *Journal of Abnormal and Social Psychology 60*: 100–110

McCormack K and Megna NL 2001 The effects of privacy walls on aggression in a captive group of rhesus macaques (*Macaca mulatta*). *American Journal of Primatology 54* (Supplement): 50–51

McDermott J and Hauser MD 2007 Nonhuman primates prefer slow tempos but dislike music overall. *Cognition 104*: 654–668

McGrew WC, Brennan JA and Russell J 1986 An artificial 'Gum-tree' for marmosets (*Callithrix j. jacchus*). *Zoo Biology 5*: 45–50

McKinley J, Buchanan-Smith HM, Bassett L and Morris K 2003 Training common marmosets (*Callithrix jacchus*) to cooperate during routine laboratory procedures: Ease of training and time investment. *Journal of Applied Animal Welfare Science 6*: 209–220

McMillan JL, Perlman JE, Galvan A, Wichmann T and Bloomsmith MA 2014 Refining the pole-and-collar method of restraint: Emphasizing the use of positive training techniques with rhesus macaques (*Macaca mulatta*). *Journal of the American Association for Laboratory Animal Science 53*(1): 61–68

Medical Research Council 2004 *MRC Ethics Guide: Best Practice in the Accommodation and Care of Primates used in Scientific Research*. Medical Research Council: London, UK

Menzel EW 1991 Chimpanzees *(Pan troglodytes):* Problem seeking versus the bird-in-hand, least-effort strategy. *Primates 32*: 497–508

Meyerson BJ 1986 Ethology in animal quarters. *Acta Physiologica Scandinavica 554*(Supplement): 24–31

Mitchell DS, Wigodsky HS, Peel HH and McCaffrey TA 1980 Operant conditioning permits voluntary, noninvasive measurement of blood pressure in conscious, unrestrained baboons *(Papio cynocephalus)*. *Behavior Research Methods and Instrumentation 12*: 4192–498

Morton DB 1990 Adverse effects in amimals and their relevance to refining scientific procedures. *ATLA 18*: 29–39

Murchison MA 1995 Forage feeder box for single animal cages. *Laboratory Primate Newsletter 34*(1): 1–2

Murphy DE 1976 Enrichment and occupational devices for orang utans and chimpanzees. *International Zoo News 137*(23.5): 24–26 http://www.awionline.org/lab_animals/biblio/izn-mur.htm

Murray L, Hartner M and Clark LP 2002 Enhancing post-surgical recovery of pair-housed nonhuman primates (*M. fascicularis*). *Contemporary Topics in Laboratory Animal Science 41*(4): 112–113

Nadler RD, Herndon JG, Metz B, Ferrer AC and Erwin J 1992 Environmental enrichment by varied feeding strategies for individually caged young chimpanzees. In: Erwin J and Landon JC (eds) *Chimpanzee Conservation and Public Health: Environments for the Future* pp 137–145. Diagnon/Bioqual: Rockville, MD

Nagel V and Kummer H 1974 Variation in cercopithecoid aggressive behavior. In: Holloway R (ed) *Primate Aggression, Territoriality, and Xenophobia* pp 159–184. Academic Press: New York, NY

Nahon NS 1968 A device and techniques for the atraumatic handling of the sub-human primate. *Laboratory Animal Care 18*: 486–487

Nakamichi M and Asanuma K 1998 Behavioral effects of perches on group-housed adult female Japanese monkeys. *Perceptual and Motor Skills 87*: 707–714

Napier JR and Napier PH 1994 *The Natural History of the Primates*. MIT Press: Cambridge, MA

National Center for the Replacement Refinement Reduction of Animals in Research 2006 *Handling and Restraint*. National Center for the Replacement, Refinement and Reduction of Animals in Research: London

National Health and Medical Research Council Animal Welfare Committee 1997 *Policy on the Use of Nonhuman Primates in Medical Research*. National Health and Medical Research Council: Canberra, Australia

National Research Council 1985 *Guide for the Care and Use of Laboratory Animals, Sixth Edition*. National Institutes of Health: Bethesda, MD

National Research Council 1996 *Guide for the Care and Use of Laboratory Animals, Seventh Edition*. National Academy Press: Washington, DC

National Research Council 1998 *The Psychological Well-Being of Nonhuman Primates*. National Academy Press: Washington, DC

Neveu H and Deputte BL 1996 Influence of availability of perches on the behavioral well-being of captive, group-living mangabeys. *American Journal of Primatology 38*: 175–185

Niemeyer C, Eaton GG and Kelley ST 1998 Practical aspects of the program to promote psychological well-being in nonhuman primates at the Oregon Regional Primate Research Center. In: Hare VJ and Worley E (eds) *Proceedings of the International Conference on Environmental Enrichment* pp 345–354. The Shape of Enrichment: San Diego, CA

Novak MA and Bayne K 1991 Monkey behavior and laboratory issues. *Laboratory Animal Science 41*: 306–307

O'Connor E and Reinhardt V 1994 Caged stumptailed macaques voluntarily work for ordinary food. *In Touch 1*(1): 10–11

O'Neill PL, Lauter AC and Weed JL 1997 Curious response of three monkey species to mirrors. *American Zoo and Aquarium Association Regional Conference Proceedings*: 95–101

Ochiai OT and Matsuzawa T 2001 Introduction of two wooden climbing frames as environmental enrichment for captive chimpanzees (*Pan troglodytes*) and its assessment. *Japanese Journal of Animal Psychology 51*(1): 1–9

Öbrink KJ and Rehbinder C 1999 Animal definition: a necessity for the validity of animal experiments? *Laboratory Animals 22*: 121–130

Ott JN 1974 The importance of laboratory lighting as an experimental variable. In: Magalhaes H (ed) *Environmental Variables in Animal Experimentation* pp 39–57. Bucknell University: Lewisburg, PA

Panneton M, Alleyn S and Kelly N 2001 Chair restraint for squirrel monkeys. *AALAS 52nd National Meeting Official Program*: 92

Paquette D and Prescott J 1988 Use of novel objects to enhance environments of captive chimpanzees. *Zoo Biology 7*: 15–23

Pearson BL, Judge P and Reeder DM 2008 Effectiveness of saliva collection and enzyme-immunoassay for the quantification of cortisol in socially housed baboons. *American Journal of Primatology 70*: 1145–1151

Perlman J, Guhad FA, Lambeth S, Fleming T, Lee D, Martino M and Schapiro SJ 2001 Using positive reinforcement training techniques to facilitate the assessment of parasites in captive chimpanzees. *American Journal of Primatology 54*(Supplement): 56

Perlman JE, Bowsher TR, Braccini SN, Kuehl TJ and Schaprio SJ 2003 Using positive reinforcement training techniques to facilitate the collection of semen in chimanzees (*Pan troglodytes*). *American Journal of Primatology* 60(Supplement): 77–78

Peterson G, Kelly K and Miller L 1988 Use of an artificial gum-tree feeder for marmosets. *Animal Keepers Forum 15*: 396–401

Phillippi-Falkenstein K and Clarke MR 1992 Procedure for training corral-living rhesus monkeys for fecal and blood-sample collection. *Laboratory Animal Science 42*: 83-85

Platt MM and Thompson RL 1985 Mirror responses in a Japanese macaque troop (Arashiyama West). *Primates 26*: 300–314

Poole TB 1990 Environmental enrichment for marmosets. *Animal Technology 41*: 81–86

Poole TB, Hubrecht R and Kirkwood JK 1999 Marmosets and Tamarins. In: Poole T and English P (eds) *The UFAW Handbook on the Care and Management of Laboratory Animals, Seventh Edition* pp. 558-573. Blackwell Science: Oxford, UK

Pranger LA, Maier A, Coleman K, Lambeth SP, Perlman JE, Thiele E, McMillam JL and Schapiro SJ 2006 Venipuncture training using positive reinforcement training techniques: a comparison of chimpanzee and rhesus macaques. *American Journal of Primatology 68*(Supplement): 61–62

Priest GM 1990 The use of operant conditioning in training husbandry behavior with captive animals. *Proceedings of the National American Association of Zoo Keepers Conference 16*: 94–108

Priest GM 1991 Training a diabetic drill (*Mandrillus leucophaeus*) to accept insulin injections and venipuncture. *Laboratory Primate Newsletter 30*(1): 1–4

Primate Research Institute 2003 *Guide of the Care and Use of Laboratory Primates (Second Edition).* Primate Research Institute of Kyoto University: Kyoto, Japan

Pruetz JD and Bloomsmith MA 1992 Comparing two manipulable objects as enrichment for captive chimpanzees. *Animal Welfare 1*: 127–137

Rawlins J 2005 Stock tanks for yearlong primate enrichment. *Tech Talk 10*(3): 1–2

Reasinger DJ and Rogers JR 2001 Ideas of improving living conditions of nonhuman primates by improving cage design. *Contemporary Topics in Laboratory Animal Science 40*(4): 89

Reinhardt V 1989 Evaluation of the long-term effectiveness of two environmental enrichment objects for singly caged rhesus macaques. *Lab Animal 18*(6): 31–33

Reinhardt V 1990a Time budget of caged rhesus monkeys exposed to a companion, a PVC perch and a piece of wood for an extended time. *American Journal of Primatology 20*: 51–56

Reinhardt V 1990b Social enrichment for laboratory primates: A critical review. *Laboratory Primate Newsletter 29*(3): 7–11

Reinhardt V 1990c Avoiding undue stress: Catching individual animals in groups of rhesus monkeys. *Lab Animal 19*(6): 52–53

Reinhardt V 1990d Environmental enrichment program for caged stump-tailed macaques (*Macaca arctoides*). *Laboratory Primate Newsletter 29*(2): 10–11

Reinhardt V 1991a Group formation of previously single-caged adult rhesus macaques for the purpose of environmental enrichment. *Journal of Experimental Animal Science .34*: 110–115

Reinhardt V 1991b Training adult male rhesus monkeys to actively cooperate during in-homecage venipuncture. *Animal Technology 42*: 11–17

Reinhardt V 1992 Environmental enrichment branches that do not clog drains. *Laboratory Primate Newsletter 31* (2): 8

Reinhardt V 1993a Using the mesh ceiling as a food puzzle to encourage foraging behaviour in caged rhesus macaques (*Macaca mulatta*). *Animal Welfare 2*: 165–172

Reinhardt V 1993b Enticing nonhuman primates to forage for their standard biscuit ration. *Zoo Biology 12*: 307–312

Reinhardt V 1993c Promoting increased foraging behaviour in caged stumptailed macaques. *Folia Primatologica 61*: 47–51

Reinhardt V 1994a Caged rhesus macaques voluntarily work for ordinary food. *Primates 35*: 95–98

Reinhardt V 1994b Pair-housing rather than single-housing for laboratory rhesus macaques. *Journal of Medical Primatology 23*: 426–431

Reinhardt V 1994c Social enrichment for previously single-caged stumptail macaques. *Animal Technology 5*: 37–41

Reinhardt V 1997a The Wisconsin Gnawing Stick. *Animal Welfare Information Center Newsletter 7*(3–4): 11–12

Reinhardt V 1997b Lighting conditions for laboratory monkeys: Are they adequate? *Animal Welfare Information Center Newsletter 8*(2): 3–6

Reinhardt V 1998 Housing and handling of nonhuman primates. In: Bekoff M and Meaney C (eds) *Encyclopedia of Animal Rights and Animal Welfare* pp 217–222. Greenwood Press: Westport, CT

Reinhardt V 1999 Pair-housing overcomes self-biting behavior in macaques. *Laboratory Primate Newsletter 38*(1): 4

Reinhardt, V and Dodsworth, R 1989 *Facilitated Socialization of Previously Single Caged Adult Rhesus Macaques (Videotape with accompanying text)* Wisconsin Regional Primate Research Center: Madison, WI

Reinhardt V and Cowley D 1990 Training stumptailed monkeys to cooperate during in-homecage treatment. *Laboratory Primate Newsletter 29*(4): 9–10

Reinhardt V and Reinhardt A 1991 Impact of a privacy panel on the behavior of caged female rhesus monkeys living in pairs. *Journal of Experimental Animal Science 34*: 55–58

Reinhardt V and Cowley D 1992 In-homecage blood collection from conscious stump-tailed macaques. *Animal Welfare 1*: 249–255

Reinhardt V and Hurwitz S 1993 Evaluation of social enrichment for aged rhesus macaques. *Animal Technology 44*: 53–57

Reinhardt V and Reinhardt A 2002 Introduction. In: Reinhardt V and Reinhardt A (eds) *Comfortable Quarters for Laboratory Animals, Ninth Edition* ii-iv. Animal Welfare Institute: Washington, DC

Reinhardt V, Houser WD, Eisele S, Cowley D and Vertein R 1988 Behavior responses of unrelated rhesus monkey females paired for the purpose of environmental enrichment. *American Journal of Primatology 14*: 135–140

Reinhardt V, Liss C and Stevens C 1995 Restraint methods of laboratory nonhuman primates: A critical review. *Animal Welfare 4*: 221–238

Richmond J 2002 Refinement, reduction, and replacement of animal use for regulatory testing: Future improvements and implementation within the regulatory framework. *ILAR Journal 43*(Supplement): 63–68

Robbins DQ, Zwick H, Leedy M and Stearns G 1986 Acute restraint device for rhesus monkeys. *Laboratory Animal Science 36*: 68–70

Roberts RL, Roytburd LA and Newman JD 1999 Puzzle feeders and gum feeders as environmental enrichment for common marmosets. *Contemporary Topics in Laboratory Animal Science 38*(5): 27–31

Roberts SJ and Platt ML 2004 Pair-housing macaques with biomedical implants: a safe and practical alternative to single-housing. *American Journal of Primatology 62*(Supplement): 96–97

Roberts SJ and Platt ML 2005 Effects of isosexual pair-housing on biomedical implants and study participation in male macaques. *Contemporary Topics in Laboratory Animal Science 44*(5): 13–18

Rolland RM 1991 A prescription for psychological well-being. In: Novak MA and Petto AJ (eds) *Through the Looking Glass. Issues of Psychological Well-being in Captive Nonhuman Primates* pp 129–134. American Psychological Association: Washington DC

Rosenberg DP and Kesel ML 1994 Old-World monkeys. In: Rollin BE and Kesel ML (eds) *The Experimental Animal in Biomedical Research. Volume II, Care, Husbandry, and Well-Being - An Overview by Species* pp 457–483. CPR Press: Boca Raton, FL

Ross PW and Everitt JI 1988 A nylon ball device for primate environmental enrichment. *Laboratory Animal Science 38*(4): 481–483

Rowell TE 1967 A quantitative comparison of the behaviour of a wild and a caged baboon group. *Animal Behaviour 15*: 499–509

Russell JL, Taglialatela JP and Hopkins WD 2006 The use of positive reinforcement training in chimpanzees (*Pan troglodytes*) for voluntary presentation for IM injections. *American Journal of Primatology 68*(Supplement): 122

Russell WMS 2002 The ill-effects of uncomfortable quarters. In: Reinhardt V and Reinhardt A (eds) *Comfortable Quarters for Laboratory Animals, Ninth Edition* pp 1–5. Animal Welfare Institute: Washington, DC

Sackett D, Oswald M and Erwin J 1975 Aggression among captive female pigtail monkeys in all-female and harem groups. *Journal of Biological Psychology 17*: 17–20

Savane S 2008 Use of flashlights in Old World nonhuman primate health monitoring. *American Association for Laboratory Animal Science Meeting Official Program*: 103

Savastano G, Hanson A and McCann C 2003 The development of an operant conditioning training program for New World priamtes at the Bronx Zoo. *Journal of Applied Animal Welfare Science 6*: 247–261

Schapiro S. J. 2000 A few new developments in primate housing and husbandry. *Scandinavian Journal of Laboratory Animal Science 27*: 103–110

Schapiro SJ 2005 Chimpanzees used in research: Voluntary blood samples differ from anesthetized samples. *Animal Welfare Institute (AWI) Quarterly 54*(3): 15–16

Schapiro SJ and Bushong D 1994 Effects of enrichment on veterinary treatment of laboratory rhesus macaques (*Macaca mulatta*). *Animal Welfare 3*: 25–36

Schapiro SJ and Bloomsmith MA 1995 Behavioral effects of enrichment on singly-housed, yearling rhesus monkeys: An analysis including three enrichment conditions and a control group. *American Journal of Primatology 35*: 89–101

Schapiro SJ, Lee-Parritz DE, Taylor LL, Watson LM, Bloomsmith MA and Petto AJ 1994 Behavioral management of specific pathogen-free rhesus macaques: Group formation, reproduction, and parental competence. *Laboratory Animal Science 44*: 229–234

Schapiro SJ, Perlman JE, Thiele E and Lambeth S 2005 Training nonhuman primates to perform behaviors useful in biomedical research. *Lab Animal 34*(5): 37–42

Schapiro SJ, Laule G and Seelig D 2007 "Applied Behavior" panel discussion. *Journal of Applied Animal Welfare Science 10*: 79–81

Schnell CR and Gerber P 1997 Training and remote monitoring of cardiovascular parameters in nonhuman primates. *Primate Report 49*: 61–70

Schultz P 2006 I see myself. *AWI Quarterly 55*(3): 6

Schwenk B 1992 Bungee jumping monkeys. *Animal Keepers' Forum 19*(12): 437–438

Schwindaman D 1991 The 1985 animal welfare act amendments. In: Novak MA and Petto AJ (eds) *Through the Looking Glass. Issues of Psychological Well-being in Captive Nonhuman Primates* pp 26–32. American Psychological Association: Washington DC

Scientists Center for Animal Welfare 1987 Consensus recommendations on effective Institutional Animal Care and Use Committees. *Laboratory Animal Science 37*: 11–13

Segerson L and Laule GE 1995 Initiating a training program with gorillas at the North Carolina Zoological Park. *American Zoo and Aquarium Association (AZA) Annual Conference Proceedings* : 488–489

Shefferly N, Fritz J and Howell S 1993 Toys as environmental enrichment for captive juvenile chimpanzees (*Pan troglodytes). Laboratory Primate Newsletter 32*(2): 7–9

Shideler SE, Savage A, Ortuño AM, Moorman EA and Lasley BL 1994 Monitoring female reproductive function by measurement of fecal estrogen and progesterone metabolites in the white-faced saki (*Pithecia pithecia). American Journal of Primatology 32*: 95–108

Shimoji M, Bowers CL and Crockett CM 1993 Initial response to introduction of a PVC perch by singly caged *Macaca fascicularis. Laboratory Primate Newsletter 32*(4): 8–11

Shively CA 2001 *Psychological Well-Being of Laboratory Primates at Oregon Regional Primate Research Center.* Willamette Week Online, March 21, 2001: Portland, OR

Skoumbourdis EK 2008 Pole-and-collar-and-chair training. *Laboratory Animal Refinement & Enrichment Forum* [electronic discussion group]: January 24, 2008

Smith TE, McCallister JM, Gordon SJ and Whittikar M 2004 Quantitative data on training new world primates to urinate. *American Journal of Primatology 64*(1): 83–93

Snowdon CT, Savage A and McConnell PB 1985 A breeding colony of cotton-top tamarins (*Saguinus oedipus*). *Laboratory Animal Science 35*: 477–480

Southwick CH 1967 An experimental study of intragroup agonistic behaviour in rhesus monkeys *(Macaca mulatta). Behaviour 28*: 182–209

Spector MR and Bennett BT 1988 The use of naturally occurring manipulanda can reduce the frequency of cage stereotypy in solitary-housed primates. *The Psychological Well-Being of Captive Primates Conference*: 44–45

Spragg SDS 1940 Morphine addiction in chimpanzees. *Comparative Psychology Monographs 15*: 1–132

Stahl D, Herrmann F and Kaumanns W 2001 Group formation of a captive all-male group of lion-tailed macaques (*Macaca silenus*). *Primate Report 59*: 93–108

Stringfield CE and McNary JK 1998 Operant conditioning for diabetic primates to accept insulin injections. *AAZV/AAWV Joint Conference Proceedings*: 396–397

Taylor TD 2002 Feeding enrichment for red-handed tamarins. *The Shape of Enrichment 11*(2): 1–3

Taylor WJ, Brown DA, Lucas-Awad J and Laudenslager ML 1997 Response to temporally distributed feeding schedules in a group of bonnet macaques (*Macaca radiata*). *Laboratory Primate Newsletter 36*(3): 1–3

Tiefenbacher S, Lee B, Meyer JS and Spealman RD 2003 Noninvasive technique for the repeated sampling of salivary free cortisol in awake, unrestrained squirrel monkeys. *American Journal of Primatology 60*: 69–75

Truelove M 2009 Social housing of nonhuman primates with cranial implants: A discussion. *Laboratory Primate Newsletter 48*(2): 1–2

Turkkan JS, Ator NA, Brady JV and Craven KA 1989 Beyond chronic catheterization in laboratory primates. In: Segal EF (ed) *Housing, Care and Psychological Wellbeing of Captive and Laboratory Primates* pp 305–322. Noyes Publications: Park Ridge, NJ

United States Department of Agriculture 1991 Title 9, CFR (Code of Federal Register), Part 3. Animal Welfare; Standards; Final Rule. *Federal Register 56*(32): 6426–6505

University of Arizona - IACUC Certification Coordinator 2008 Restraint. IACUC Learning Module - Primates [Web site]: Accessed 02/11/2008

University of Minnesota - Investigators, and Animal Husbandry and Veterinary Staff 2008 Nonhuman Primates. Restraint and Handling of Animals [Web site]: Accessed 02/11/2008

Valerio, DA, Miller, RL, Innes, JRM, Courntey, KD, Pallotta, AJ and Guttmacher, RM 1969 *Macaca mulatta. Management of a Laboratory Breeding Colony*. Academic Press: New York, NY

Vertein R and Reinhardt V 1989 Training female rhesus monkeys to cooperate during in-homecage venipuncture. *Laboratory Primate Newsletter 28*(2): 1–3

Vertein R and Reinhardt V 1993 Empirical use of liquid supplemental nutrition for aged macaques. *Laboratory Primate Newsletter 32*(1): 3

Videan EN, Fritz J, Schwandt ML, Smith HF and Howell S 2005a Controllability in environmental enrichment for captive chimpanzees (*Pan troglodytes*). *Journal of Applied Animal Welfare Science 8*: 117–130

Videan EN, Fritz J, Murphy J, Borman R, Smith HF and Howell S 2005b Training captive chimpanzees to cooperate for an anesthetic injection. *Lab Animal 34*(5): 43–48

Visalberghi E and Anderson JR 1993 Reasons and risks associated with manipulating captive primates' social environments. *Animal Welfare 2*: 3–15

Visalberghi E and Anderson JR 1999 Capuchin Monkeys. In: Poole T and English P (eds) *The UFAW Handbook on the Care and Management of Laboratory Animals Seventh Edition* pp 601–610. Blackwell Science: Oxford, UK

Washburn DA and Rumbaugh DM 1992 Investigations of rhesus monkey video-task performance: Evidence for enrichment. *Contemporary Topics in Laboratory Animal Science 31*(5): 6–10

Waugh, C 2002 Coconuts as enrichment item for macaques. *Primate Enrichment Forum (electronic discussion group)*: October 24, 2002

Weick BG, Perkins SE, Burnett DE, Rice TR and Staley EC 1991 Environmental enrichment objects and singly housed rhesus monkeys: Individual preferences and the restoration of novelty. *Contemporary Topics in Laboratory Animal Science 30*(5): 18

Westlund K 2002 Preference of the vertical dimension of cyno pairs living in high cages. *Laboratory Animal Refinement and Enrichment Forum* [electronic discussion group]: November 28, 2002

Whitney RA, Johnson, DJ and Cole, WC 1973 *Laboratory Primate Handbook*. Academic Press: New York, NY

Wickings EJ and Nieschlag E 1980 Pituitary response to LRH and TRH stimulation and peripheral steroid hormones in conscious and anaesthetized adult male rhesus monkeys (*Macaca mulatta*). *Acta Endocrinologica 93*: 287–293

Williams LE, Palughi PJ, Cushman A and Gibson SV 1992 Vegetables as dietary enrichment for *Saimiri*. *American Journal of Primatology 27*: 63–64

Winterborn A 2007 Cooperation counts. *AWI Quarterly 56*(3): 16

Wolfensohn SE and Lloyd M 1994 *Handbook of Laboratory Animal Management and Welfare*. Oxford University Press: New York, NY

Wolfle TL 1987 Control of stress using non-drug approaches. *Journal of the American Veterinary Medical Association 191*: 1219–1221

Woodbeck T and Reinhardt V 1991 Perch use by *Macaca mulatta* in relation to cage location. *Laboratory Primate Newsletter 30*(4): 11–12

Yerkes RM and Yerkes AW 1929 *The Great Apes. A Study of Anthropoid Life*. Yale University Press: New Haven

Extraneous Variables

Viktor Reinhardt, DVM, PhD

"A GOOD MANAGEMENT PROGRAM [for animals in biomedical research institutions] provides the environment, housing, and care that ... minimizes variations that can affect research" (National Research Council, 1996, p 21). Indeed, it would be naïve to rely on data collected from animals who experience fear and anxiety when they are approached by an investigator, from animals who suffer from depression and frustration resulting from the inability to show species-typical behaviors, and from animals who experience fear and anxiety resulting from involuntary restraint during procedures. In the context of biomedical research, these distressing experiences constitute unaccounted-for variables that affect the animals' physiological homeostasis. Sound scientific methodology requires that research data are not influenced by uncontrolled variables. Unless this requirement is met, an experiment is not considered scientifically valid (American Medical Association, 1992).

This chapter reviews common extraneous variables, discusses their potential impact on research data, and elaborates briefly on refinement options that minimize, eliminate, or avoid them.

Investigator

It seems to be a widely accepted practice among biomedical investigators to *use* animals as tools to promote their professional careers (Rollin, 1995). For the presumed sake of objectivity, animals are regarded as standardized test objects/models/systems rather than sentient beings endowed with feelings (Hummer, 1965; Arluke, 1988).

It is not uncommon that investigators have little or no direct contact with the animals of their research programs (Traystman, 1987; Arluke, 1993; Herzog, 2002; Baumans et al., 2007; Reinhardt, 2013); they are not aware when uncontrolled variables related to animal husbandry, care and handling are affecting the data obtained (Reese, 1991; Öbrink & Rehbinder, 1999; Baldwin, et al., 2007) and, therefore, require an unnecessarily large number of animals in order to achieve statistical significance of the research results (Russell & Burch, 1959; Home Office, 1989; Brockway et al., 1993; Baldwin et al., 2007; Reinhardt, 2013). Many investigators do not realize the influence of environmental variables on experimental results or at least do not adequately describe the environmental history of the animals used for experimentation (Davis et al., 1973; Claassen, 1994; Reinhardt & Reinhardt, 2000a; Reinhardt & Reinhardt, 2000b).

When handling animals during a research procedure, many investigators do it with little or no consideration that these animals experience anxiety, fear and distress—feelings that change their physiological equilibrium and, hence, influence the research data collected from them. Not giving much thought about how an animal feels during an uncomfortable and life-threatening procedure, it is perhaps not surprising that investigators often fail to describe how the animals were handled and/or how they reacted during the data collection procedure, even if stress-dependent parameters were measured (Reinhardt & Reinhardt, 2000a).

As a result of adverse conditioning, animals in research laboratories often show intense fear and physiological stress reactions when the researcher enters their room (Tatoyan & Cherkovich, 1972; Döhler et al., 1977; Manuck et al., 1983; Schnell & Wood, 1993; Späni et al., 2003; Reinhardt, 2013). The investigator may be completely unaware that he or she constitutes an overlooked extraneous variable in the research endeavor.

"Whether an investigator maintains a high personal respect for the well-being of the individual animal or holds classic concepts of animals as being experimental models, it should be more widely recognized that there is typically a scientific necessity to have animals at ease with their environments if studies are to remain objective" (Warwick, 1990, p 363). "Stressed animals do not make good research subjects" (American Medical Association, 1992, p 18).

Noise

Animals in research facilities are very often exposed to artificial noise without the option of retreating to a quiet location within their living quarters.

The noise environment of animals in research labs is usually not mentioned in the methodology section of scientific articles, even though it is likely to have important implications for the validity of data collected from such animals (Gamble, 1979; Clough, 1982; Milligan et al., 1993; Claassen, 1994). Noise as a potential data-confounding variable and a potential animal welfare concern has been given little consideration in the published literature. The few studies published do suggest that common sources of noise in the research laboratory are stressors that activate the hypothalamic-pituitary-adrenal axis.

Construction noise is an uncontrolled stressor for animals confined in cages. Mice and rats show a significant increase of plasma ACTH and corticosterone secretion along with a disruption of energy balance when they are exposed to construction noise and associated vibrations (Dallman et al., 1999; Raff et al., 2011). Exposure to construction noise can decrease the reproductive efficiency of guinea pigs (Anthony & Harclerode, 1959), mice (Rasmussen et al., 2009; Reinhardt, 2010), rats (Schipper et al., 2011), rabbits (Reinhardt, 2010) and zebrafish (Reinhardt, 2010).

The banging of metal cages in animal rooms can produce bursts of intense noise, triggering a significant endocrine stress response not only in rats (Barrett & Stockham, 1963) but probably also in other animal species found in research labs. Anecdotal evidence indicates that long-tailed and rhesus macaques get very agitated and resort to behavioral pathologies such as hair-pulling and self-biting when noisy construction work occurs nearby (Reinhardt, 2010). Marmosets show a significant rise in saliva cortisol concentrations during periods of loud noise caused by routine human activities (Cross et al., 2004) or by noisy construction work (Pines et al., 2004). A significant increase in fecal cortisol concentration was noticed in long-tailed macaques who were exposed to construction noise (Westlund et al., 2012).

Investigators may be unaware that exposure to common noises in animal holding areas affects the physiological equilibrium of animals and hence the quality of research data obtained from them (Jain & Baldwin, 2003; Baldwin et al., 2007). Noise in research animal facilities could be buffered with readily available industrial and architectural sound absorbing panels (Carlton & Richards, 2002; Jeans et al., 2008) along with sound buffering epoxy flooring (Johnson et al., 2005); noise levels could be reduced with considerate and correct working methods (Voipio et al., 2006); some noise sources—such as loud radio, loud talking, squeaky carts and squeaky doors—could be eliminated altogether.

Understimulation

The living quarters of animals in research facilities are often intentionally not provisioned with species-adequate stimuli, so that they are the same for all research subjects. The easiest way to create such *standardized* living quarters is to keep them empty of anything that is not necessary for the tenants' survival. It is not correct that "barren environments *may* not meet the species-specific needs of an animal [emphasis added]" (Committee on Recognition and Alleviation of Distress in Laboratory Animals, National Research Council, 2008, p 55); barren living quarters *cannot* meet these needs.

Animals kept in an impoverished environment with no or insufficient opportunities to perform genetically programmed, species-typical behaviors are unlikely to be good models for research (Hockly et al., 2002; Baldwin, 2010) because their physiological and behavioral response to chronic understimulation (boredom) will not be uniform. Some will cope reasonably well, others will get so frustrated that they get distressed and develop behavioral pathologies. A boring environment, therefore, bears the risk that not all subjects will respond to a test situation in a normative manner.

The most serious condition of understimulation is single housing of gregarious animals. In mice, distress resulting from social deprivation can decrease their resistance to spontaneously developing tumors (Andervont, 1944; Muhlbock, 1951), increase the variance of sensitivity to drugs (Mackintosh, 1962) and toxic agents (Hatch et al., 1965), alter immunological responses, brain neurochemistry, learning ability, and pain thresholds (Valzelli, 1973), produce abnormal behaviors and changes in body weight, lead to alterations of organ weights, blood cell counts, and adrenal function (Baer 1971), increase heart rate (Einstein et al., 2000), and disrupt the normal circadian sleep pattern (Späni et al. 2003). Individually caged rats show a compromised immuncompetence and, hence, an impaired resistance to disease (Ader & Friedman, 1964); they also exhibit a greater incidence of stereotypical tail manipulation and pawing (Baenninger, 1967; Hurst et al., 1997), a higher heart rate and blood pressure (Sharp et al., 2002a), higher levels of corticosterone and prolactin (Cambardella et al., 1994), higher variations in certain biochemical parameters (Pérez et al., 1997) and a lower survival rate than rats housed in groups (Shaw & Gallagher, 1984). When they are housed alone, guinea pigs markedly lose weight and reduce their water intake (Fenske, 1992).

Rabbits are prone to develop serious behavioral pathologies when housed alone (Gunn & Morton, 1995; Krohn et al., 1999; Held et al., 2001). Trichophagy is one such behavioral disorder that often results in the formation of gastric trichobezoars (Wagner et al., 1974), a clinical problem that may lead to an alarmingly high incidence of mortality associated with intestinal stasis (Jackson, 1991).

Single housing is likely to be stressful for dogs and is associated with an increased incidence of bizarre movements and barking (Hetts et al., 1992; Hubrecht et al., 1992).

In cats, single housing can result in extreme boredom manifesting in pathological behaviors such as psychogenic alopecia and polyphagia with resultant obesity (Buffington, 1991; DeLuca & Kranda, 1992).

When sheep are kept alone, they are prone to developing stereotypical pulling and chewing of wool and gnawing at wooden structures of their enclosure. "The nature and extent of the abnormal behaviour found may be indicative of a differing background physiological state"

(Marsden & Wood-Gush, 1986, p 159) which can make the extrapolation of experimental results rather problematic (Done-Currie et al., 1984). Single-housed sheep show an increase in heart, respiration and metabolic rates (Baldock & Sibly, 1990; Van Adrichem, 1993; Cockram et al. 1994; Carbajal & Orihuela, 2001; McLean & Swanson, 2004), plus an increase in adrenal and thyroid activity (Bobek et al., 1986; Bowers et al., 1993).

In goats, plasma norephinephrine levels increase when they are kept without contact with other goats (Carbonaro et al., 1992).

The deleterious effect of social deprivation is particularly pronounced in nonhuman primates. It is, therefore, a legislative imperative in the United States that:

> *Research facilities must develop, document, and follow an appropriate plan for environmental enhancement. … The plan must include specific provisions to address the social needs of nonhuman primates* (United States Department of Agriculture, 2002, p 94).

European regulations also make it clear that: "Because the common laboratory non-human primates are social animals, they should be housed with one or more compatible conspecifics. … Single housing should only occur if there is justification on veterinary or welfare grounds. Single housing on experimental grounds should be determined in consultation with the animal technician and with the competent person charged with advisory duties in relation to the well-being of the animals (Council of Europe, 2006, p 44 & 14). In rhesus macaques, the species most commonly found in laboratories, single caging can produce "long-term features of immunosuppression and significant increases in plasma prolactin concentrations, indicative of stress-induced anxiety" (Lilly et al., 1999, p 197). A major problem among individually caged primates is apathy and depression (Erwin & Deni, 1979). Clinical records and immunological data indicate that single-caged primates are more susceptible to health problems than animals living in compatible social settings (Shively et al., 1989; Schapiro & Bushong, 1994; Schapiro et al., 1997). Behavioral assessments of individually caged rhesus macaques of a Primate Research Center revealed that "of the 362 animals surveyed, 321 [88.7%] exhibited at least one abnormal behavior" (Lutz et al., 2003, p 1). Many single-caged individuals spend more than 20% of the day engrossed in abnormal behaviors (Bayne et al., 1991; Bayne et al., 1992; Kessel & Brent, 1996; Bellanca et al., 1999; Bourgeois & Brent, 2003). A particularly alarming abnormal behavior is self-injurious biting. "Monkeys with SIB [self-injurious behavior] bite their own bodies frequently, occasionally inflicting wounds" (Novak et al., 1998, p 213) that can be life-threatening (Author's unpublished observations). "Research has shown that approximately 10% of captive, individually housed monkeys have had some veterinary record of self-injurious behavior within their life-time" (Jorgensen et al., 1998, p 187). In humans, self-injurious behavior is classified as a major psychotic disorder that occurs not only in mentally handicapped individuals (Simeon et al., 1992) but also in socially isolated prisoners (Yaroshevsky, 1975). In large colonies of single-caged rhesus macaques, the incidence of this behavioral pathology may be as high as 14% (Novak, 2003) or even 39% (Alexander & Fontenot, 2003).

It is not necessary to keep animals in barren living quarters in research facilities. The provision of species-appropriate stimuli (environmental enrichment) "may reduce variability between animals and produce animals that are better models of normal function" (Garner, 2002, p 95).

Inanimate enrichment does not cure animals of behavioral pathologies but it temporarily reduces the frequency of their occurrence during the time the animals are actively engaged with the enrichment. The distraction derived from the availability of the enrichment is likely to ameliorate the overall stress attendant to boredom. Rats kept in enriched cages show longer durations of sleep and lower levels of agonistic behavior (Abou-Ismail et al., 2010), they are less timid and less afraid (Denenberg & Morton, 1962; Klein et al., 1994; Eskola & Kaliste-Korhonen, 1998) and show significantly lower baseline adrenocorticotropic hormone and corticosterone concentrations compared to rats housed without enrichment (Belz et al., 2003). Mice kept in enriched versus barren standard cages are less emotional/fearful/excitable (Manosevitz & Joel, 1973; Chamove, 1989; Scharmann, 1994; Roy et al., 2001; Van de Weerd et al. 2002; Benaroya-Milshtein et al., 2004; Õkva et al., 2010). They show lower plasma corticosterone levels (Hennesy & Foy, 1987; Roy et al., 2001) and a less variable, better regulated immune response (Kingston & Hoffman-Goetz, 1996), and are less aggressive among each other (Ambrose & Morton, 2000). Access to appropriate inanimate enrichment can diminish aggressive interactions among hamsters (Arnold & Westbrook, 1997/1998), rabbits (Mis & Warren, 2003), chickens (Yasutomi & Adachi, 1987; Gvaryahu et al., 1994; Baroli et al., 1997), pigs (Schaefer et al., 1990; Blackshaw et al., 1997; Durrell et al., 1997; O'Connell & Bettie, 1999) and primates (Neveu, 1994; Nakamichi & Asanuma, 1998; Honess & Marin, 2006). In chickens, environmental enrichment also reduces the incidence of cannibalism (Yasutomi & Adachi, 1987). In mice (Engellenner et al., 1982; Van de Weerd et al., 2002), rats (Deacon, 2001; Morrison, 2001), hamsters (Arnold & Gillaspy, 1994), rabbits (Mis & Warren, 2003), pigs (Grandin, 1986; Pearce et al., 1989; Moore et al., 1994; Rodarte et al., 2004) and chickens (Brake, 1987; Reed et al., 1993) proper enrichment buffers the animals' fear and aggressive defense response to personnel, thereby diminishing stress reactions to being captured for procedures.

Social enrichment in the form of compatible conspecific companionship avoids physiological, stress-related imbalances in single-caged gregarious animals; this has been demonstrated in mice (Goldsmith et al., 1976), rats (Ader & Friedman, 1964; Gardiner & Bennet, 1978; Fagin et al., 1983; Cambardella et al., 1994; Baldwin et al., 1995; Nyska et al., 1998), squirrel monkeys (Gonzalez et al., 1982), baboons (Coelho et al., 1991), rhesus macaques (Schaprio et al., 2000a; Doyle et al., 2008) and chimpanzees (Reimers et al., 2007). Social companionship also attenuates behavioral disorders. It actually prevents the development of stereotypies and trichophagy in rabbits (Poderscek et al., 1991; Love, 1994; Krohn et al., 1999; Held et al., 2001) and ameliorates or eliminates the behavioral pathology of self-biting (Fritz, 1989; Line et al., 1990; Reinhardt, 1999; Weed et al., 2003) as well as self-directed hair pulling (Hartner et al., 2000; Reinhardt, 2010) in chimpanzees and macaques.

Optional visual seclusion is an essential environmental enrichment/refinement, fostering the well-being and physiological equilibrium of confined animals. Individually housed animals may seek seclusion for undisturbed resting, while socially housed animals may seek cover to diffuse social tension and avoid aggressive conflicts. Access to a covered retreat area is particularly important during alarming situations—for example, when fear-inducing personnel enter the room. Furnishing the tanks of frogs with hollow structures for hiding significantly decreases both aggression and mortality rate (Hedge et al., 2002; Torreilles & Green, 2007); it also diminishes startle responses, suggesting that the frogs get less stressed when personnel are around (Archard, 2012). In groups of male mice, access to soft paper material decreases fecal corticosterone levels and reduces aggressive interactions by offering subordinates cover and escape routes (Armstrong et al., 1998; Van Loo et al., 2002; Niu et al., 2011). Pair-housed male guinea pigs show significantly lower fecal cortisol concentration when they are provided with a hut serving as a buffer against social tension (Walters et al., 2012). Common rats with access to shelters are less timid and engage in stereotyped backflipping less often than those in barren cages (Townsend, 1997; Callard et al., 2000). Cotton rats become less aggressive among each other and against personnel when they are provisioned with tubes serving them as refuge places (Neubauer & Buckmaster, 2011). Hamsters often develop bizarre aggressive behaviors when housed individually in suspended wire cages. Providing a piece of polyvinyl chloride pipe as a place for seclusion can resolve this problem (McClure & Thomson, 1992). Aggression is reduced among rabbits when sections of their pens are screened, tubes are placed on the floor or shelves installed so that the animals can withdraw or escape from each other as needed (Howard et al., 1999; Stauffacher, 2000). Visual barriers reduce aggression among pigs (Waran & Broom, 1993), cattle (Bouissou, 1970) and primates (Erwin, 1977; Reinhardt & Reinhardt, 1991; Maninger et al., 1998; McCormack & Megna, 2001). Placement of vertical panels in the middle of the enclosure assures that chickens—and presumably also rodents—show less disturbance reactions and make use of the available floor space more evenly by no longer aggregating at the peripheral walls and shunning the otherwise unprotected central area (Newberry & Shackleton, 1997; Cornetto & Estevez, 2001; Cornetto et al., 2002).

Music and/or talk radio: Attending care personnel often listen to music or talk radio while doing their routine work in the animal rooms. The type of music and the type of talk radio listened to will differ from person to person, and there will be some caregivers who do their work in silence.

Music can have a calming effect in dogs, as exposure to new age music decreases the amount and intensity of their barking (Kilcullen-Steiner & Mitchell, 2001). Radio music can make laying hens more productive (Jones & Rayner, 1999), suggesting that music affects their endocrine system.

Exposure to gentle music has a calming effect on mice (Chikahisa et al., 2007; Li et al., 2010) and boost their immune system (Núñez et al., 2002).

Rats are able to distinguish between different radio sound patterns; they show a clear preference for silence to anything else, which may be taken as an indication that they feel disturbed by the sound from the radio (Krohn et al., 2011). Hearing music modifies their cardiovascular functions by either increasing or decreasing heart rate and blood pressure, depending on the tempo, rhythm, pitch and tonality of the music (Lemmer, 2008; Akiyama & Sutoo, 2011).

Low-volume music is likely to reduce the intensity of the startle response that rabbits typically exhibit when fear-inducing personnel enter their room (Reinhardt, 2010). Music can have a very calming effect on pigs; they will quietly lie in body contact with each other and barely move when personnel enter their room (Reinhardt, 2013).

In primates, certain types of music have no noticeable effect (e.g., harp music, vervets: Hinds et al., 2007), or have a calming effect as reflected in lowered heart rate and increased resting behavior (e.g., oldies, baboons and chimpanzees: Brent & Weaver, 1996; Howell et al., 2003), while other types of music can make the animals more restless (e.g., high-beat, chimpanzees: Harvey et al., 2000). A study with tamarins and marmosets showed that the animals prefer slow-tempo to fast-tempo music, but when allowed to choose between slow-tempo music and silence, they prefer silence (McDermott & Hauser, 2007). Salivary cortisol concentrations doubled in marmosets following 30 minutes of exposure to radio music/talk (Pines et al., 2004), suggesting that the radio is a source of stress rather than entertainment for these animals.

Music and/or talk radio may be a pleasant distraction for attending animal care personnel (Reinhardt, 2010) but a stressful disturbance for the confined animals, especially those who are nocturnal (e.g., mice and rats) and want to sleep during the day. Before animals are exposed to regular noise from radios, a simple preference test should be conducted to find out if the animals actually prefer the noise over silence. If they prefer silence, no music and/or radio talk should be played in their holding areas, not only for the sake of animal well-being but also for the sake of sound scientific methodology.

Removal from the home environment

Animals in research institutions are regularly caught and transferred to treatment, test or experimental areas for procedures that entail involuntary restraint. The stress/distress resulting from enforced restraint is augmented by the stress resulting from the transfer to an unfamiliar treatment area. For example, in rats and mice, even moving animals in their familiar cages to an unfamiliar room in the same building increases corticosteroid levels for several hours (Kvetnansky et al., 1978; Ursin & Murison, 1986; Drozdowicz et al., 1990; Tuli et al., 1995a). "Exposure to a new environment, or novelty, may contribute significantly to the adrenocortical response often attributed to the effects of noxious or painful stimulation" (Friedman & Ader, 1967). Dobrakovavá & Jurcovicova (1984) tried to adapt male Wistar rats, caged in groups of four, to being transferred in their home cage to another room, stay there for 10 minutes, and then return to the original location. This procedure was repeated once a day for a 15-day period. The animals showed a significant increase in plasma corticosterone and prolactin concentration in response to the transfer procedure on day 15, to levels that were not lower than those observed on day 1.

> Wherever possible, every effort should be made to design in ways that bring the treatment to the animal, instead of the reverse. Removal for any purpose exposes the animal to overly novel, frequently noxious, and always stressful stimuli (Lindburg & Coe, 1995, p 565).

Novelty of the environment activates the pituitary-adrenal axis not only in rodents but also in nonhuman primates (Friedman & Ader, 1967; Brown & Martin, 1974; Pfister & King, 1976; Line et al., 1987; Cabib et al., 1990). "Removing an animal from its home cage prior to monitoring anything biological will probably affect the event being monitored" (Mitchell & Gomber, 1976, p 546). Single-caged rhesus macaques show an increase in self-biting behavior, evidence of sleep disturbance, and elevated cortisol levels in saliva and serum after being removed from their familiar living quarters to unfamiliar living quarters (Davenport et al., 2008). The magnitude of cortisol response to blood collection is significant when the procedure occurs in the hallways but not when it occurs in the home cage (Herndon et al., 1984; Reinhardt et al., 1990; Reinhardt et al., 1991). Macaques show significantly higher cortisol and catecholamine levels when they are chair-restrained in an unfamiliar versus a familiar environment (Mason 1972; Mason et al., 1973).

Investigators often fail to mention in scientific articles if their research subjects were removed or if they were allowed to stay in the familiar home environment during procedures (Reinhardt & Reinhardt, 2000b). To ignore this variable while ascertaining *basal* or *normal* values of stress-sensitive parameters would contravene basic scientific rules.

Removal from social companion(s)

Experiments and tests are often a source of stress or distress; in addition, they usually entail the subject being removed not only from the familiar home environment but also from familiar conspecifics. Any gregarious animal who is unwillingly removed from familiar social companions experiences separation stress/anxiety/depression, accompanied by significant physiological and biochemical reactions that are bound to influence subsequently collected physiological data (*rats*: Ehlers et al., 1993; *mice*: Pibiri et al., 2008; *pigs*: Ruis et al., 1997; *sheep*: Apple et al., 1993; *cattle*: Hopster & Blokhuis, 1993; *chickens*: Jones & Merry, 1988; *primates*: Pearson et al., 2008).

There is ample evidence indicating that the presence of one or several companions has a stress-buffering effect, reducing both the magnitude and frequency of physiological stress reactions to aversive circumstances. When rodents are exposed to a stressful situation, physiological stress reactivity is buffered by the presence of another conspecific (*rats*: Davitz & Mason, 1955; Conger, 1957; Latané & Glass 1968; Taylor 1981; Giralt & Armario, 1989; Heath, 1999; de Jong et al., 2005; *mice*: Goldsmith et al., 1978; Van Loo et al., 2007; *guinea pigs*: Sachser et al., 1998; Kaiser et al., 2003; Hennessy et al., 2006; *hamsters*: Detillion et al., 2004).

In ruminants, the stress associated with experimental conditions is buffered by a familiar conspecific (*sheep*: Pearson & Mellor, 1976; Baldock & Sibly, 1990; Fraser, 1995; *cattle*: Veissier & Le Neindre, 1992).

In nonhuman primates, the presence of a companion reduces signs of behavioral disturbance and the magnitude of cortisol increase during a fear-provoking situation (Vogt et al., 1981; Coe et al., 1982; Gonzalez et al., 1982; Stanton et al., 1985). Individuals recover from the stress of being transferred to a novel environment significantly faster when a companion is present than when they are alone (Gust et al., 1994). Being tethered during an experiment is an extremely disturbing situation (Kaplan et al., 1983; Adams et al., 1988; Crockett et al., 1993). The cardiovascular stress response is significantly lower when tethered subjects are allowed to keep visual, tactile and auditory contact with other conspecifics than when they are kept alone (Coelho et al., 1991). "Social stimuli may function as a source of security and a means of mitigating emotional distress" (Mason, 1960, p 110).

For common husbandry- and research-related procedures there is often no need to remove animals from their familiar living quarters and separate them from social companions. This applies particularly to nonhuman primates.

It has been documented that socially housed marmosets readily learn to (a) step, one at a time, on a scale and remain still during weighing (McKinley et al., 2003) and (b) cooperate during urine sample collection (Anzenberger & Gossweiler, 1993; McKinley et al., 2003). Group-housed tamarins, vervet monkeys, and chimpanzees can also be trained to reliably produce urine samples in their home enclosure (Kelley & Bramblett, 1981; Snowdon et al., 1985; Stone et al., 1994; Lambeth et al., 2000). Chimpanzees living in a social setting can be trained to provide semen samples, present for subcutaneous and intramuscular injection, and cooperate during blood collection without needing to be removed from companions (Perlman et al., 2003; Schapiro et al., 2005; Coleman et al., 2008). Group-housed stump-tailed macaques can readily learn to present for vaginal swabbing within their home enclosure (Bunyak et al., 1982). Pair-housed stump-tailed macaques learn easily to cooperate during topical drug application in their home cages (Reinhardt & Cowley, 1990). Pair-housed stump-tailed and pair-housed rhesus macaques have been trained to voluntarily present for blood collection in their familiar living quarters (Reinhardt et al., 2002). It has been shown in pair- and group-housed marmosets and baboons that saliva samples can be obtained without difficulties from individual animals without having to remove them from their group (Cross et al., 2004; Pearson et al., 2008).

Group-housed rats learn quickly to cooperate during oral drug delivery in their home cages (Huang-Brown & Guhad, 2002; Rourke & Pemberton, 2007).

Multi-tier caging

Small and medium-size animals are traditionally kept in rows of cages that are stacked on top of each other so that a maximum number of animals can be accommodated per room. This creates different living environments in the cages of different racks in terms of:
1. distance from the light source and
2. distance from the floor of the room.

A bright upper-row cage may provide a species-suitable environment for a diurnal arboreal animal such as a primate, but it would be unsuitable for a nocturnal terrestrial animal such as a rodent (Ader et al., 1991). Yet, many of the caged primates live in the crepuscular environment of lower rows while many of the caged rodents live in the bright environment of upper rows.

It should be noted here that US animal welfare regulations pertaining to dogs, cats, guinea pigs, hamsters, rabbits and primates explicitly stipulate that:

> *Lighting must be uniformly diffused throughout animal facilities* (United States Department of Agriculture, 2002, pp 58, 72, 80, 90).

This unequivocal legal requirement is a safeguard for reliable scientific methodology, as light influences physiological systems, metabolism, general activity, behavior and emotionality (Marshall, 1940; Ross et al., 1966; Wurtman, 1967; Weihe et al., 1969, Martinez, 1972; Hauntzinger & Piacsek, 1973; Vriend & Lauber, 1973; Weihe, 1976; Newton, 1978; Saltarelli & Coppola, 1979; Clough, 1984; Heger et al., 1986; Martin, 1991). "The illumination conditions existing around the animals in their cages need to be considered for each species and for a much wider range of functions than has previously been thought necessary" (Weihe, 1976, p 74). A difference in light quantity and light quality between locations on a cage rack constitutes an extraneous variable that has to be taken into account if the validity of scientific research methodology is to be warranted (Ott, 1974; Canadian Council on Animal Care, 1993). Surprisingly, it is rarely mentioned in scientific articles whether the animals of the research project were *all* caged at the *same* level of the room (cf., Davis et al., 1973; Claassen, 1994).

With multi-tier caging it is difficult to provide uniform lighting for all animals since upper rows will always cast shade on cages in lower rows. Not only the quantity but also the quality of light in lower rows differs from that in upper rows. The light that lower-row-caged animals are receiving is not direct but is reflected from the walls of the room, thereby changing its spectral distribution, depending on the colors of the walls. It is probably not an overstatement that:

> *The intensity of light in animal cages is likely to be the most variable environmental factor in the average animal room* (Clough, 1982, p 512).

"What we basically have done to date is to provide lighting suitable to our needs and assumed it was all right for the animal" (Bellhorn, 1980, p 441)—and for research. In standard multi-tier rodent cages and in standard double-tier primate cages, variation in light intensities far exceeds a two-fold difference between bottom- and top-row cages (Clough & Donnelly, 1984; Reinhardt & Reinhardt, 1999; Schapiro et al., 2000b).

The International Primatological Society (2007, p 12) notes that "A two-tiered system is not recommended as these cages are usually too small. The lower tiers do not allow primates to engage in their vertical flight response, are often darker, and animals in the lower cages tend to receive less attention from attending personnel." The bottom-row cages are often so dark that it is difficult to identify and check individual animals without the use of flashlights (Reinhardt, 1997; Reasinger & Rogers, 2001; Savane, 2008). This situation not only introduces an uncontrolled variable into research data but also undermines good housekeeping. Surprisingly, it is mentioned in primatological research articles only rarely—2% of 96 articles surveyed—whether the research subjects were housed at different levels or if all of them were housed at the same level of the multi-tier cage rack (Reinhardt & Reinhardt, 2000b).

Multi-tier caging systems also make it impossible to provide all animals of a room the same feeling of security when personnel approach them. Especially the investigator constitutes a potential stressor, often triggering pronounced physiological and behavioral disturbance in caged subjects (*mice*: Kramer et al., 2001; *rats*: Döhler et al., 1977; *monkeys*: Tatoyan & Cherkovich, 1972, Malinow et al., 1974; Manuck et al., 1983; Hassler et al., 1989; Line et al., 1989; Schnell & Wood, 1993; Bowers et al., 1998; Boinski et al., 1999; Crockett & Gough, 2002), presumably due to aversive conditioning (Robbins et al., 1986). Empirical evidence indicates that the fear response to personnel differs in animals kept in upper rows versus those kept in lower rows. In chickens, fear reactions are moderate when they are kept in bottom-row cages but intense when they are kept in upper-row cages (Sefton, 1976; Hemsworth et al., 1993). "Under natural conditions, many primates spend much of their lives above ground and escape upward to avoid terrestrial threats. Therefore, these animals might perceive the presence of humans above them as particularly threatening" (National Research Council, 1998, p 118). A monkey who is kept in a bottom-row cage is practically cornered when being approached by a person, while a monkey in an upper-row cage can retreat to a relatively safe place above the person. The different emotional reactions of upper-row versus lower-row caged animals are likely to impact differently the subjects' physiological responses during subsequent experiments.

Rotating cages through different positions on a rack (Ross & Everitt, 1988; Canadian Council on Animal Care, 1993; National Research Council, 1996) rotates the methodological problems arising from the multi-tier caging but it does not solve them. Cage-specific light sources could avoid differences in illumination (MacLean et al., 2009; Baumans et al., 2013) but not differences in cage distance from the floor of the room. Presently, there is no other alternative than single-tier caging that can provide all animals of a room a uniform housing environment, both in terms of distance from the light source and distance from the floor.

Restraint

It is common practice in research laboratories to restrain animals during drug administration and sample collection procedures (Wolfensohn & Lloyd, 1994; Fowler, 1995; Hrapkiewicz et al., 1998). Being caught and subsequently immobilized by a predator—such as a human—is a life-threatening, hence distressing experience for any animal, purpose-bred beagles probably being the exception.

> *Because restraint itself is a stressor that affects the physiological functioning of the subject, measurement error and variability are introduced into the data* (Brockway et al., 1993, p 57).

"Since the purpose of physiological recording should be to obtain a record that is an exact facsimile or analog of the events under investigation, stress induced by restraint and handling, even when these are of minor nature and performed by skilled staff, is one of the major problems encountered in biomedical investigations" (Schnell & Gerber, 1997, p 68). "It is only common sense ... that an animal will not respond normally if it is stressed" (Schwindaman, 1991, p 30).

Manual restraint is usually applied with rats and mice. The animals show a variety of stress reactions that may include:
» increase in adrenocortical activity (Tuli et al., 1995b),
» activation of adrenal-medullary discharge of epinephrine and a sympathetic neuronal release of norepinephrine (Kvetnansky et al., 1978),
» increase in heart rate (Harkin et al., 2002; Sharp et al., 2003),
» changes in plasma prolactin levels (Krulich et al., 1974; Lenox et al., 1980; Gala & Haisenleder, 1986),
» increase in plasma glucose level (Besch & Chou, 1971),
» increase in core temperature (Berkey et al., 1990; Harkin et al., 2002), and
» alterations of metabolism and excretion of drugs (Kissinger et al., 2001).

Even witnessing the restraint of another conspecific can be enough to trigger stress reactions in unrestrained animals (Pitman et al., 1988; Fuchs et al., 1987; Sharp et al., 2002b).

For hamsters, routine manual restraint is a stronger stressor—as measured in heart rate increase and core body temperature increase—than intruder confrontation, cage change, and group formation (Gattermann & Weinandy, 1996/97). Guinea pigs often exhibit signs of fear (struggling) and anxiety (squealing) when they are manually restrained for procedures (Author's unpublished observations). This suggests that they also experience stress-triggered physiological alterations when being restrained.

In sheep, restraint increases cortisol, lactate, and glutamic oxaloacetic transaminase (Apple et al., 1993).

In pigs, enforced restraint raises cortisol levels, induces prostaglandin-mediated hyperthermia, and affects the acid-base balance (Van de Wal et al., 1986; Parrott & Lloyd, 1995).

In ferrets, manual restraint results in a significant increase of plasma cortisol and adrenocorticotrophic hormone and a decrease of alpha-MSH (Schoemaker et al., 2003).

In a study with dogs, an acclimatization period of at least 4 weeks was required to notably reduce stress-related effects associated with periodic manual restraint during venipuncture. Epinephrine (adrenaline) values gradually declined, but they were still almost 50% above baseline by the end of the study on day 41 (Slaughter et al., 2002).

Cats do not like to be restrained and will take every opportunity to get free, even if this implies defensive aggression against the handling person (Author's unpublished observations). A cat's aversive reactions suggest that enforced restraint is a stressor that is likely to affect the animal's physiological equilibrium.

Involuntary restraint is an especially distressing experience for nonhuman primates. Subjects are restrained either manually or mechanically during brief sample collection or drug administration procedures; they are strapped into restraint chairs or on restraint crosses during long-term procedures (Fowler, 1995). Restrained individuals usually struggle, exhibit self-defensive aggression, and often squeal. Physiological reactions to brief restraint include:
  » increased respiration rate (Berendt & Williams, 1971),
  » metabolic acidosis (Manning et al., 1969),
  » increased heart rate (Osborne 1973; Line et al., 1991; Schnell & Wood, 1993),
  » increased blood pressure (Golub & Anderson, 1986; Schnell & Wood, 1993),
  » raised rectal temperature (Bush et al., 1977),
  » rise in SGO-T (serum glutamic-oxalacetic transaminase) (Cope & Polis, 1959),
  » rise in AST (aspartate aminotransferase) and ALT (alanine aminotransferase) (Landi et al., 1990),
  » increased plasma cortisol concentrations (Elvidge et al., 1976; Puri et al., 1981; Fuller et al.,1984; Suzuki et al., 2002),
  » leukocytosis (Ives & Dack, 1956; Loomis et al., 1980; Goosen et al., 1984),
  » increased plasma concentrations of adrenal androgens (Fuller et al., 1984),
  » elevation of plasma prolactin (Quadri et al., 1978),
  » increased glucagon levels (Myers et al., 1988),
  » impaired glucose clearance (Yasuda et al., 1988),
  » impaired testosterone release (Puri et al., 1981; Hayashi & Moberg, 1987; Torii et al., 1993),
  » baseline variability in growth hormone levels (Mason et al., 1968), and
  » alterations of the electrocorticogram (Bouyer et al., 1978).

Even after repeated (12 times) exposure to brief restraint in the familiar home cage, rhesus macaques continue to show a pronounced heart-rate response, indicating that they do not habituate to this common procedure (Line et al., 1991).

The sedative ketamine is often injected with the assumption that it will reduce the overall stress that primates experience during the most common procedure, namely blood collection (Laudenslager & Worlein, 2003). Traditionally, the subject is forcibly restrained during the injection, a circumstance that introduces restraint-stress as an extraneous variable even before the blood has been drawn (Aidara et al., 1981; Bentson et al., 2003). Ketamine sedates the animal for the subsequent blood collection but it does not modify the cortisol response to the initial enforced injection of the ketamine (Loomis et al., 1980; Puri et al., 1981; Fuller et al.1984; Crockett et al., 2000). It is questionable that reliable control values of stress-sensitive blood parameters can be obtained under such conditions. The statement that "simple procedures such as injections of relatively harmless substances and blood sampling ... are expected to produce little or no discomfort" (Scientists Center for Animal Welfare, 1987, p 12) is valid only under the condition that the subject is *not* forcibly restrained during these procedures.

> [Possibly data-influencing] *physiological, biochemical and hormonal changes occur in any restrained animal. ... Restraint procedures should [therefore] only be invoked after all other less stressful procedures have been rejected as alternatives* (Canadian Council on Animal Care, 1993, p 95).

Alternative "procedures that reduce reliance on forced restraint ... are less stressful for animals and staff, safer for both, and generally more efficient" (National Research Council, 1998, p 45).

Biotelemetry offers alternative means of obtaining physiological measurements from freely moving animals, without introducing stress artifacts resulting from restraint. This system consists of a radio transmitter placed in a jacket or implanted subcutaneously or intraperitoneally, depending on the species and size of the subject. It has been used successfully in mice, rats, gerbils, guinea pigs, hamsters, rabbits, dogs, cats, primates, chickens, goats, sheep, amphibians, reptiles and fishes to monitor activity, body temperature, heart rate, blood pressure, electrocardiogram, and electroencephalogram (Malinow et al., 1974; Laburn et al., 1992; Kramer et al., 1993; Schnell & Wood, 1993; Depasquale et al., 1994; Christian & Bedford, 1995; Sato et al., 1995; Truett & West, 1995; Colbourne et al., 1996; Dejardins et al., 1996; Savory & Kostal, 1996; Brown et al., 1997; Heybring et al., 1997; Seebacher & Alford, 2002; Bridger & Booth, 2003; Krohn et al., 2003; Morton et al., 2003; Van Ginneken et al., 2004).

Positive reinforcement training is an alternative to involuntary restraint during procedures that necessitate the direct handling of the subject.

> *The least distressing method of handling is to train the animal to co-operate in routine procedures. Advantage should be taken of the animal's ability to learn* (Home Office, 1989, p 18).

Training animals to cooperate rather than resist during procedures not only refines research methodology but also improves personnel safety; a cooperative animal who works *with* the handling person has no reason to resort to self-defensive aggression.

In rats and rabbits, the stress and risk associated with gastric intubation can be avoided by training the animals to accept and swallow test drugs (e.g., indomethacin, celecoxib, tosufloxacin) that are masked with chocolate or sucrose (Marr et al., 1993; Huang-Brown & Guhad, 2002). Eight of ten rabbits cooperated within 2 days. They "would stand with their paws on the front of the cages, protrude their faces from between the bars, and appear to beg for the syringe containing the antibiotic" (Marr et al., 1993, p 46). Rats have successfully been trained to cooperate during saliva collection in their familiar home environment (Guhad & Hau, 1996).

Sheep can easily be trained to voluntarily enter a tilt table and accept brief immobilization for a grain reward (Grandin, 1989). It takes pigs just 2 weeks of training before they voluntarily run down the hallway to get onto a platform scale, where they stand still and enjoy a food reward while their weights are being recorded (Neubauer et al., 2011). Göttinger minipigs learn easily to voluntarily come forward, step onto a box, and hold still during dosing in one nostril (Brodersen et al., 2010); they also learn to follow a target stick, walk onto a scale for weighing, and stand still for physical examinations, electrocardiography, and dermal dosing (Blye et al., 2006).

Primates readily learn to cooperate during injection (Spragg, 1940; Levison et al., 1964; Priest, 1991; Nelms et al., 2001; Bentson et al., 2003), blood collection (Wall et al., 1985; Hein et al., 1989; Priest, 1990; Reinhardt, 1991; Reinhardt & Cowley, 1992; Priest, 1998), saliva collection (Bettinger et al., 1998; Tiefenbacher et al., 2003), urine collection (Kelley & Bramblett, 1981; Anzenberger & Gossweiler, 1993; Schnell & Gerber, 1997; Lambeth et al., 2000; McKinley et al., 2003), vaginal swabbing (Bunyak et al., 1982), oral drug administration (Turkkan et al., 1989; Klaiber-Schuh & Welker, 1997; Schnell & Gerber, 1997; Crouthamel & Sackett, 2004; Reinhardt, 2013) and topical drug application (Reinhardt & Cowley, 1990). Trained subjects show no behavioral signs of fear or distress and the physiological stress response to the procedure is considerably reduced or eliminated altogether (Michael et al., 1974; Elvidge et al., 1976; Reinhardt & Cowley, 1992; Schnell & Gerber, 1997; Reinhardt, 2003; Bentson et al., 2003). Macaques can be trained to cooperate for pole-and-collar transfer to a restraint chair (Down et al., 2005; Bliss-Moreau et al., 2013; McMillan et al., 2014).

Conclusions

The variables summarized in this chapter are not the only ones that have the potential to increase variance and reduce the reliability of research data, but they are the most obvious ones that are often overlooked in research protocols.

It is a prerequisite of truly scientific biomedical research methodology to take extraneous variables into account and investigate *prior* to the experiment if one or several of them have the potential to confound the effects induced by the experimental manipulation. Eliminating or avoiding variables that can interfere with the research subject's response to a test situation is a safeguard that the variance of the collected data will be minimal. This, in turn, will enable investigators to assess their findings statistically with a minimum number of research subjects. It is unethical to use more animals in order to increase statistical power and achieve significance of research findings, rather than make an effort to assess husbandry-related variables and eliminate or avoid them if they alter the research subjects' responses to given experiments or tests (Öbrink & Rehbinder, 1999). At a very minimum all husbandry-, housing-, and handling-related variables should be adequately described in any scientific publication so that other investigators can repeat the experiment/test or carry out comparative studies (Morton, 1992; Smith et al., 1997).

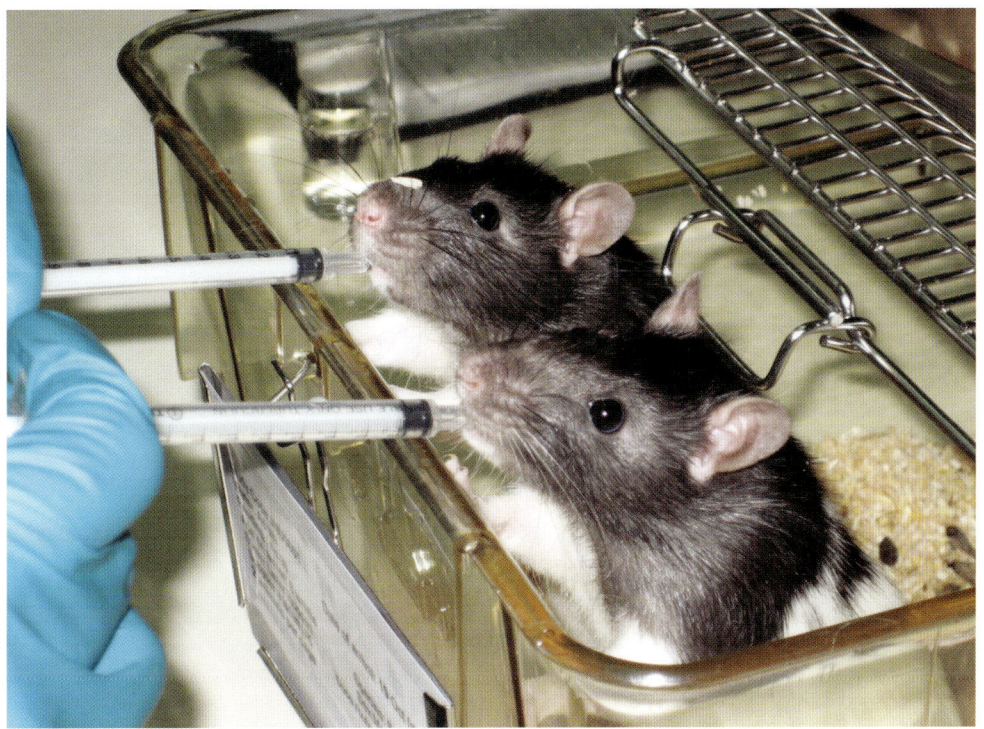

REFERENCES

Abou-Ismail UA, Burman OHP, Nicol CJ and Mendl M 2010 The effects of enhancing cage complexity on the behaviour and welfare of laboratory rats. *Behavioural Processes 85*: 172–180

Adams MR, Kaplan JR, Manuck SB, Uberseder B and Larkin KT 1988 Persistent sympathetic nervous system arousal associated with tethering in cynomolgus macaques. *Laboratory Animal Science 38*: 279–282

Ader DN, Johnson SB, Huang SW and Riley WJ 1991 Group–size, cage shelf level, and emotionality in non-obese diabetic mice—Impact on onset and incidence of IDDM. *Psychosomatic Medicine 53*: 313–321

Ader R and Friedman SB 1964 Social factors affecting emotionality and resistance to disease in animals: IV. Differential housing, emotionality, and Walker 256 carcinosarcoma in the rat. *Psychological Reports 15*: 535–541

Aidara D, Tahiri-Zagret C and Robyn C 1981 Serum prolactin concentrations in mangabey (*Cercocebus atys lunulatus*) and patas (*Erythrocebus patas*) monkeys in response to stress, ketamine, TRH, sulpiride and levodopa. *Journal of Reproduction and Fertility 62*: 165–172

Akiyama K and Sutoo D 2011 Effect of different frequencies of music on blood pressure regulation in spontaneously hypertensive rats. *Neuroscience Letters 487*(1): 58–60

Alexander SN and Fontenot MB 2003 Isosexual social group formation for environmental enrichment in adult male *Macaca mulatta*. *American Association for Laboratory Animal Science Meeting Official Program:* 141 (Abstract)

Ambrose N and Morton DB 2000 The use of cage enrichment to reduce male mouse aggression. *Journal of Applied Animal Welfare Science 3*: 117–125

American Association for Laboratory Animal Science 2001 *Cost of Caring: Recognizing Human Emotions in the Care of Laboratory Animals.* American Association for Laboratory Animal Science: Memphis, TN

American Medical Association 1992 *Use of Animals in Biomedical Research – The Challenge and Response – An American Medical Association White Paper.* AMA Group on Science and Technology: Chicago, IL

Andervont HB 1944 Influence of environment on mammary cancer in mice. *Journal of the National Cancer Institute 4*: 579–581

Anthony A and Harclerode JE 1959 Noise stress in laboratory rodents. II: Effects of chronic noise exposure on sexual performance and reproductive function of guinea pigs. *Journal of the Acoustical Society of America 31*: 1437–1440

Anzenberger G and Gossweiler H 1993 How to obtain individual urine samples from undisturbed marmoset families. *American Journal of Primatology 31*: 223–230

Apple JK, Minton JE, Parson KM and Unruh JA 1993 Influence of repeated restraint and isolation stress and electrolyte administration on pituitary–adrenal secretions, electrolytes, and other blood constituents of sheep. *Journal of Animal Science 71*: 71–77

Archard GA 2012 Effect of enrichment on the behaviour and growth of juvenile *Xenopus laevis*. *Applied Animal Behaviour Science 39*: 264–270

Arluke AB 1988 Sacrificial symbolism in animal experimentation: Object or pet? *Anthrozoös 2*: 98–117

Arluke AB 1993 Trapped in a guilt cage. *Animal Welfare Information Center Newsletter 4*(2): 1–2 & 7–8

Armstrong KR, Clark TR and Peterson MR 1998 Use of cornhusk nesting material to reduce aggression in caged mice. *Contemporary Topics in Laboratory Animal Science 37*(4): 64–66

Arnold CE and Gillaspy S 1994 Assessing laboratory life for Golden Hamsters: Social preference, caging selection, and human interaction. *Lab Animal 23*(2): 34–37

Arnold CE and Westbrook RD 1997/1998 Enrichment in group–housed laboratory golden hamsters. *Animal Welfare Information Center Newsletter 8*(3/4): 22–24

Baenninger LP 1967 Comparison of behavioural development in socially isolated and grouped rats. *Animal Behaviour 15*: 312–323

Baer H 1971 Long–term isolation stress and its effects on drug response in rodents. *Laboratory Animal Science 21*: 341–349

Baldock NM and Sibly RM 1990 Effects of handling and transportation on the heart rate and behaviour of sheep. *Applied Animal Behaviour Science 28*: 15–39

Baldwin AL 2010 Does lack of enrichment invalidate scientific data obtained from rodents? *Enrichment Record 5*: 10–12

Baldwin AL, Schwartz GE and Hopp DH 2007 Are investigators aware of environmental noise in animal facilities and that this noise may affect experimental data? *Journal of the American Association for Laboratory Animal Science 46*(1): 45–51

Baldwin DR, Wilcox ZC and Baylosis RC 1995 Impact of differential housing on humoral immunity following exposure to an acute stressor in rats. *Physiology and Behavior 57*: 649–653

Baroli D, Ghiandoni S, Mantovani C and Cavalchini LG 1997 Environmental enrichment device and egg production in laying hens. In: Koene P and Blokhuis HJ (eds) *Proceedings of the European Symposium on Poultry Welfare* pp. 107–108. Ponsen and Looyen: Wageningen, Netherlands

Barrett AM and Stockham MA 1963 The effect of housing conditions and simple experimental procedures upon the corticosterone level in the plasma of rats. *Journal of Endocrinology 26*: 97–105

Baumans V, Bennett K, Donnelly M, Andrews-Kelly G, Nelsen S, Rappaport K, Reiffer A and Reinhardt V 2013 The problem with the dark bottom–row cage for non–human primates: A discussion by the Laboratory Animal Refinement & Enrichment Forum. *Animal Technology and Welfare 12*: 111–114

Baumans V, Coke CS, Green J, Moreau E, Morton D, Patterson-Kane E, Reinhardt A, Reinhardt V and Van Loo P (eds) 2007 *Making Lives Easier for Animals in Research Labs: Discussions by the Laboratory Animal Refinement & Enrichment Forum.* Animal Welfare Institute: Washington, DC

Bayne K, Mainzer H, Dexter SL, Campbell G, Yamada F and Suomi SJ 1991 The reduction of abnormal behaviors in individually housed rhesus monkeys (*Macaca mulatta*) with a foraging/grooming board. *American Journal of Primatology 23*: 23–35

Bayne K, Dexter SL, Mainzer H, McCully C, Campbell G and Yamada F 1992 The use of artificial turf as a foraging substrate for individually housed rhesus monkeys (*Macaca mulatta*). *Animal Welfare 1*: 39–53

Bellanca RU, Heffernan KS, Grabber JE and Crockett CM 1999 Behavioral profiles of laboratory monkeys referred to a Regional Primate Research Center's psychological well–being program. *American Journal of Primatology 49*: 33 (Abstract)

Bellhorn RW 1980 Lighting in the animal environment. *Laboratory Animal Science 30*: 440–450

Belz EE, Kennell JS, Czambel RK, Rubin RT and Rhodes ME 2003 Confinement lowers stress–responsive hormones in singly housed male and female rats. *Pharmacology Biochemistry and Behavior 76*: 481–486

Benaroya-Milshtein N, Hollander N, Apter A, Kukulansky T, Raz N, Wilf A, Yaniv I and Pick CG 2004 Environmental enrichment in mice decreases anxiety, attenuates stress responses and enhances natural killer cell activity. *European Journal of Neuroscience 20*: 1341–1347

Bentson KL, Capitanio JP and Mendoza SP 2003 Cortisol responses to immobilization with Telazol and ketamine in baboons (*Papio cynocephalus/anubis*) rhesus macaques (*Macaca mulatta*). *Journal of Medical Primatology 32*: 148–160

Berendt R and Williams TD 1971 The effect of restraint and position upon selected respiratory parameters of two species of *Macaca*. *Laboratory Animal Science 21*: 502–509

Besch EL and Chou BJ 1971 Physiological responses to blood collection methods in rats. *Proceedings of the Society for Experimental Biology and Medicine 138*: 1019–1021

Bettinger T, Kuhar C, Sironen A and Laudenslager M 1998 Behavior and salivary cortisol in gorillas housed in an all male group. *American Zoo and Aquarium Association Annual Conference Proceedings*: 242–246

Blackshaw JK, Thomas FJ and Lee J-A 1997 The effect of a fixed or free toy on the growth rate and aggressive behaviour of weaned pigs and the influence of hierarchy on initial investigation of the toys. *Applied Animal Behaviour Science 53*: 203–212

Bliss-Moreau E, Moadab G and Theil J 2013 Efficient cooperative chair training. *36th Meeting of the American Society of Primatologists Scientific Program*: Abstract # 118

Blye R, Burke R, James C, Fitzgerald AL and Cox ML 2006 The use of operant conditioning of Göttingen minipigs for topical safety studies. *American Association for Laboratory Animal Science Meeting Official Program*: 153–154 (Abstract)

Bobek S, Niezgoda J, Pierzchala K, Litynski P and Sechman A 1986 Changes in circulating levels of iodothyronines, cortisol and endogenous thiocyanate in sheep during emotional stress caused by isolation of animals from the flock. *Zentralblatt für Veterinärmedizin 33*: 698–705

Boinski S, Gross TS and Davis JK 1999 Terrestrial predator alarm vocalizations are a valid monitor of stress in captive brown capuchins (*Cebus apella*). *Zoo Biology 18*: 295–312

Bouissou MF 1970 Rôle du contact physique dans la manifestation des relations hierarchiques chez les bovins: conséquences pratiques. *Annales Zootechniques 19*: 279–285

Bourgeois SR and Brent L 2003 The effect of four enrichment conditions on abnormal behavior in seven singly caged baboons (*Papio hamadryas anubis*). *American Journal of Primatology 60*(Supplement): 80–81 (Abstract)

Bouyer JJ, Dedet L, Debray O and Rougeul A 1978 Restraint in primate chair may cause unusual behavior in baboons: Electrocorticographic correlates and corrective effects of diazepam. *Electroencephalic Clinical Neurophysiology 44*: 562–567

Bowd AD 1980 Ethics and animal experimentation. *American Psychologist 35*: 224–225

Bowers CL, Friend TH, Grissom KK and Lay DC 1993 Confinement of lambs (*Ovis aries*) in metabolism stalls increased adrenal function, thyroxine and motivation for movement. *Applied Animal Behaviour Science 36*: 149–158

Bowers CL, Crockett CM and Bowden DM 1998 Differences in stress reactivity of laboratory macaques measured by heart period and respiratory sinus arrhythmia. *American Journal of Primatology 45*: 245–261

Brake J 1987 Influence of presence of perches during rearing on incidence of floor laying in broiler breeders. *Poultry Science 66*: 1587–1589

Brent L and Weaver D 1996 The physiological and behavioral effects of radio music on singly housed baboons. *Journal of Medical Primatology 25*: 370–374

Bridger CJ and Booth RK 2003 The effects of biotelemetry transmitter presence and attachment procedures on fish physiology and behavior. *Reviews in Fisheries Science 11*: 13–34

Brockway BP, Hassler CR and Hicks N 1993 Minimizing stress during physiological monitoring. In: Niemi SM and Willson JE (eds) *Refinement and Reduction in Animal Testing* pp. 56–69. Scientists Center for Animal Welfare: Bethesda, MD

Brodersen T, Glerup P, Molgaard S, Andersen L and Sorensen DB 2010 The use of positive reinforcement with Göttinger minipigs. *American Association for Laboratory Animal Science Meeting Official Program*: 167–168 (Abstract)

Brown GM and Martin JB 1974 Corticosterone, prolactin, and growth hormone responses to handling and new environment in the rat. *Psychosomatic Medicine 36*: 241–247

Brown SA, Langford K and Tarver S 1997 Effects of certain vasoactive agents on the long–term pattern of blood pressure, heart rate, and motor activity in cats. *American Journal of Veterinary Research 58*: 647–652

Buffington CA 1991 Nutrition and nutrional disorders. In: Pedersin N (ed) *Feline Husbandry: Diseases and Management in the Multiple Cat Environment* p. 345. American Veterinary Publishing: Goleta, CA

Bunyak SC, Harvey NC, Rhine RJ and Wilson MI 1982 Venipuncture and vaginal swabbing in an enclosure occupied by a mixed–sex group of stumptailed macaques (*Macaca arctoides*). *American Journal of Primatology 2*: 201–204

Bush M, Custer R, Smeller J and Bush LM 1977 Physiologic measures of nonhuman primates during physical restraint and chemical immobilization. *Journal of the American Veterinary Medical Association 171*: 866–869

Callard MD, Bursten SN and Price EO 2000 Repetitive backflipping behaviour in captive roof rats (*Rattus rattus*) and the effect of cage enrichment. *Animal Welfare 9*: 139–152

Cambardella P, Greco AM, Sticchi R, Bellotti R and Di Renzo G 1994 Individual housing modulates daily rhythms of hypothalamic catecholaminergic system and circulating hormones in adult male rats. *Chronobiology International* 11: 213–221

Canadian Council on Animal Care 1993 *Guide to the Care and Use of Experimental Animals, Volume 1, 2nd Edition*. Canadian Council on Animal Care: Ottawa, Canada

Carbajal S and Orihuela A 2001 Minimal number of conspecifics needed to minimize the stress response of isolated mature ewes. *Journal of Applied Animal Welfare Science 4*: 249–255

Carbonaro DA, Friend TH, Dellmeier GR and Nuti LC 1992 Behavioral and physiological responses of dairy goats to isolation. *Physiology and Behavior 51*: 297–301

Carlton DL and Richards W 2002 Affordable noise control in a laboratory animal facility. *Lab Animal 31*(1): 47–48

Chamove AS 1989 Cage design reduces emotionality in mice. *Laboratory Animals 23*: 215–219

Chikahisa S, Sano A, Kitaoka K, Miyamoto K and Sei H 2007 Anxiolytic effect of music depends on ovarian steroid in female mice. *Behavioural Brain Research 179*: 50–59

Christian KA and Bedford G 1995 Seasonal–changes in thermoregulation by the frillneck lizard, *Chlamydosaurus–kingii*, in tropical Australia. *Ecology 76*: 124–132

Claassen, V 1994 *Neglected Factors in Pharmacology and Neuroscience Research*. Elsevier: Amsterdam, Netherlands

Clough G 1982 Environmental effects on animals used in biomedical research. *Biological Reviews 57*: 487–523

Clough G 1984 Environmental factors in relation to the comfort and well–being of laboratory rats and mice. In: The Universities Federation for Animal Welfare (ed) *Standards in Laboratory Animal Management* pp. 7–23. The Universities Federation for Animal Welfare: Potters Bar, UK

Clough G and Donnelly HT 1984 Light intensity influences the oestrous cycle of LACA mice. In: The Universities Federation for Animal Welfare (ed) *Standards in Laboratory Animal Management* p. 60. The Universities Federation for Animal Welfare: Potters Bar, UK

Cockram MS, Ranson M, Imlah P, Goddard PJ, Burrells C and Harkiss GD 1994 The behavioural, endocrine and immune response of sheep to isolation. *Animal Production 58*: 389–400

Coe CL, Franklin D, Smith ER and Levine S 1982 Hormonal responses accompanying fear and agitation in the squirrel monkey. *Physiology and Behavior 29*: 1051–1057

Coelho AM, Carey KD and Shade RE 1991 Assessing the effects of social environment on blood pressure and heart rates of baboons. *American Journal of Primatology 23*: 257–267

Colbourne F, Sutherland GR and Auer RN 1996 An automated system for regulating brain temperature in awake and freely moving rodents. *Journal of Neuroscience Methods 67*: 189–190

Coleman K, Pranger L, Maier A, Lambeth SP, Perlman JE, Thiele E and Schapiro SJ 2008 Training rhesus macaques for venipuncture using positive reinforcement techniques: A comparison with chimpanzees. *Journal of the American Association for Laboratory Animal Science 47*: 37–41

Committee on Recognition and Alleviation of Distress in Laboratory Animals, NRC 2008 *Recognition and Alleviation of Distress in Laboratory Animals*. The National Academies Press: Washington, DC

Conger JJ, Sawrey WL and Turrell ES 1957 An experimental investigation of the role of social experience in the production of gastric ulcers in hooded rats. *American Psychologist 12*: 410 (Abstract)

Cope FW and Polis BD 1959 Increased plasma glutamic–oxalacetic transaminase activity in monkeys due to nonspecific stress effect. *Journal of Aviation Medicine 30*: 90–94

Cornetto TL and Estevez I 2001 Influence of vertical panels on use of space by domestic fowl. *Applied Animal Behaviour Science 71*: 141–153

Cornetto TL, Estevez I and Douglas LW 2002 Using artificial cover to reduce aggression and disturbances in domestic fowl. *Applied Animal Behaviour Science 75*: 325–336

Council of Europe 2006 *Appendix A of the European Convention for the Protection of Vertebrate Animals Used for Experimental and Other Scientific Purposes (ETS No. 123) enacted June 15, 2007*. Council of Europe: Strasbourg, France

Crockett CM, Bowers CL, Sackett GP and Bowden DM 1993 Urinary cortisol responses of longtailed macaques to five cage sizes, tethering, sedation, and room change. *American Journal of Primatology 30*: 55–74

Crockett CM, Shimoji M and Bowden DM 2000 Behavior, appetite, and urinary cortisol responses by adult female pigtailed macaques to cage size, cage level, room change, and ketamine sedation. *American Journal of Primatology 52*: 63–80

Crockett C. M. and Gough GM 2002 Onset of aggressive toy biting by a laboratory baboon coincides with cessation of self–injurious behavior. *American Journal of Primatology 57*: 39 (Abstract)

Cross N, Pines MK and Rogers LJ 2004 Saliva sampling to assess cortisol levels in unrestrained common marmosets and the effect of behavioral stress. *American Journal of Primatology 62*: 107–114

Crouthamel B and Sackett G 2004 Oral medication administration: Training monkeys to take juice. *Laboratory Primate Newsletter 43*(1): 5–6

Döhler KD, Gärtner K, Mühlen AV and Döhler U 1977 Activation of anterior pituitary, thyroid and adrenal glands in rats after disturbance stress. *Acta Endocrinologica 86*: 489–497

Dallman MF, Akana SF, Bell ME, Bhatnager S, Choi S, Chu A, Gomez F, Laugero K, Soriano L and Viau V 1999 Warning! Nearby construction can profoundly affect your experiments. *Endocrine 11*: 111–113

Davenport MD, Lutz CK, Tiefenbacher SNMA and Meyer JS 2008 A rhesus monkey model of self–injury: effects of relocation stress on behavior and neuroendocrine function. *Biological Psychiatry 63*: 990–996

Davis DE, Bennett CL, Berkson G, Lang CM, Snyder RL and Pick JR 1973 Recommendations for a standardized minimum description of animal treatment. *ILAR Journal 16*(4): 3–4

Davitz JR and Mason DJ 1955 Socially facilitated reduction of a fear response in rats. *Journal of Comparative and Physiological Psychology 48*: 149–151

DeJong JG, van der Vegt BJ, Buwalda B and Koolhaas JM 2005 Social environment determines the long–term effects of social defeat. *Physiology and Behavior 84*: 87–95

Deacon R 2001 Play makes rats more "user–friendly". *Animal Technology 52*: 51–52

Dejardins S, Cauchy MJ and Kozliner E 1996 The running cardiomyopathic hamster with continous telemetric ECG: A new heart failure model to evaluate 'symptoms', cause and death of heart rate. *Experimental Clinical Cardiology 1*: 29–36

DeLuca AM and Kranda KC 1992 Environmental enrichment in a large animal facility. *Lab Animal 21*(1): 38–44

Denenberg VH and Morton JRC 1962 Effects of environmental complexity and social groupings upon the modification of emotional behavior. *Journal of Comparative and Physiological Psychology 55*: 242–246

Depasquale MJ, Ringer LW, Winslow RL, Buchholz RA and Fossa AA 1994 Chronic monitoring of cardiovascular function in the conscious guinea pig using radio–telemetry. *Clinical Experimentation of Hypertension 16*: 245–260

Detillion CE, Craft TK, Glasper ER, Prendergast BJ and de Vries AC 2004 Social facilitation of wound healing. *Psychoneuroendocrinology 29*: 1004–1011

Dobrakovavá J and Jurcovicova J 1984 Corticosterone and prolactin response to repeated handling and transfer of male rats. *Experimental Clinical Endocrinology 5*: 21–27

Done-Currie JR, Hecker JF and Wodzicka-Tomaszewka M 1984 Behaviour of sheep transferred from pasture to an animal house. *Applied Animal Behaviour Science 12*: 121–130

Down N, Skoumbourdis E, Walsh M, Francis R, Buckmaster C and Reinhardt V 2005 Pole–and–collar training: A discussion by the Laboratory Animal Refinement and Enrichment Forum. *Animal Technology and Welfare 4*: 157–161

Doyle LA, Baker KC and Cox LD 2008 Physiological and behavioral effects of social introduction on adult male rhesus macaques. *American Journal of Primatology 70*(6): 542–550

Drozdowicz CK, Bowman TA, Webb ML and Lang CM 1990 Effect of in–house transport on murine plasma corticosterone concentration and blood lymphocyte population. *American Journal of Veterinary Research 51*: 1841–1846

Ehlers CL, Kaneko WM, Owens MJ and Nemeroff CB 1993 Effects of gender and social isolation on electroencephalogram and neuroendocrine parameters in rats. *Biological Psychiatry 33*: 358–366

Einstein R, Rowan C, Billing R and Lavidis N 2000 The use of telemetry to refine experimental technique. In: Balls M, Van Zeller AM and Halder ME (eds) *Progress in the Reduction, Refinement and Replacement of Animal Experimentation* pp. 1187–1197. Elsevier: Amsterdam, Netherlands

Elvidge H, Challis JRG, Robinson JS, Roper C and Thorburn GD 1976 Influence of handling and sedation on plasma cortisol in rhesus monkeys (*Macaca mulatta*). *Journal of Endocrinology 70*: 325–326

Engellenner WJ, Goodlett CR, Burright RG and Donovick PJ 1982 Environmental enrichment and restriction: Effects on reactivity, exploration and maze learning in mice with septal lesions. *Physiology and Behavior 29*: 885–893

Erwin J 1977 Factors influencing aggressive behavior and risk of trauma in the pigtail macaque (*Macaca nemestrina*). *Laboratory Animal Science 27*: 541–547

Erwin J and Deni R 1979 Strangers in a strange land: Abnormal behavior or abnormal environments? In: Erwin J, Maple T and Mitchell G (eds) *Captivity and Behavior* pp. 1–28. Van Nostrand Reinhold: New York, NY

Eskola S and Kaliste-Korhonen E 1998 Effects of cage type and gnawing blocks on weight gain, organ weights and open–field behaviour in Wistar rats. *Scandinavian Journal of Laboratory Animal Science 25*: 180–193

Fagin KD, Shinsako J and Dallman MF 1983 Effects of housing and chronic cannulation on plasma ACTH and corticosterone in the rat. *American Journal of Physiology 245*: E515–E520

Fenske M 1992 Body weight and water intake of guinea pigs: influence of single caging and an unfamiliar new room. *Journal of Experimental Animal Science 35*: 71–79

Fowler, ME 1995 *Restraint and Handling of Wild and Domestic Animals (Second Edition)*. Iowa State University Press: Ames, IA

Fraser AF 1995 Sheep. In: Rollin BE and Kesel ML (eds) *The Experimental Animal in Biomedical Research, Volume II – Care, Husbandry, and Well–Being* pp. 87–118. CRC Press: Boca Raton, FL

Friedman SB and Ader R 1967 Adrenocortical response to novelty and noxious stimulation. *Neuroendocrinology 2*: 209–212

Fritz J 1989 Resocialization of captive chimpanzees: An amelioration procedure. *American Journal of Primatology 19*(Supplement): 79–86

Fuchs E, Fluegge G and Hutzelmeyer HD 1987 Response of rats to the presence of stressed conspecifics as a function of day time. *Hormones and Behavior 21*: 245–252

Fuller GB, Hobson WC, Reyes FI, Winter JSD and Faiman C 1984 Influence of restraint and ketamine anesthesia on adrenal steroids, progesterone, and gonadotropins in rhesus monkeys. *Proceedings of the Society for Experimental Biology and Medicine 175*: 487–490

Gamble, MR 1979 *Effects of noise on laboratory animals.* Ph.D. Thesis, University of London: London, UK

Gardiner SM and Bennet T 1978 Factors affecting the development of isolation–induced hypertension in rats. *Medical Biology 56*: 277–281

Garner JP 2002 Why every scientist should care about animal welfare: Abnormal repetitive behavior and brain function in captive animals. *Proceedings of the World Congress on Alternatives and Animal Use in the Life Sciences*: 95 (Abstract)

Gattermann R and Weinandy R 1996/97 Time of day and stress response to different stressors in experimental animals. Part I: Golden hamster (*Mesocricetus auratus* Waterhouse, 1839). *Journal of Experimental Animal Science 38*: 66–76

Giralt M and Armario A 1989 Individual housing does not influence the adaptation of the pituitary–adrenal axis and other physiological variables to chronic stress in adult male rats. *Physiology and Behavior 45*: 477–481

Goldsmith JF, Brain PF and Benton D 1976 Effects of age at differential housing and the duration of individual housing/grouping on intermale fighting behavior and adrenocortical activity in TO strain mice. *Aggressive Behavior* 2: 307–323

Goldsmith JF, Brain PF and Benton D 1978 Effects of the duration of individual or group housing on behavioral and adrenocortical reactivity in male mice. *Physiological Psychology* 21: 757–760

Golub MS and Anderson JH 1986 Adaptation of pregnant rhesus monkeys to short–term chair restraint. *Laboratory Animal Science* 36: 507–511

Gonzalez CA, Coe CL and Levine S 1982 Cortisol responses under different housing conditions in female squirrel monkeys. *Psychoneuroendocrinology* 7: 209–216

Goosen DJ, Davies JH, Maree M and Dormehl IC 1984 The influence of physical and chemical restraint on the physiology of the chacma baboon (*Papio ursinus*). *Journal of Medical Primatology* 13: 339-351

Grandin T 1986 Minimizing stress in pig handling. *Lab Animal* 15(3): 15-20 Grandin T 1989 Voluntary acceptance of restraint by sheep. *Applied Animal Behaviour Science* 23: 257–261

Guhad FA and Hau J 1996 Salivary IgA as a marker of social stress in rats. *Neuroscience Letters* 27: 137–140

Gunn D and Morton DB 1995 Inventory of the behaviour of New Zealand white rabbits in laboratory cages. *Applied Animal Behaviour Science* 45: 277–292

Gust DA, Gordon TP, Brodie AR and McClure HM 1994 Effect of a preferred companion in modulating stress in adult female rhesus monkeys. *Physiology and Behavior* 55: 681–684

Gvaryahu G, Ararat E, Asaf E, Lev M, Weller JI, Robinzon B and Snapir N 1994 An enrichment object that reduces aggressiveness and mortality in caged laying hens. *Psychology and Behavior 55*: 313–316

Harkin A, Connor TJ, O'Donnell JM and Kelly JP 2002 Physiological and behavioral responses to stress: What does a rat find stressful? *Lab Animal* 31(4): 42–49

Hartner MK, Hall J, Penderghest J, White E, Watson S and Clark L 2000 A novel approach to group–housing male cynomolgus macaques in a pharmaceutical environment. *Contemporary Topics in Laboratory Animal Science* 39: 67 (Abstract)

Harvey H, Rice T, Kayhart R and Torres C 2000 The effects of specific types of music on the activity levels of singly housed chimpanzees (*Pan troglodytes*). *American Journal of Primatology 51*(Supplement): 60 (Abstract)

Hatch AM, Wiberg GS, Zawidzka Z, Cann M, Airth JM and Grice HC 1965 Isolation syndrome in the rat. *Toxicology and Applied Pharmacology 7*: 737–745

Hauntzinger GM and Piacsek BE 1973 Influence of duration, intensity and spectrum of light exposure on sexual maturation of female rats. *Federation Proceedings* 32: 213

Hayashi KT and Moberg GP 1987 Influence of acute stress and the adrenal axis on regulation of LH and testosterone in the male rhesus monkey (*Macaca mulatta*). *American Journal of Primatology* 12: 263–273

Heath M 1999 Preliminary behaviour data for single and pair housed rats. *Animal Technology 50*: 47–48

Hedge TA, Saunders KE and Ross CA 2002 Innovative housing and environmental enrichment for bullfrogs (*Rana catesbiana*). *Contemporary Topics in Laboratory Animal Science 41*(4): 120–121 (Abstract)

Heger W, Merker H-J and Neubert D 1986 Low light intensity decreases the fertility of *Callithrix jacchus*. *Primate Report 14*: 260 (Abstract)

Hein PR, Schatorje JS, Frencken HJ, Segers MF and Thomas CM 1989 Serum hormone levels in pregnant cynomolgus monkeys. *Journal of Medical Primatology 133–142*: 133–142

Held SDE, Turner RJ and Wootton RJ 2001 The behavioural repertoire of non–breeding group–housed female laboratory rabbits (*Oryctolagus cuniculus*). *Animal Welfare 10*: 437–443

Hemsworth PH, Barnett JL and Jones RB 1993 Situational factors that influence the level of fear of human by laying hens. *Applied Animal Behaviour Science 36*: 197–210

Hennesy MB and Foy T 1987 Non–edible material elicits chewing and reduces the plasma corticosterone response during novelty exposure in mice. *Behavioral Neuroscience 101*: 237–245

Hennessy MB, Hornschuh G, Kaiser S and Sachser N 2006 Cortisol responses and social buffering: a study throughout the life span. *Hormones and Behavior 49*(3): 383–390

Herndon JG, Turner JJ, Perachio AA, Blank MS and Collins DC 1984 Endocrine changes induced by venipuncture in rhesus monkeys. *Physiology and Behavior 32*: 673–676

Herzog H 2002 Ethical aspects of relationships between humans and research animals. *ILAR Journal 43*(1): 27–32

Hetts S, Clark JD, Calpin JP, Arnold CE and Mateo JM 1992 Influence of housing conditions on beagle behaviour. *Applied Animal Behaviour Science 34*: 137–155

Heybring E, Macdonald E and Olsson K 1997 Radiotelemetrically recorded blood pressure and heart rate changes in relation to plasma catecholamine levels during parturition in the conscious, unrestrained goat. *Acta Physiologica Scandinavica 161*: 295–302

Hinds SB, Raimond S and Purcell BK 2007 The effect of harp music on heart rate, mean blood pressure, respiratory rate, and body temperature in the African green monkey. *Journal of Medical Primatology 36*: 95–100

Hockly E, Cordery PM, Woodman B, Mahal A, van Dellen A, Blakemore C, Lewis CM, Hannan AJ and Bates GP 2002 Enrichment slows disease progression in R6/2 Huntington's disease mice. *Annals of Neurology 51*: 235–242

Home Office 1989 *Animals (Scientific Procedures) Act 1986. Code of Practice for the Housing and Care of Animals Used in Scientific Procedures.* Her Majesty's Stationery Office: London, UK

Honess PE and Marin CM 2006 Enrichment and aggression in primates. *Neuroscience and Biobehavioral Reviews 30*: 413–346

Hopster H and Blokhuis HJ 1993 Consistent stress response of individual dairy cows to social isolation. *Proceedings of the International Congress on Applied Ethology*: 123–126

Howard B, Wortley M and Kay R 1999 Rabbit enclosures – structure and space. *Animal Technology 50*: 156–157

Howell S, Schwandt M, Fritz J, Roeder E and Nelson C 2003 A stereo music system as environmental enrichment for captive chimpanzees. *Lab Animal 32*(10): 31–36

Hrapkiewicz, K, Medina, L and Holmes, DD 1998 *Clinical Medicine of Small Mammals and Primates, Second Edition.* Manson Publishing: London, UK

Huang-Brown KM and Guhad FA 2002 Chocolate, an effective means of oral drug delivery in rats. *Lab Animal 31*(10): 34–36

Hubrecht RC, Serpell JA and Poole TB 1992 Correlates of pen size and housing conditions on the behaviour of kennelled dogs. *Applied Animal behaviour Science 34*: 365–383

Hummer RL 1965 Principles of public health importance in the management of a subhuman primate colony. *Journal of the American Veterinary Medical Association 147*: 1063 1067

Hurst JL, Barnard CJ, Nevison CM and West CD 1997 Housing and welfare in laboratory rats: Welfare implications of isolation and social contact among caged males. *Animal Welfare 6*: 327–347

International Primatological Society 2007 *IPS International Guidelines for the Acquisition, Care and Breeding of Nonhuman Primates.* International Primatological Society: Bronx, NY

Ives M and Dack GM 1956 "Alarm reaction" and normal blood picture in *Macaca mulatta. Journal of Laboratory Clinical Medicine 47*: 723–729

Jackson G 1991 Intestinal stasis and rupture in rabbits. *The Veterinary Record 129*: 287–289

Jain M and Baldwin AL 2003 Are laboratory animals stressed by their housing environment and are investigators aware that this stress can affect physiological data? *Medical Hypotheses 60*: 284–289

Jeans GL, Fryer DA, Casey KW, Bohne CR, Calihan GL and Fix AS 2008 Enhanced indoor housing for research dogs. *American Association for Laboratory Animal Science Meeting Official Program:* 91–92 (Abstract)

Johnson CV, Wismer M, Francis J, Seeburger G, Cunningham P, Parlapiano A, Hora D, Kath G, Hayden C, Gardener M, Benyak C and Feeney W 2005 Noise assessment of multiple floor surfaces within an animal facility. *American Association for Laboratory Animal Science Meeting Official Program:* 102 (Abstract)

Jones RB and Merry BJ 1988 Individual or paired exposure of domestic chicks to an open field: some behavioural and adrenocortical consequences. *Behavioural Processes 16*: 75–86

Jones RB and Rayner S 1999 Music in the hen house: a survey of its incidence and perceived benefits. *Poultry Science 78*(Supplement): 110 (Abstract)

Jorgensen MJ, Kinsey JH and Novak MA 1998 Risk factors for self–injurious behavior in captive rhesus monkeys (*Macaca mulatta*). *American Journal of Primatology 45*: 187 (Abstract)

Kaplan JR, Adam MR and Bumsted P 1983 Heart rate changes associated with tethering of cynomolgus monkeys. *Laboratory Animal Science 38*: 493

Kelley TM and Bramblett CA 1981 Urine collection from vervet monkeys by instrumental conditioning. *American Journal of Primatology 1*: 95–97

Kessel AL and Brent L 1996 The effectiveness of cage toys in reducing abnormal behavior in individually housed pigtail macaques. *Proceedings of the Congress of the International Primatological Society and Conference of the American Society of Primatologists:* 519 (Abstract)

Kilcullen-Steiner C and Mitchell A 2001 Quiet those barking dogs. *American Association for Laboratory Animal Science Meeting Official Program:* 103 (Abstract)

Kingston SG and Hoffman-Goetz L 1996 Effect of environmental enrichment and housing density on immune system reactivity to acute exercise stress. *Physiology and Behavior 60*: 145–150

Klaiber-Schuh A and Welker C 1997 Crab–eating monkeys (*Macaca fascicularis*) can be trained to cooperate in non–invasive oral medication without stress. *Primate Report 47*: 11–30

Klein SL, Lambert KG, Durr D, Schaefer T and Waring RE 1994 Influence of environmental enrichment and sex on predator stress response in rats. *Physiology and Behavior 56*: 291–297

Kramer K, Mulder A, van de Weerd H, Baumans V, van Heijningen C, Remie R, Voss H-P and van Zutphen B 2001 Does conditioning influence the increase of heart rate and body temperature as provoked by handling in the mouse? *Contemporary Topics in Laboratory Animal Science 40*(4): 92 (Abstract)

Kramer K, van Acker SABE, Voss H-P, Grimbergen JA, Van der Vijgh WJF and Bast A 1993 Use of telemetry to record electrocardiogram and heart rate in freely moving mice. *Journal of Pharmacological and Toxicological Methods 30*: 209–215

Krohn TC, Ritskes-Hoitinga J and Svendsen P 1999 The effect of feeding and housing on the behaviour of the laboratory rabbit. *Laboratory Animals 33*: 101–107

Krohn TC, Hansen AK and Dragsted N 2003 Telemetry as a method for measuring the impact of housing conditions on rats' welfare. *Animal Welfare 12*: 53–62

Krohn TC, Salling B and Kornerup Hansen A 2011 How do rats respond to playing radio in the animal facility? *Laboratory Animals 45*: 141–144

Kvetnansky R, Sun CL, Lake CR, Thoa N, Torda T and Kopin IJ 1978 Effect of handling and forced immobilization on rat plasma levels of epinephrine, norepinephrine, and dopamine–beta–hydroxylase. *Endocrinology 103*: 1868–1874

Laburn HP, Mitchell D and Goelst K 1992 Fetal and maternal body temperatures measured by radiotelemetry in near–term sheep during thermal stress. *Journal of Applied Physiology 72*: 894–900

Lambeth SP, Perlman JE and Schapiro SJ 2000 Positive reinforcement training paired with videotape exposure decreases training time investment for a complicated task in female chimpanzees. *American Journal of Primatology 51*(Supplement): 79–80 (Abstract)

Landi MS, Kissinger JT, Campbell SA, Kenney CA and Jenkins EL 1990 The effects of four types of restraint on serum alanine aminotransferase and asparate aminotransferase in the *Macaca fascicularis. Journal of the American College of Toxicology 9*: 517 523

Latané B and Glass D 1968 Social and nonsocial attraction in rats. *Journal of Personality and Social Psychology 9*: 142–146

Laudenslager M and Worlein J 2003 Saliva collection as a means of avoiding the stress of phlebotomy or ketamine anesthesia. *American Journal of Primatology 60*(Supplement): 90–91(Abstract)

Lemmer B 2008 Effects of music composed by Mozart and Ligeti on blood pressure and heart rate circadian rhythms in normotensive and hypertensive rats. *Chronobiology International 25*: 971–986

Lenox RH, Kant GJ, Sessions GR, Pennington LL, Mougey EH and Meyerhoff JL 1980 Specific hormonal and neurochemical responses to different stressors. *Neuroendocrinology 30*: 300–308

Levison PK, Fester CB, Nieman WH and Findley JD 1964 A method for training unrestrained primates to receive drug injection. *Journal of the Experimental Analysis of Behavior 7*: 253–254

Li WJ, Yu H, Yang JM, Gao J, Jiang H, Feng M, Zhao YX and Chen ZY 2010 Anxiolytic effect of music exposure on BDNFMet/Met transgenic mice. *Brain Research 1347*: 71–79

Lilly AA, Mehlman PT and Higley J 1999 Trait–like immunological and hematological measures in female rhesus across varied environmental conditions. *American Journal of Primatology 48*: 197–223

Lindburg DG and Coe J 1995 Ark design update: Primate needs and requirements. In: Gibbons EF, Durrant BS and Demarest AJ (eds) *Conservation of Endangered Species in Captivity* pp. 553–570. SUNY Press: Albany, NY

Line SW, Clarke AS and Markowitz H 1987 Plasma cortisol of female rhesus monkeys in response to acute restraint. *Laboratory Primate Newsletter 26*(4): 1–3

Line SW, Markowitz H, Morgan KN and Strong S 1991 Effect of cage size and environmental enrichment on behavioral and physiological responses of rhesus macaques to the stress of daily events. In: Novak MA and Petto AJ (eds) *Through the Looking Glass. Issues of Psychological Well–being in Captive Nonhuman Primates* pp. 160–179. American Psychological Association: Washington DC

Line SW, Morgan KN, Markowitz H and Strong S 1989 Heart rate and activity of rhesus monkeys in response to routine events. *Laboratory Primate Newsletter 28*(2): 9–12

Line SW, Morgan KN, Markowitz H, Roberts J and Riddell M 1990 Behavioral responses of female long–tailed macaques (*Macaca fascicularis*) to pair formation. *Laboratory Primate Newsletter 29*(4): 1–5

Loomis MR, Henrickson RV and Anderson JH 1980 Effects of ketamine hydrochloride on the hemogram of rhesus monkeys (*Macaca mulatta*). *Laboratory Animal Science 30*: 851–853

Love JA 1994 Group housing: Meeting the physical and social needs of the laboratory rabbits. *Laboratory Animal Science 44*: 5–11

Lutz CK, Well A and Novak M 2003 Stereotypic and self–injurious behavior in rhesus macaques: A survey and retrospective analysis of environment and early experience. *American Journal of Primatology 60*: 1–15

Mackintosh JH 1962 Effect of strain and group size on the response of mice to "sconal" anaesthesia. *Nature 194*: 1304

MacLean EL, Roberts Prior S, Platt ML and Brannon EM 2009 Primate location preference in a double–tier cage: The effects of illumination and cage height. *Journal of Applied Animal Welfare Science 12*: 73–81

Malinow MR, Hill JD and Ochsner AJ 1974 Heart rate in caged rhesus monkeys (*Macaca mulatta*). *Laboratory Animal Science 24*: 537–540

Maninger N, Kim JH and Ruppenthal GC 1998 The presence of visual barriers decreases antagonism in group housed pigtail macaques (*Macaca nemestrina*). *American Journal of Primatology 45*: 193–194 (Abstract)

Manning PJ, Lehner NDM, Feldner MA and Bullock BC 1969 Selected hematologic, serum chemical, and arterial blood gas characteristics of squirrel monkeys (*Saimiri sciureus*). *Laboratory Animal Care 19*: 831–837

Manosevitz M and Joel U 1973 Behavioral effects of environmental enrichment in randomly bred mice. *Journal of Comparative and Physiological Psychology 85*: 373–382

Manuck SB, Kaplan JR and Clarkson TB 1983 Behavioral induced heart rate reactivity and atherosclerosis in cynomolgus monkeys. *Psychosomatic Medicine 45*: 95–108

Marr JM, Gnam EC, Calhoun J and Mader JT 1993 A non–stressful alternative to gastric gavage for oral administration of antibiotics in rabbits. *Lab Animal 22*(2): 47–49

Marsden MD and Wood-Gush DGM 1986 A note on the behaviour of individually–penned sheep regarding their use for research purposes. *Animal Production 42*: 157–159

Marshall FHA 1940 The experimental modification of the oestrous cycle in the ferret by different intensities of light irradiations and other methods. *Journal of Experimental Biology 17*: 139–147

Martin G 1991 Ecological aspects of chicken husbandry – Interaction between environmental condition, behavioural activity of hens and quality of deep litter. In: Boehnke E and Mokenthin V (eds) *Alternatives in Animal Husbandry* pp. 87–94. University of Kassel: Witzenhausen, Germany

Martinez JL 1972 Effects of selected illumination levels on circadian periodicity in the rhesus monkey (*Macaca mulatta*). *Journal of Interdisciplinary Cycle Research 3*: 47–59

Mason JW 1972 Corticosteroid response to chair restraint in the monkey. *American Journal of Physiology 222*: 1291–1294

Mason JW, Wool MS, Wherry FE, Pennington LL, Brady JV and Beer B 1968 Plasma growth hormone response to avoidance in the monkey. *Psychosomatic Medicine 30*: 760–773

Mason JW, Mougey EH and Kenion CC 1973 Urinary epinephrine and norepinephrine responses to chair restraint in the monkey. *Physiology and Behavior 10*: 801–803

Mason WA 1960 Socially mediated reduction in emotional responses of young rhesus monkeys. *Journal of Abnormal and Social Psychology 60*: 100–110

McClure DE and Thomson JI 1992 Cage enrichment for hamsters housed in suspended wire cages. *Contemporary Topics in Laboratory Animal Science 31*(4): 33 (Abstract)

McCormack K and Megna NL 2001 The effects of privacy walls on aggression in a captive group of rhesus macaques (*Macaca mulatta*). *American Journal of Primatology 54* (Supplement 1): 50–51 (Abstract)

McDermott J and Hauser MD 2007 Nonhuman primates prefer slow tempos but dislike music overall. *Cognition 104* : 654–668

McKinley J, Buchanan-Smith HM, Bassett L and Morris K 2003 Training common marmosets (*Callithrix jacchus*) to cooperate during routine laboratory procedures: Ease of training and time investment. *Journal of Applied Animal Welfare Science 6*: 209–220

McLean CB and Swanson LE 2004 Reducing stress in individually housed sheep. *American Association for Laboratory Animal Science Meeting Official Program:* 144 (Abstract)

McMillan JL, Perlman JE, Galvan A, Wichmann T and Bloomsmith MA 2014 Refining the pole–and–collar method of restraint: Emphasizing the use of positive training techniques with rhesus macaques (*Macaca mulatta*). *Journal of the American Association for Laboratory Animal Science 53*(1): 61–68

Michael RP, Setchell KDR and Plant TM 1974 Diurnal changes in plasma testosterone and studies on plasma corticosteroids in non–anaesthetized male rhesus monkeys (*Macaca mulatta*). *Journal of Endocrinology 63*: 325–335

Milligan SR, Sales GD and Khirnykh K 1993 Sound levels in rooms housing laboratory animals: An uncontrolled daily variable. *Physiology and Behavior 53*: 1067–1076

Mis J and Warren F 2003 A novel and cost–effective approach to New Zealand White Rabbit enrichment. *Tech Talk 8*(6): 4

Mitchell G and Gomber J 1976 Short– and long–term attachments in adult heterosexual pairs of rhesus monkeys. *Primates 12*: 543–547

Moore EA, Broom DM and Simmins PH 1994 Environmental enrichment in flatdeck accomodation for exploratory behaviour in early–weaned piglets. *Applied Animal Behaviour Science 41*: 277–278 (Abstract)

Morrison P 2001 The rat floor pen: Fact or fantasy? *Animal Technology 52*: 33–34

Morton DB 1992 A fair press for animals. *New Scientist 134*(1816): 28–30

Morton DB, Hawkins P, Bevan R, Heath K, Kirkwood J, Pearce P, Scott L, Whelan G and Webb A 2003 Refinements in telemtry procedures—Sevenths report of the BVAAWF/FRAME/RSPCA/UFAW Joint Working Group on Refinment, Part A. *Laboratory Animals 37*: 261–299

Muhlbock O 1951 Influence of environment on mammary tumors in mice. *Acta Unio Internationalis Contra Cancrum 7*: 351–353

Myers BA, Mendoza SP and Cornelius CE 1988 Elevation of plasma glucagon levels in response to stress in squirrel monkeys: Comparison of two subspecies (*Saimiri sciureus boliviensis* and *Saimiri sciureus* sciureus). *Journal of Medical Primatology 17*: 205–214

Núñez MJ, Mañá P, Liñares D, Riveiro MP, Balboa J, Suárez-Quintanilla J, Maracchi M, Méndez MR and López JMF-GM 2002 Music, immunity and cancer. *Liefe Science 71*: 1047–1057

Nakamichi M and Asanuma K 1998 Behavioral effects of perches on group–housed adult female Japanese monkeys. *Perceptual and Motor Skills 87*: 707–714

National Research Council 1996 *Guide for the Care and Use of Laboratory Animals, 7th Edition*. National Academy Press: Washington, DC

National Research Council 1998 *The Psychological Well–Being of Nonhuman Primates*. National Academy Press: Washington, DC

Nelms R, Davis BK, Tansey G and Raber JM 2001 Utilization of training techniques to minimize distress and facilitate the treatment of a chronically ill macaque. *American Association for Laboratory Animal Science Meeting Official Program:* 97–98 (Abstract)

Neubauer T and Buckmaster C 2011 Decline in aggression in cotton rats through enrichment. *Tech Talk 16*(4): 2–3

Neubauer T, Betts T and Evans C 2011 The use of enrichment to facilitate data collection in a pig study. *American Association for Laboratory Animal Science Meeting – Abstracts of Poster Sessions:* 21 (Abstract)

Neveu H 1994 Influence of two changes in the physical environment on the well–being of a group of mangabeys. *Folia Primatologica 62*: 206 (Abstract)

Newberry RC and Shackleton DM 1997 Use of visual cover by domestic fowl: a Venetian blind effect? *Animal Behaviour 54*: 387–395

Newton WM 1978 Environmental impact on laboratory animals. *Advances in Veterinary Science and Comparative Medicine 22*: 1–28

Niu Y, Zhang M and Liu J 2011 The effect of cage enrichment onf fluctuating asymmetry and fecal corticosterone of group–housed laboratory mice. *American Association for Laboratory Animal Science Meeting Official Program:* 150 (Abstract)

Novak MA 2003 Self–injurious behavior in rhesus monkeys: New insights into its etiology, physiology, and treatment. *American Journal of Primatology 59*: 3–19

Novak MA, Kinsey JH, Jorgensen MJ and Hazen TJ 1998 Effects of puzzle feeders on pathological behavior in individually housed rhesus monkeys. *American Journal of Primatology 46*: 213–227

Nyska A, Leininger JR and Maronpot RR 1998 Effect of individual versus group caging on the incidence of pituitary and Leydig cell tumors in F344 rats: proposed mechanism. *Medical Hypotheses 50*: 525–529

O'Connell NE and Beattie VE 1999 Influence of environmental enrichment on aggressive behaviour and dominance relationships in growing pigs. *Animal Welfare 8*: 269–279

Öbrink KJ and Rehbinder C 1999 Animal definition: a necessity for the validity of animal experiments? *Laboratory Animals 22*: 121–130

Ökva K, Nevlalainen T, Mauranen K and Pokk P 2010 The effect of three different items of cage furniture on the behaviour of male C57BL/6J mice in the plus–maze test. *Animal Welfare 19*: 401–409

Osborne BE 1973 A restraining device for use when recording electrocardiograms in monkeys. *Laboratory Animals 7*: 289–292

Pérez C, Canal JR, Dominguez E, Campillo JE and Guillén M 1997 Individual housing influences certain biochemical parameters in the rat. *Laboratory Animals 31*: 357–361

Parrott RF and Lloyd DM 1995 Restraint, but not frustration, induces prostaglandin–mediated hyperthermia in pigs. *Physiology and Behavior 57*: 1051–1055

Pearson BL, Judge P and Reeder DM 2008 Effectiveness of saliva collection and enzyme–immunoassay for the quantification of cortisol in socially housed baboons. *American Journal of Primatology 70*: 1145–1151

Pearson RA and Mellor DJ 1976 Some behavioral and physiological changes in pregnant goats and sheep during adaptation to laboratory conditions. *Research in Veterinary Science 20*: 215–217

Perlman JE, Bowsher TR, Braccini SN, Kuehl TJ and Schaprio SJ 2003 Using positive reinforcement training techniques to facilitate the collection of semen in chimpanzees (*Pan troglodytes*). *American Journal of Primatology 60*(Supplement): 77–78 (Abstract)

Pfister HP and King MG 1976 Adaptation of the glucocorticosterone response to novelty. *Physiology and Behavior 17*: 43–46

Pibiri F, Nelson M, Guidotti A, Costa E and Pinna R 2008 Decreased corticolimbic allopregnanolone expression during social isolation enhances contextual fear: A model relevant for posttraumatic stress disorder. *Proceedings of The National Academy of Sciences 105*: 5567–5572

Pines MK, Kaplan G and Rogers LJ 2004 Stressors of common marmosets (*Callithrix jacchus*) in the captive environment: Effects on behaviour and cortisol levels. *Folia Primatologica 75*(Supplement 1): 317–318

Pitman DL, Ottenweller JE and Natelson BH 1988 Plasma corticosterone levels during repeated presentation of two intensities of restraint stress: Chronic stress and habituation. *Physiology and Behavior 43*: 47–55

Poderscek AL, Blackshaw JK and Beattie AW 1991 The behaviour of group penned and individually caged laboratory rabbits. *Applied Animal Behaviour Science 21*: 353–363

Priest GM 1998 *New Frontiers in Animal Behavior Management (Videotape with Commentary)*. San Diego Zoo: San Diego, CA

Priest GM 1991 Training a diabetic drill (*Mandrillus leucophaeus*) to accept insulin injections and venipuncture. *Laboratory Primate Newsletter 30*(1): 1–4

Priest GM 1990 The use of operant conditioning in training husbandry behavior with captive exotic animals. *Proceedings of the National American Association of Zoo Keepers Conference 16*: 94–108

Puri CP, Puri V and Anand-Kumar TC 1981 Serum levels of testosterone, cortisol, prolactin and bioactive luteinizing hormone in adult male rhesus monkeys following cage–restraint or anaesthetizing with ketamine hydrochloride. *Acta Endocrinologica 97*: 118–124

Quadri SK, Pierson C and Spies HP 1978 Effects of centrally acting drugs on serum levels in rhesus monkeys. *Neuroendocrinology 27*: 136–147

Raff H, Bruder ED, Cullinan WE, Ziegler DR and Cohen EP 2011 Effect of animal facility construction on basal hypothalamic–pituitray–adrenal and renin–aldosterone activity in the rat. *Endocrinology 152*: 1218–1221

Rasmussen S, Glickman GNR, Quimby FW and Tolwani RJ 2009 Construction noise decreases reproductive efficiency in mice. *Journal of the American Association for Laboratory Animal Science 48*(4): 363–370

Reasinger DJ and Rogers JR 2001 Ideas of improving living conditions of non–human primates by improving cage design. *Contemporary Topics in Laboratory Animal Science 40*(4): 89 (Abstract)

Reed HJ, Wilkins LJ, Austin SD and Gregory NG 1993 The effect of environmental enrichment during rearing on fear reactions and depopulation trauma in adult caged hens. *Applied Animal Behaviour Science 36*: 39–46

Reese EP 1991 The role of husbandry in promoting the welfare of laboratory animals. In: Hendriksen CFM and Koeter HBWM (eds) *Animals in Biomedical Research* pp. 155–192. Elsevier: Amsterdam, Netherlands

Reimers M, Schwarzenberger F and Preuschoft S 2007 Rehabilitation of research chimpanzees: Stress and coping after long–term isolation. *Hormones and Behavior 51*: 428–435

Reinhardt V 1991 Training adult male rhesus monkeys to actively cooperate during in–homecage venipuncture. *Animal Technology 42*: 11–17

Reinhardt V 1997 Lighting conditions for laboratory monkeys: Are they adequate? *Animal Welfare Information Center Newsletter 8*(2): 3–6

Reinhardt V 1999 Pair–housing overcomes self–biting behavior in macaques. *Laboratory Primate Newsletter 38*(1): 4

Reinhardt V 2003 Working with rather than against macaques during blood collection. *Journal of Applied Animal Welfare Science 6*: 189–197

Reinhardt, V (ed) 2010 *Caring Hands – Discussions by the Laboratory Animal Refinement & Enrichment Forum, Volume II*. Animal Welfare Institute: Washington, DC

Reinhardt, V (ed) 2013 *Compassion Makes a Difference – Discussions by the Laboratory Animal Refinement & Enrichment Forum, Volume III*. Animal Welfare Institute: Washington, DC

Reinhardt V and Cowley D 1990 Training stumptailed monkeys to cooperate during in–homecage treatment. *Laboratory Primate Newsletter 29*(4): 9–10

Reinhardt V, Cowley D, Scheffler J, Vertein R and Wegner F 1990 Cortisol response of female rhesus monkeys to venipuncture in homecage versus venipuncture in restraint apparatus. *Journal of Medical Primatology 19*: 601–606

Reinhardt V and Reinhardt A 1991 Impact of a privacy panel on the behavior of caged female rhesus monkeys living in pairs. *Journal of Experimental Animal Science 34*: 55–58

Reinhardt V, Cowley D, Eisele S and Scheffler J 1991 Avoiding undue cortisol responses to venipuncture in adult male rhesus macaques. *Animal Technology 42*: 83–86

Reinhardt V and Cowley D 1992 In–homecage blood collection from conscious stumptailed macaques. *Animal Welfare 1*: 249–255

Reinhardt V and Reinhardt A 1999 The monkey cave: The dark lower–row cage. *Laboratory Primate Newsletter 38*(3): 8–9

Reinhardt V and Reinhardt A 2000a Blood collection procedure of laboratory primates: A neglected variable in biomedical research. *Journal of Applied Animal Welfare Science 3*: 321–333

Reinhardt V and Reinhardt A 2000b The lower row monkey cage: An overlooked variable in biomedical research. *Journal of Applied Animal Welfare Science 3*: 141–149

Reinhardt V, Buchanan-Smith HM and Prescott MJ 2002 Training macaques to voluntarily co–operate during two common procedures: Blood collection and capture of group–housed animals. *Congress of the International Primatological Society* pp. 182–183 (Abstract). Mammalogical Society of China: Beijing, China

Robbins DQ, Zwick H, Leedy M and Stearns G 1986 Acute restraint device for rhesus monkeys. *Laboratory Animal Science 36*: 68–70

Rodarte LF, Ducoing A and Galindo F 2004 The effect of environmental manipulation on behavior, salivary cortisol, and growth of piglets weaned at 14 days of age. *Journal of Applied Animal Welfare Science 7*: 171–179

Rollin BE 1995 Laws relevant to animal research in the United States. In: Tuffery AA (ed) *Laboratory Animals – An Introduction for Experimenters, Second Edition* pp. 67–86. John Wiley & Sons: New York, NY

Ross PW and Everitt JI 1988 A nylon ball device for primate environmental enrichment. *Laboratory Animal Science 38*: 481–483

Ross S, Nagy ZM, Kessler C and Scott JP 1966 Effects of illumination on wall–leaving behavior and activity in three inbred mouse strains. *Journal of Comparative and Physiological Psychology 62*: 338–340

Rourke C and Pemberton DJ 2007 Investigation of a novel refined oral dosing method. *Animal Technology and Welfare 6*: 15–17

Roy V, Belzung C, Delarue C and Chapillon P 2001 Environmental enrichment in BALB/c mice: effects in classical tests of anxiety and exposure to a predatory odor. *Physiology and Behavior 74*: 313–320

Ruis MAW, Te Brake JHA, Engel B, Ekkel ED, Buist WG, Blokhuis HJ and Koolhaas JM 1997 The circadian rhythm of salivary cortisol in growing pigs: Effects of age, gender, and stress. *Physiology and Behaviour 62*: 623–630

Russell WMS and Burch RL 1959 *The Principles of Humane Experimental Technique*. Methuen & Co.: London, UK

Sachser N, Dürschlag M and Hirzel D 1998 Social relationships and the management of stress. *Psychoneuroendocrinology 23*: 891–904

Saltarelli CG and Coppola CP 1979 Influence of visible light on organ weights of mice. *Laboratory Animal Science 29*: 319–322

Sato K, Chatani F and Sato S 1995 Circadian and short–term variabilities in blood pressure and heart rate measured by telemetry in rabbits and rats. *Journal of the Autonomous Nervous System 54*: 235–246

Savane S 2008 Use of flashlights in Old World nonhuman primate health monitoring. *American Association for Laboratory Animal Science Meeting Official Program:* 103 (Abstract)

Savory CJ and Kostal L 1996 Application of a radiotelemetry system for chronic measurement of blood pressure, heart rate, EEG, and activity in the chicken. *Physiology and Behavior 61*: 963–969

Schaefer AL, Salomons MO, Tong AKW, Sather AP and Lepage P 1990 The effect of environmental enrichment on aggression in newly weaned pigs. *Applied Animal Behaviour Science 27*: 41–52

Schapiro SJ and Bushong D 1994 Effects of enrichment on veterinary treatment of laboratory rhesus macaques (*Macaca mulatta*). *Animal Welfare 3*: 25–36

Schapiro SJ, Nehete PN, Perlman JE and Sastry KJ 1997 Social housing condition affects cell–mediated immune responses in adult rhesus macaques. *American Journal of Primatology 42*: 147 (Abstract)

Schapiro SJ, Nehete PN, Perlman JE and Sastry KJ 2000a A comparison of cell–mediated immune responses in rhesus macaques housed singly, in pairs, or in groups . *Applied Animal Behaviour Science 68*: 67–84

Schapiro SJ, Stavisky R and Hook M 2000b The lower–row cage may be dark, but behavior does not appear to be affected. *Laboratory Primate Newsletter 39*(1): 4–6

Schapiro SJ, Perlman JE, Thiele E and Lambeth S 2005 Training nonhuman primates to perform behaviors useful in biomedical research. *Lab Animal 34*(5): 37–42

Scharmann W 1994 Housing of mice in an enriched environment. In: Bunyan J (ed) *Proceedings of the Federation of European Laboratory Animal Science Associations Symposium* pp. 335–337. Royal Society of Medicine Press: London, UK

Schipper P, Nonkes LJ, Karel P, Kiliaan AJ and Homberg JR 2011 Serotonin transporter genotype x construction stress interaction in rats. *Behavioural Brain Research 223*: 169–175

Schnell CR and Wood JM 1993 Measurement of blood pressure, heart rate, body temperature, ECG and activity by telemetry in conscious unrestrained marmosets. *Proceedings of the Federation of European Laboratory Animal Science Associations Symposium*: 107–111

Schnell CR and Gerber P 1997 Training and remote monitoring of cardiovascular parameters in non–human primates. *Primate Report 49*: 61–70

Schoemaker NJ, Mol JA, Lumeij JT, Thijssen JH and Rijnberk A 2003 Effects of anaesthesia and manual restraint on the plasma concentrations of pituitary and adrenocortical hormones in ferrets. *Veterinary Record 10*: 591–595

Schwindaman D 1991 The 1985 animal welfare act amendments. In: Novak MA and Petto AJ (eds) *Through the Looking Glass. Issues of Psychological Well–being in Captive Nonhuman Primates* pp. 26–32. American Psychological Association: Washington, DC

Scientists Center for Animal Welfare 1987 Consensus recommendations on effective Institutional Animal Care and Use Committees. *Laboratory Animal Science 37*: 11–13

Seebacher F and Alford RA 2002 Shelter microhabitats determine body temperature and dehydration rates of a terrestrial amphibian (*Bufomarinus*). *Journal of Herpetology 36*: 69–75

Sharp JL, Zammit TG, Azar TA and Lawson DM 2002a Stress–like responses to common procedures in male rats housed alone or with other rats. *Contemporary Topics in Laboratory Animal Science 41* (4): 8–14

Sharp JL, Zammit T, Azar TA and Lawson DM 2002b Does witnessing experimental procedures produce stress in male rats? *Contemporary Topics in Laboratory Animal Science 41*(5): 8–12

Sharp JL, Zammit T, Azar TA and Lawson DM 2003 Stress–like responses to common procedures in individually and group–housed female rats. *Contemporary Topics in Laboratory Animal Science 42*(1): 9–18

Shaw DC and Gallagher RH 1984 Group or singly housed rats? In: The Universities Federation for Animal Welfare *Standards in Laboratory Animal Management* pp. 65–70. The Universities Federation for Animal Welfare: Potters Bar, UK

Shively CA, Clarkson TB and Kaplan JR 1989 Social deprivation and coronary artery atherosclerosis in female cynomolgus monkeys. *Atherosclerosis 77*: 69–76

Simeon D, Stanley B and Frances A 1992 Self–mutilation in personality disorders: Psychological and biological correlates. *American Journal of Psychiatry 149*: 221–226

Slaughter MR, Birmingham JM, Patel B, Whelan GA, Krebs-Brown AJ, Hockings PD and Osborne JA 2002 Extended acclimatization is required to eliminate stress effects of periodic blood–sampling procedures on vasoactive hormones and blood volume in beagle dogs. *Laboratory Animals 36*: 403–410

Smith JA, Birke L and Sadler D 1997 Reporting animal use in scientific papers. *Laboratory Animals 31*: 312–317

Snowdon CT, Savage A and McConnell PB 1985 A breeding colony of cotton–top tamarins (*Saguinus oedipus*). *Laboratory Animal Science 35*: 477–480

Späni D, Arras M, König B and Rülicke T 2003 Higher heart rate of laboratory mice housed individually vs in pairs. *Laboratory Animals 37*: 54–62

Spragg SDS 1940 Morphine addiction in chimpanzees. *Comparative Psychology Monographs 15*: 1–132

Stanton ME, Patterson JM and Levine S 1985 Social influences on conditioned cortisol secretion in the squirrel monkey. *Psychoneuroendocrinology 10*: 125–134

Stauffacher M 2000 Refinement in rabbit housing and husbandry. In: Balls M, Van Zeller AM and Halder M (eds) *Progress in the Reduction, Refinement and Replacement of Animal Experimentation* pp. 1269–1277. Elsevier: Amsterdam, Netherlands

Stone AM, Bloomsmith MA, Laule GE and Alford PL 1994 Documenting positive reinforcement training for chimpanzee urine collection. *American Journal of Primatology 33*: 242 (Abstract)

Suzuki J, Ohkura S and Terao S 2002 Baseline and stress levels of cortisol in conscious and unrestrained Japanese macaques (*Macaca fuscata*). *Journal of Medical Primatology 31*: 340–344

Tatoyan SK and Cherkovich GM 1972 The heart rate in monkeys (Baboons and Macaques) in different physiological states recorded by radiotelemetry. *Folia Primatologica 17*: 255–266

Taylor GT 1981 Fear and affiliation in domesticated male rats. *Journal of Comparative and Physiological Psychology 95*: 685–693

Tiefenbacher S, Lee B, Meyer JS and Spealman RD 2003 Noninvasive technique for the repeated sampling of salivary free cortisol in awake, unrestrained squirrel monkeys. *American Journal of Primatology 60*: 69–75

Torii R, Kitagawa N, Nigi H and Ohsawa N 1993 Effects of repeated restraint stress at 30–minute intervals during 24–hours on serum testosterone, LH and glucocorticoids levels in male Japanese monkeys (*Macaca fuscata*). *Experimental Animal 42*: 67–73

Torreilles SL and Green SL 2007 Refuge cover decreases the incidence of bite wounds in laboratory south african clawed frogs (Xenopus laevis). *Journal of the American Association for Laboratory Animal Science 46*(5): 33–36

Townsend P 1997 Use of in–cage shelters by laboratory rats. *Animal Welfare 6*: 95–103

Traystman RJ 1987 ACUC, who needs it? The investigator's viewpoint. *Laboratory Animal Science 37*(Special Issue): 108–110

Truett A and West D 1995 Validation of a radiotelemetry system for continuous blood pressure and heart rate monitoring in dogs. *Laboratory Animal Science 45*: 299–302

Tuli J, Smith JA and Morton DB 1995a Stress measurements in mice after transportation. *Laboratory Animals 29*: 132–138

Tuli J, Smith JA and Morton DB 1995b Effects of acute and chronic restraint on the adrenal gland weight and serum corticosterone concentration of mice and their faecal output of oocysts after infection with *Eimeria apionodes*. *Research in Veterinary Science 59*: 82–86

Turkkan JS, Ator NA, Brady JV and Craven KA 1989 Beyond chronic catheterization in laboratory primates. In: Segal EF (ed) *Housing, Care and Psychological Wellbeing of Captive and Laboratory Primates* pp. 305–322. Noyes Publications: Park Ridge, NJ

United States Department of Agriculture 2002 *Animal Welfare Regulations Revised as of January 1, 2002 – Code of Federal Regulations, Title 9, Chapter 1, Parts 1–4*. U.S. Government Printing Office: Washington, DC

Ursin H and Murison R 1986 Facts, fiction and rational decisions. *Acta Physiologica Scandinavica 554*(Supplement 1): 234–242

Valzelli L 1973 The "isolation syndrome" in mice. *Psychopharmacologia 31*: 305–320

van Adrichem PWM and Vogt JE 1993 The effect of isolation and separation on the metabolism of sheep. *Livestock Production Science 33*: 151–159

van de Wal PG, Engel BEG and Huslhof HG 1986 Changes in blood acid–base characteristics, haemoglobin and lactate concentrations due to increasing moderate stress in pigs. *Netherlands Journal of Agricultural Science 34*: 108–111

van de Weerd HA, Aarsen EL, Mulder A, Kruitwagen CLJJ, Hendriksen CFM and Baumans V 2002 Effects of environmental enrichment for mice: Variation in experimental results. *Journal of Applied Animal Welfare Science 5*: 87–109

van Ginneken VJT, Snelderwaard P, van der Linden R, Van den Reijden N, Van den Thillart GEEJM and Kramer K 2004 Coupling of heart rate with metabolic depression in fish: a radiotelemetric and calorimetric study. *Thermochimica Acta 414*: 1–10

van Loo PLP, Kruitwagen CLJJ, Koolhaas JM, van de Weerd HA, Van Zutphen LFM and Baumans V 2002 Influence of cage enrichment on aggressive behaviour and physiological parameters in male mice. *Applied Animal Behaviour Science 76*: 65–81

van Loo PLP, Kuin N, Sommer R, Avsaroglu H, Pham T and Baumans V 2007 Impact of 'living apart together' on postoperative recovery of mice compared with social and individual housing. *Laboratory Animals 41*: 441–455

Veissier I and Le Neindre P 1992 Reactivity of Aubrac heifers exposed to a novel environment alone or in groups of four. *Applied Animal Behaviour Science 33*: 11–15

Voipio H-M, Nevalainen T, Halonen P, Hakumäki M and Björk E 2006 Role of cage material, working style and hearing sensitivity in perception of animal care noise. *Laboratory Animals 40*: 400–409

Vriend J and Lauber JK 1973 Effects of light intensity, wavelength and quanta on gonads and spleen of the deer–mouse. *Nature 244*: 37–38

Wagner JL, Hackel DB and Samsell AG 1974 Spontaneous death in rabbits resulting from gastric trichobezoars. *Laboratory Animal Science 24*: 826–830

Wall HS, Worthman C and Else JG 1985 Effects of ketamine anaesthesia, stress and repeated bleeding on the haematology of vervet monkeys. *Laboratory Animals 19*: 138–144

Walters SL, Torres-Urbano CJ, Chichester L and Rose RE 2012 The impact of huts on physiological stress: a refinement in post–transport housing of male guineapigs (*Cavia porcellus*). *Laboratory Animals 46*: 220–224

Waran NK and Broom DM 1993 The influence of a barrier on the behaviour and growth of early–weaned piglets. *Animal Production 56*: 115–119

Warwick C 1990 Important ethological and other considerations of the study and maintenance of reptiles in captivity. *Applied Animal Behaviour Science 27*: 363–366

Weed JL, Wagner PO, Byrum R, Parrish S, Knezevich M and Powell DA 2003 Treatment of persistent self–injurious behavior in rhesus monkeys through socialization: A preliminary report. *Contemporary Topics in Laboratory Animal Science 42*(5): 21–23

Weihe WH 1976 The effect of light on animals. *Laboratory Animal Handbooks 7*: 63–76

Weihe WH, Schidlow J and Strittmatter J 1969 The effect of light intensity on the breeding and development of rats and golden hamsters. *International Journal of Biometeorology 13*: 69–79

Westlund K, Fernström A-L, Wergård E-M, Fredlund H, Hau J and Spångberg M 2012 Physiological and behavioural stress responses in cynomolgus macaques (*Macaca fascicularis*) to noise associated with construction work . *Laboratory Animals 46*: 51–58

Wolfensohn S and Lloyd M 1994 *Handbook of Laboratory Animal Management and Welfare – Chapter on small species: Gerbils and Hamsters*. Oxford University Press: Oxford, UK

Wurtman RL 1967 Effects of light and visual stimuli on endocrine function. In: Martini L. and Ganong WF (eds) *Neuroendocrinology, Vol. II* pp. 19–59. Academic Press: New York

Yaroshevsky F 1975 Self–mutilation in Soviet prisons. *Canadian Psychiatric Association Journal 20*: 443–446

Yasuda M, Wolff J and Howard CF 1988 Effects of physical and chemical restraint on intravenous glucose tolerance test in crested black macaques (*Macaca nigra*). *American Journal of Primatology 15*: 171–180

Yasutomi M and Adachi N 1987 Effects of playthings on prevention of cannibalism in rearing chickens. *Japanese Poultry Science 24*: 372–373

Human-Animal Bond

Joanna Cruden, Bsc (Hons) FIAT RAnTech

FEW WOULD QUESTION the strength of the bond that people have with their pets. Yet, to suggest a similar bond between a caretaker and a laboratory animal seems incongruous to many, even within the research setting. However, such bonds can exist and can enrich the lives of the animals and the people caring for them. Furthermore, close bonds can help science, as caretakers are able to more readily observe changes in behavior signifying potential health and welfare issues.

Over the thousands of years that humans have interacted with domesticated animals, our relationship with them has taken many different forms, including: worship, food, labor, and sport. We often use anthropomorphism when describing animals and their behavior. In some cases, anthropomorphism can be taken to the extreme, such as in the case of performing a marriage ceremony for two dogs (Podberscek et al., 2000). However, it is difficult to judge these cases, as the interpretation of the human-animal relationship varies greatly between and within societies.

When we are working with animals it is extremely important to have knowledge of the species and, in some cases, the individual animal. One may argue that what makes a great animal caretaker is someone who not only has knowledge, but also empathy with the animals in his or her charge. But how can we be sure we can empathize with animals, if we are uncertain of the causes of their behavior? Social psychologist Lauren Wispe (1987) offers the following observations:

» The object of empathy is understanding.
» The object of sympathy is the other [individual's] well-being.

One might argue that to be a compassionate caretaker, one must not only empathize, but sympathize with the animal.

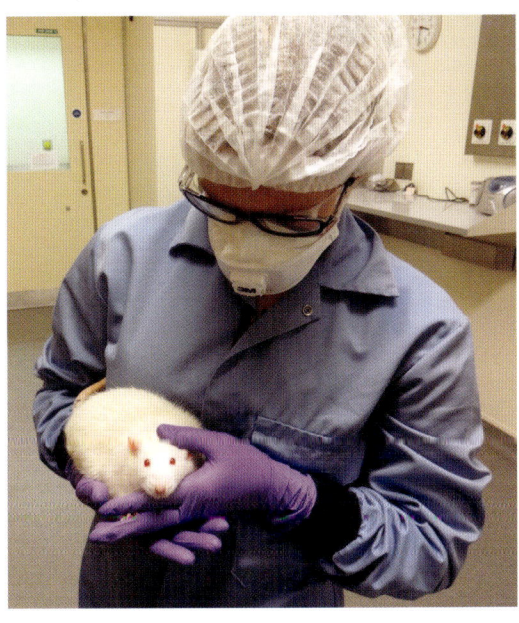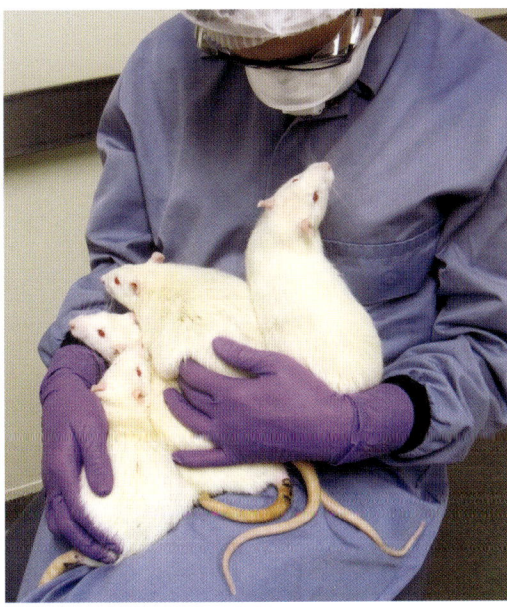

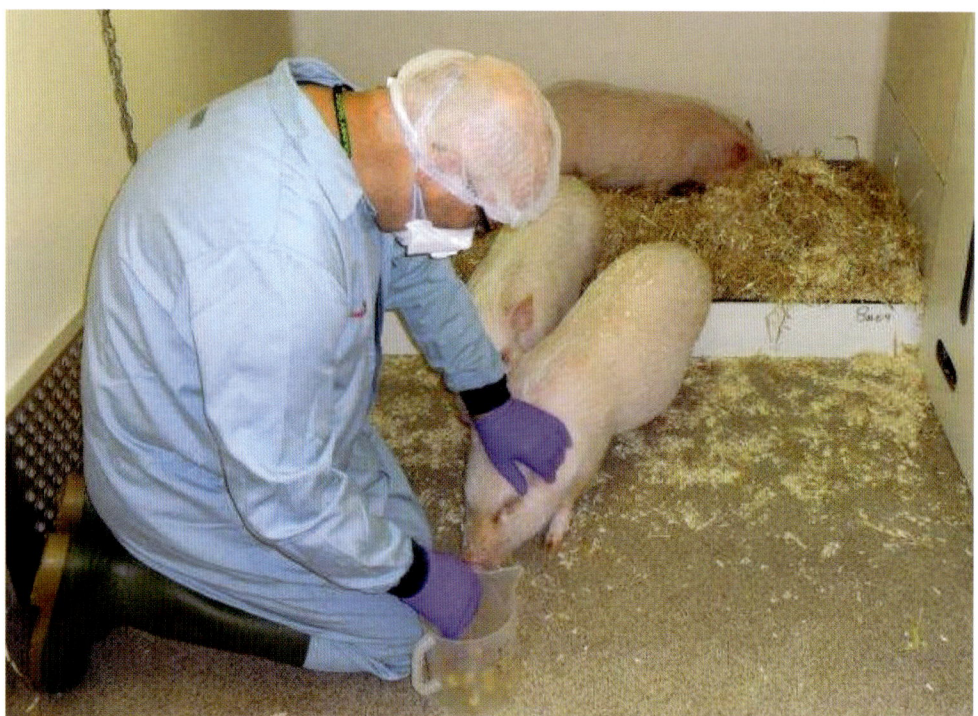

The Farm Animal Welfare Council developed fundamental requirements for farm animals, known as the "five freedoms," which have become prominent in all industries that use animals (Farm Animal Welfare Council, 1997). These are—

» Freedom from thirst, hunger and malnutrition;
» Freedom from discomfort;
» Freedom from pain, injury and disease;
» Freedom to express normal behavior; and
» Freedom from fear and distress.

The following question was raised on the Laboratory Animal Refinement & Enrichment Forum (LAREF) (Anonymous, 2003): "Should animal care personnel be encouraged to establish affectionate, rather than neutral, relationships with the animals in their charge?" Most forum participants who responded to the question felt that the development of an affectionate relationship with the animals in their charge was almost unavoidable. The consensus was that emotional attachment provides an assurance that the animals would receive optimal care, both physically and behaviorally. As one respondent observed,

Having a close relationship with your animals is necessary to regard them as living beings, rather than biological test tubes. As such, you are more careful and patient, and will think more about what the procedures mean to the animals. You will become more creative in finding animal friendly alternatives for the procedures you need to do on the animals. You will thus increase the well-being of your animals and, by doing so, make them better research subjects and increase the validity of the test results.

One individual, however, expressed concern that establishing an affectionate relationship with experimental animals and knowing them as individuals might hamper her impartiality and capacity to be objective when observing and registering their behavior. Another person countered:

> It seems to me that we get hung up on trying to divorce our emotions from what we hope to be our objectivity. I do not think that any normally functioning human being in the world does anything for any reason other than emotional. Sure, research is done to answer questions, but isn't the premise of all research to make human (or animal) lives better? If you want to make lives better, it's because of emotion, not because you are logically attached to life.

While there was disagreement about whether it was more difficult to establish a relationship with some animal species, it was postulated that working closely with individual animals or small groups and observing them for extended periods of time was a more important bonding factor than evolutionary relatedness with our own species. Such close contact provides insight into the personalities of individual animals and allows the caretaker to give more personalized care. An example of this was described by a colleague (personal communication, August 21, 2014):

> I spent 6 fantastic years working within a Small Primate Unit and during this time it was amazing to see how staff and marmosets built relationships with each other. Each marmoset had a unique personality, which had to be taken into account during the daily routines. Some animals would gradually build a greater trust and bond with individual members of staff and this would not always be replicated with other members of the team. For example, Happy initially was quite a timid animal, though over time built a close bond with me. He would sit on my shoulder and was content to be escorted around the unit—we had an understanding that he was safe on my shoulder and he should not leave there unless told to do so. However, even after several years he still showed some distrust to most other members of staff.

Barriers to human-animal bonding in the research setting

Within the research setting there are many barriers to forming a meaningful human-animal bond. This section discusses some of the most prominent.

For many animals in the research setting, their only interactions with human beings are when they are handled. Subtle differences in how the animals are handled can have significant effects on how they view human beings, in turn affecting the opportunity for a bond to form. For example, the traditional way laboratory mice are handled (i.e., picking up by the scruff of the neck or the base of the tail) is not conducive to providing the animals with a sense of security—the first step towards bonding with their caretaker (Baumann et al., 2007). Hurst & West (2013) found that using a handling tube or gently cupping mice reduced their anxiety in behavioral studies. Hurst suggested that picking mice up by the tail was similar to how a predator would catch the mice and thus, to reduce their fear this traditional method should be avoided.

Bonding with mice may seem unrealistic due to the vast number of these animals typically housed in a facility. However, if caretakers took just 1 minute to really observe an occupied mouse cage, they would notice that each mouse has a unique personality and is reacting differently to the observer's presence (Baumann et al., 2007). Seeing each mouse as an individual improves the likelihood of forming a sympathetic bond with them.

Other significant barriers to the formation of bonds are the cages (particularly for rodent species) and personal protective equipment (PPE) requirements (particularly for nonhuman primates). This barrier is particularly prominent with the use of Individually Ventilated Cages (IVCs). Among the reasons IVCs are used:

» They foster a cleaner working environment.
» They remove most of the odors associated with rodents.
» There is less disturbance to the animals when people enter the housing room.
» They can effectively isolate health issues.
» They minimize human exposure to rodent allergens.
» They provide very efficient use of space, allowing far greater numbers of animals to be housed in a single room when compared to conventional housing options.

Unfortunately, these otherwise positive aspects create an almost unbreakable barrier to forming any bond with the cage inhabitants. Caretakers are encouraged to minimize contact with the animals, often only looking at them briefly each day, through the cage side. In larger facilities, the time pressures of changing several hundred cages daily prevent more than cursory contact with the animals. This production-line mentality can lead to an emotional detachment from the animals, creating a negative or disinterested view of them. When the focus shifts to volume-driven factors, the first things that tend to be lost are personal touches that improve animal welfare. Consider the contrast to what this caretaker describes (personal communication, September 8, 2014):

> When I started working with laboratory animals we could walk from room to room and when I had spare time at the end of the day I could wander round the facility and handle the animals. I used to make a bee line for the rats. I used to just sit there with a whole cage-worth either in my pockets or on my lap exploring because in those days we had time to devote to animal playtime.

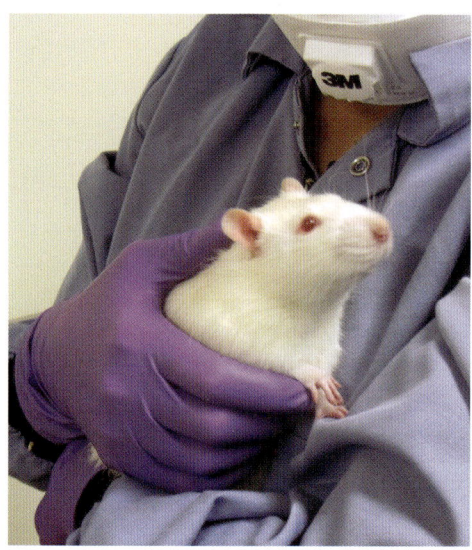 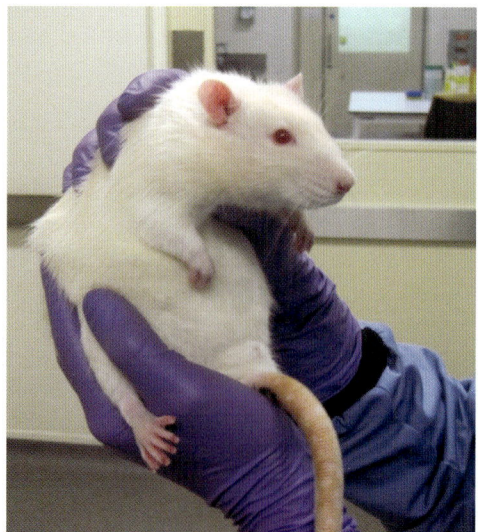

If it is difficult or impossible for us to establish positive relationships with the animals in our care, we may become disengaged from their well-being. "Researchers must continue to question the barriers that have traditionally been erected against forming HABs [human-animal bonds] in the name of objectivity and to investigate seriously the ways in which fostering the formation of HABs can promote animal welfare without compromising the scientific respectability of research" (Russow, 2002).

In 2002, the Institute of Animal Technology commissioned a survey of all its 2,000-plus members, asking a number of questions regarding their professional motivations; the Institute received 511 responses (Institute of Animal Technology, 2002). Some of the reasons people choose to work with laboratory animals include the following:

» They are fascinated by working with the animals, and enjoy observing them.
» They enjoy handling and interacting with the animals in their charge.
» They like attending to the animals' needs and improving their quality of life.

Addressing these reasons can be very difficult within a modern research facility, where rodents are primarily housed in IVCs. The remainder of the chapter will discuss ways to promote the human-animal bond, focusing on the most commonly used animals in research.

Building a trust relationship with animals

Any relationship between a human and an animal is based on mutual trust that the relationship will not result in harm. Without that trust, a bond can never develop. For example, in describing his behavioral work with bison, Reinhardt (personal communication, August 2014) explained:

> When studying the behavior of bison, I did bond with several animals. You can imagine that getting close to a one-ton bison bull implied that the bull had learned that he could trust that I was not going to harm him. Concurrently, I learned that I could trust that he wouldn't harm me. So there was no reason to be afraid; we both felt safe.

As noted above, handling is the primary interaction between animals and their caretakers. Proper handling is perhaps the most important step to building the trust relationship required to form an actual bond. For example, even minimal, gentle handling can help rats become very friendly towards handlers. They appear to perceive human interactions as a positive experience. This was demonstrated in a study by Cloutier et al. (2014) where one handler tickled rats for 2 minutes a day, whereas a second handler restrained them. When the animals were placed in an environment with access to both handlers' hands they tended to interact more with the handler who had tickled them, engaging in gentle nibbling of the hand, which was interpreted as a friendly, playful behavior. Davis & Perusse (1988) showed that rats will actually work to be petted by a preferred human, even in the absence of a reward such as food. Studies like these demonstrate how easily the animals will form a bond with people, when given a positive environment.

Turning the focus to another research animal: laboratory rabbits maintain many traits of their wild counterparts and can be very nervous around people. (Mykytowycz & Hesterman, 1975). Several studies have shown that gentle, compassionate handling helps rabbits to overcome their fear of humans and makes them more compliant during handling procedures (Podberscek et al., 1991; Swennes et al., 2011). Spending a few minutes a day talking to rabbits, encouraging them

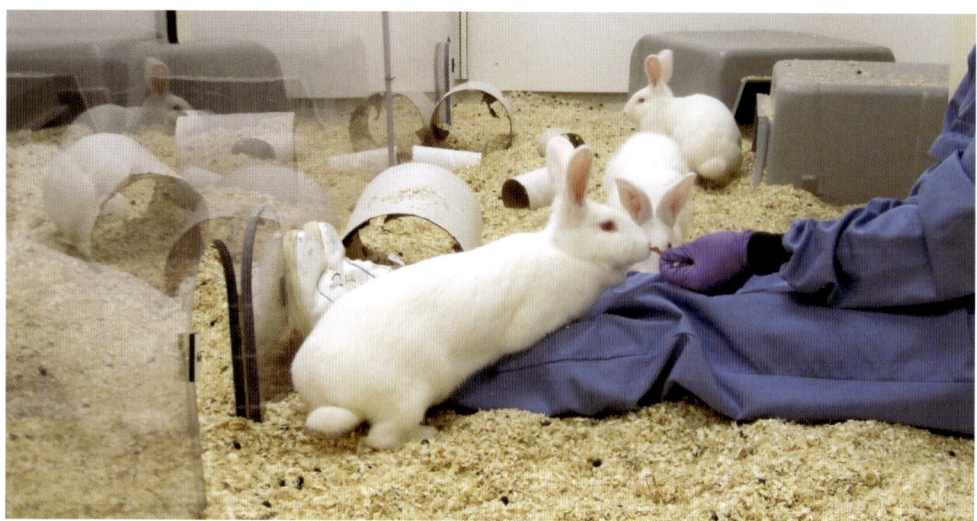

to come forward, and stroking them, decreases their fear of humans. Over time, rabbits can even get quite attached to caregivers and may hop up to the front of the cage for attention.

Again, spending time with the animals allows the caregiver time to discover the animals' distinct personalities and preferences. The following anecdote comes from another caregiver (personal communication, September 8, 2014):

> *Several years ago I worked with 3 NZW* [New Zealand White] *rabbits; I called them each a name that I thought related to their personality to help me remember any individual rabbit's quirks. Gemma used to grunt when I took her hay away to give fresh. It's nothing special but* [giving her a "G" name] *helped me remember what was her quirk and not abnormal for her. Madison (Mad) always tried to bite me when I opened the cage, so one day I just opened the cage* [and] *laid my arms across the floor. I wasn't sure what he would do to be honest; I just stayed there to see if he would come forward. He slowly did and, to my surprise, started rubbing his chin on my arms. This continued and he eventually trusted me to remove his hay and give him fresh hay without trying to bite me. He wouldn't let me stroke him. I think this was just his preference because as soon as I opened the cage he would run forward for me to put my arms down so he could scent mark me! I felt like we both reached an understanding. On the other hand, Rodger loved being stroked. As soon as you went near the cage he would run to the front and lay his head down for you to stroke him.*

Consistent handling of rabbits not only decreases their fear of humans but it will also make experimental results more reliable (Verwer et al., 2009) by avoiding physiological stress reactions.

Establishing mutual trust, while bonding with animals, provides the foundation for successful cooperation rather than resistance during the daily husbandry and research procedures. It has been documented in the literature that virtually all animals in research, including nonhuman primates (Turkkan, 1990), rats (Rourke & Pemberton, 2007), rabbits (Marr et al., 1993), dogs (Roddis, 2005), cats (Albertin, 1990), pigs (Grandin, 1986), goats (Lager, 1998), and sheep (Mellor, 2004) can be readily taught to cooperate during certain procedures that would otherwise trigger data-confounding stress reactions.

Comfortable Quarters for Laboratory Animals

Do animals remember us?

It will be of no surprise to many caretakers that rats recognize familiar people. One caretaker remarked: "When I go on holiday, the new caretaker often tells me my rats are not as calm as I make them out to be." In a study investigating whether rats remember handlers, Davis et al. (1997) found that after one 10-minute handling session, 24 of 26 Long Evans rats correctly chose the handler with whom they had previous contact. Rats prefer contact with a person with whom they are familiar and with whom they had positive experience versus an unfamiliar individual.

Primates are an excellent example of how well an animal remembers us once we have built a trust relationship with them. Augusto Vitale (2011) shares his story about a strong bond he formed with a tufted capuchin monkey, named Cammello. Cammello was a very eager monkey, who readily solved many experimental tasks. Augusto had built a bond with this monkey, but didn't realize that the monkey had also bonded with him, until he returned to the monkey colony after a 20-year absence: "Cammello rushed to greet, embrace and groom me for about 5 minutes."

Naming animals

Naming animals raises many emotions in people; a name can strengthen a bond we feel towards an animal, but naming animals in a scientific context may be considered inappropriate (Baumans et al., 2007), as it personalizes them. Many care staff will strongly disagree with this attitude and firmly believe people should have deep feelings for the animals in their care (Baumans et al., 2007, Cruden, 2010). In the LAREF discussion summarized earlier, several participants mentioned that they give names to the animals in their charge or to the animals they study. Giving names to animals can be a reflection of empathy and is a useful tool to quickly recognize individuals.

Sullivan (pictured below) was one of three friendly male ferrets at a facility where the author used to work. Sullivan and the others (Felix and Gilbert) loved to come out of their cages, chase toys, and snuggle with the caretakers. This was both great enrichment for the ferrets and the staff, who would call for them by name.

Naming animals in the research laboratory is possibly one of the most powerful tools to create a bond. It personalizes them and creates a persona. An animal with just a number or no identification can be viewed dispassionately, as a mere receptacle, whereas a named animal is viewed as a living, sentient being who must be treated with compassion.

Does the bond with animals affect research data obtained from them?

Developing a bond with animals in research laboratories and knowing their fate can be an extremely difficult situation, emotionally, for the caretakers to come to terms with (American Association for Laboratory Animal Science, 2001). In spite of this emotional cost, in a survey of 81 UK-based animal technicians, 70% agreed that people had to love animals in order to do their often emotionally draining work (Cruden, 2010). It is undeniable that, at the very least, kindness and concern for animals are desirable characteristics of anyone involved in animal research (American Association for Laboratory Animal Science, 2001). It is important to note that a strong bond can be formed with rodents just as much as it can be formed with larger species such as primates and dogs. Often this fact is overlooked when people observe the paradox of caring about animals while working with them in a research environment.

Wolfle (1996) made it very clear that stress leads to profound physiological and behavioral changes that can increase the variability of the data and decrease the reliability of the results. In order to control stress, the caretaker must strive to develop an affectionate bond with all animals in his or her charge (Wolfle, 1987). This bond conveys to the animal a quiet sense of assurance, from which coping strategies can be developed for dealing with other stressful aspects of the laboratory (Wolfle, 1987). Rather than compromising research, these human-animal bonds should be considered the very foundation of scientifically sound research methodology. If we achieve this aim we will refine the very core of laboratory animal experimental design.

Bayne (2002) elaborates that the human-animal bond is not just a benefit for the animals but also for the staff involved; it can be an enriching and rewarding experience for both. Coppola et al. (2006) found that the a greater amount of human interaction with shelter dogs led to a lower cortisol level in the saliva, suggesting that more quality time spent with laboratory dogs will lead to less stressed animals. The European Guidelines recommend that frequent contact should be maintained so that the animals become familiar with human presence and activity (European Economic Community, 1986). Where appropriate, time should be set aside for talking with, handling, and grooming the animals.

With ever-increasing understanding about the value of the human-animal bond, one can hope there may come a time when its importance will be recognized across all research institutes, and we will all be able set aside quality time to spend with the animals in order to foster trust-based relationships—and just for the pleasure of being with them.

REFERENCES

Albertin SV 1990 An alternative to distressful methods of animal immobilization. *Humane Innovations and Alternatives in Animal Experimentation 4*: 202–204

Anonymous 2003 Personnel/animal relationships: Affectionate or neutral? A discussion. *Laboratory Primate Newsletter 42*(1): 14–15. http://www.brown.edu/Research/Primate/42-1.pdf

American Association for Laboratory Animal Science 2001 *Cost of Caring: Recognizing Human Emotions in the Care of Laboratory Animals*. American Association for Laboratory Animal Science: Memphis, TN. http://www.aalas.org/pdf/06-00006.pdf

Bayne K 2002 Development of the human-research animal bond and its impact on animal well-being. *Institute for Laboratory Animal Research Journal 43*(1): 4–9. http://dels-old.nas.edu/ilar_n/ilarjournal/43_1/Development.shtml

Baumans V, Coke CS, Green J, Moreau E, Morton D Patterson-Kane E, ... Van Loo P (eds) 2007 *Making Lives Easier for Animals in Research Labs: Discussions by the Laboratory Animal Refinement & Enrichment Forum*. Animal Welfare Institute: Washington, DC. http://www.awionline.org/pubs/LAREF/LAREF-bk.html

Cloutier S, Wahl K, Baker C and Newberry RC 2014 The social buffering effect of playful handling on responses to repeated intraperitoneal injections in laboratory rats. *Journal of the American Association for Laboratory Animal Science 53*(2) 168–173

Coppola CL, Grandin T and Enns RM 2006 Human interaction and cortisol: can human contact reduce stress for shelter dogs? *Physiology and Behavior 87*: 537–541

Cruden J 2010 It is more than just a job; it is a way of life. *Animal Technology and Welfare 9*: 7–24. http://xa.yimg.com/kq/groups/8138622/1246405467/name/Cruden.pdf

Davis H, Taylor AA, and Norris C 1997 Preference for familiar humans by rats. *Psychonomic Bulletin & Review 4*: 118–120

Davis H and Perusse R 1988 Human-based social interaction can reward a rat's behavior. *Animal Learning and Behavior 16*: 89–92

European Economic Community 1986 Council Directive 86/609 on the Approximation of Laws, Regulations, and Administrative Provisions Regarding the Protection of Animals Used for Experimental and Other Scientific Purposes, Annex II Guidelines for Accommodation and Care of Animals. *Official Journal of the European Communities*: 7–28. http://eur-lex.europa.eu/LexUriServ/LexUriServ.do?uri=CELEX:31986L0609:EN:HTML

Farm Animal Welfare Council 1997 *Report on the Welfare of Laying Hens*. Farm Animal Welfare Council: Tolworth, UK

Grandin T 1986 Minimizing stress in pig handling. *Lab Animal 15*(3): 15–20

Hurst JL and West RS 2013 Taming anxiety in laboratory mice. *Nature Methods 7*: 825–828

Institute of Animal Technology 2002 Survey of membership to help assess Institute's strategic plan. Presented at the Annual General Meeting.

Lager K 1998 Apparatus and technique for conditioning goats to repeated blood collection. *Lab Animal 27*(3): 38–42

Marr JM, Gnam EC, Calhoun J and Mader JT 1993 A non-stressful alternative to gastric gavage for oral administration of antibiotics in rabbits. *Lab Animal 22*(2): 47–49

Mellor DJ 2004 Taming and training of pregnant sheep and goats and of newborn lambs, kids and calves before experimentation. *Alternatives to Laboratory Animals 32*(Supplement): 143–146. http://www.frame.org.uk/atla_issue.php?iss_id=70

Mykytowycz R and Hesterman ER 1975 An experimental study of aggression in captive European rabbits. *Behaviour 52*: 104–117

Podberscek AL, Paul ES and Serpell JA 2000 *Companion Animals and Us: Exploring the Relationships between People and Pets*. Cambridge University Press: Cambridge, UK

Podberscek AL, Blackshaw JK and Beattie AW 1991 The effects of repeated handling by familiar and unfamiliar people on rabbits in individual cages and group pens. *Applied Animal Behaviour Science 28*: 365–373

Roddis D 2005 How to teach an old dog new tricks. *Animal Technology and Welfare 4*: 181–184

Rourke C and Pemberton DJ 2007 Investigation of a novel refined oral dosing method. *Animal Technology and Welfare 6*: 15–17

Swennes AG, Alworth LC, Harvey SB, Jones CA, King CS and Crowell-Davis SL 2011 Human handling promotes compliant behavior in adult laboratory rabbits. *Journal of the American Association for Laboratory Animal Science 50*: 41–45

Russow L-M 2002 Ethical implications of the human-animal bond. *Institute for Laboratory Animal Research Journal 43*(1): 33–37. http://dels-old.nas.edu/ilar_n/ilarjournal/43_1/Implications.shtml

Turkkan JS 1990 New methodology for measuring blood pressure in awake baboons with use of behavioral training techniques. *Journal of Medical Primatology 19*: 455–466. http://www.awionline.org/lab_animals/biblio/jmp19-4.htm

Verwer CM, van der Ark A, van Amerogen, G, van den Bos R and Hendriksen CFM 2009 Reducing variation in a rabbit vaccine safety study with particular emphasis on housing conditions and handling. *Laboratory Animals 43*: 155–164

Vitale A 2011 Primatology between feelings and science: A personal experience perspective. *American Journal of Primatology 73*: 214–219

Wispe L 1987 History of the concept of empathy. In: Eisenberg N and Strayer J (eds) *Empathy and Its Development* pp 17–37. Cambridge University Press: Cambridge, UK

Wolfle TL 1987 Control of stress using non-drug approaches. *Journal of the American Veterinary Medical Association 191*: 1219–1221

Wolfle TL 1996 How different species affect the relationship. In: Krulisch L, Mayer S and Simmonds RC (eds) *The Human/Research Animal Relationship* pp 85–91. Scientists Center for Animal Welfare: Bethesda, MD

Animal Welfare Institute Policy on Research and Testing with Animals

Research must not be conducted on animals unless, at minimum, the methodology fulfills the three "Rs" of Russell and Burch, including the following:

1. The animals are maintained in an optimum, species-appropriate environment.

2. The animals are under the care of professionally trained, compassionate personnel.

3. The animals' pain, physical discomfort, maladaptive behaviors, fear and anxiety are prevented or at least minimized by considerate and scientifically sound experimental design and appropriate use of anesthetic, analgesic or tranquilizing drugs.

Detailed policy

1. All institutions that conduct research and testing with animals must refine the research methodology and reduce and seek to ultimately replace animals wherever indicated and possible. These efforts should be supported and funded by both the research-funding agencies and the research institution's administration.

2. If alternative yet equally effective methods of experimentation or testing are available, they must be used in preference to any experiment conducted with an animal, particularly an experiment that is likely to cause pain, fear or distress.

3. Any experiment or test that inflicts trauma should be conducted with a fully anesthetized animal. If the procedure causes life-threatening injury, the animal should be euthanized following the procedure and before regaining consciousness.

4. If an animal is subjected to surgery from which he or she is expected to survive, a pre-planned pain evaluation and pain management schedule must be developed. This schedule must account for overnight and weekend hours. The pain evaluation must contain specific signs, behaviors or physical parameters to be measured in the animal. Staff must ensure adequate and timely administration of pain-relieving medications until the animal has recovered and is no longer in observable discomfort.

5. Professional staff must be available at all times—day and night, weekends and holidays—to care for the animals. The staff must make rounds for the purpose of ascertaining the state of each animal's health and well-being. The staff must be trained and authorized to dispense pain-relieving or tranquilizing drugs as may be necessary. While it may be a standard operating procedure to phone the investigator or director regarding such events, this action must not delay the provision of relief for the animal. Nursing care must be provided to all animals following surgery or other injurious interventions and to animals with chronic pathological conditions.

6. Staff must be compassionate and well trained. Ongoing training regarding best practices must be provided. The staff must be observant and empowered to make its observations known to the director of the laboratory, veterinarian or another trained individual duly authorized to make animal welfare or humane endpoint decisions. For example, a moribund animal should be euthanized. An animal who is suffering should be—depending on the situation and the nature of the work—anesthetized or sedated and given supportive care such as fluids, soft food, and custom bedding and/or otherwise treated to alleviate suffering, or the animal should be euthanized.

7. Housing for animals in research must provide sufficient space and materials to permit the expression of basic species-specific behaviors, including species-typical walking and stretching, foraging, retreating to a safe/sheltered place, burrowing and gnawing (rodents), climbing, perching and swinging (nonhuman primates), perching and scratching (birds), and rooting and wallowing (pigs). Social animals must be housed with one or several compatible conspecifics to address their biological need for companionship.

8. The great majority of animals in experimentation and testing are purposely bred for sale to research facilities. This is the preferred method of acquiring animals. These animals must be raised in facilities whose standards of housing and care are equal to or better than those described herein for research laboratories. Following the legally mandated waiting period, dogs and cats at municipal pounds may be donated to veterinary schools where surgical training to conduct spays and neuters are done or other treatments are performed that are intended to facilitate adoption of the animals. After recovery, the dogs and cats should be returned to the pound where it is hoped that adoptive homes will be found for them.

9. Only noninvasive research of direct benefit to the species' own survival may be conducted on threatened or endangered species.

10. Euthanasia must be considered a major responsibility. Staff carrying out euthanasia must be well trained, efficient in performing the procedure, and empathetic to the animals. The primary concern must be the animals. The location for conducting the euthanasia should be selected so as not to increase anxiety and fear. The method of euthanasia that is selected should ensure the quickest death possible. No animal should be discarded without monitoring him or her long enough after death to ascertain rigor mortis.

11. Journals should expand the materials and methods section to include information regarding animal housing conditions, bedding type, enrichment, refinement, and details of supportive or analgesic care. The only information in most published articles is the species (or strain of rodent), sex and age—thus making it impossible for concerned scientists to find details needed to confirm sound methodology and trustworthiness of the research data and statistical results.

12. Animals should be permitted to retire after termination of their assignment(s) to research, testing and education. The funding agency and research institution should earmark funds for the life-long retirement of these animals.

PHOTO CREDITS

Cover
Dr. Brianna Gaskill

Mice
2 » iStock
3 » Brianna Gaskill
5 all » Heleen A. Van de Weerd, Utrecht University
10 bottom » iStock
13 all » Prof Jane Hurst, University of Liverpool
14 » T.P. Rooijmans, Utrecht University

Rats
21 » Courtesy of Dr. Tony Prescott (Prescott et al., 2011), University of Sheffield
22, 28 bottom, 30 » KPC, Harvard University
27 » Courtesy of Charles River
28 top, 29 » Courtesy of Dr. Debra Hickman and Jessica Peveler, Indiana University School of Medicine

Guinea Pigs
40 » Marcie Donnely
41 » Phillie Casablanca
42 top » Patrick & Preston, middle » Marcie Donnely, bottom » Cat Wendt
43 » Marcie Donnely
44 left » Ryan Owens, right » Martin/Flickr

Hamsters
51, 52, 54, 56, 57, 62 » Michele Cunneen
58 » Necrocake/Flickr

Rabbits
66 » Cheryl Dimof
67, 74 » University of Michigan
68, 71, 72 » Michele Cunneen
69 » Keith Survell
70 » Pehpsii Altemark
73 left » Keith Survell, right » University of Michigan

Ferrets
79 » Yale Photo & Design
81, 82, 84 » Jodi Scholz

Zebrafish
89, 96 » Novartis
91, 92, 97 » Christian Lawrence

Frogs
101 » Kristin Shoemaker
102 top » Tom Brandt, bottom » Susan Adams
103, 104 » Russell Yothers
105 » gcmenezes/Flickr

Cattle
109 » Razvan Antonescu
110, 112, 113, 115 » David Cawson
111 top » David Cawson, left » Michelle Bradley, right » Hornet Photography
114 top » QUOI Media Group, bottom » Compassion in World Farming

Pigs
118 » Jim Champion
119 » Dave Kav
120 left » Canolais
122 » Charles Roffey
123 left » Evelyn Skoumbourdis

Sheep
127 » Stazebla/Flickr
129 top » Carolynn Bernard-Harwell
129 bottom, 132 » Louis DiVincenti

Dogs
138 » Don Burkett
139, 140 » Michele Cunneen
142 » Carol Vinzant

Cats
146, 147, 152 » Michele Cunneen
149, 151, 154, 155 » The WALTHAM Centre for Pet Nutrition (© Mars, Incorporated. All Rights Reserved)

Primates
188 » Angelika Rehrig

Extraneous Variables
213 » Dorcas O'Rourke